Amphibian Metamorphosis

Amphibian Metamorphosis
From Morphology to Molecular Biology

Yun-Bo Shi, Ph.D.
Unit of Molecular Morphogenesis
National Institute of Child Health and Human Development
National Institutes of Health
Bethesda, MD 20892

A John Wiley & Sons, Inc., Publication

New York · Chichester · Weinheim · Brisbane · Singapore · Toronto

This book is printed on acid-free paper. ☉

Copyright © 2000 by John Wiley & Sons, Inc. All rights reserved.

Published simultaneously in Canada.

No part of this publication may be reproduced, stored in a retrieval system or transmitted in any form or by any means, electronic, mechanical, photocopying, recording, scanning or otherwise, except as permitted under Sections 107 or 108 of the 1976 United States Copyright Act, without either the prior written permission of the Publisher, or authorization through payment of the appropriate per-copy fee to the Copyright Clearance Center, 222 Rosewood Drive, Danvers, MA 01923, (978) 750-8400 fax (978) 750-4744. Requests to the Publisher for permission should be addressed to the Permissions Department, John Wiley & Sons, Inc., 605 Third Avenue, New York, NY 10158-0012, (212) 850-6011, fax (212) 850-6008, E-Mail: PERMREQ @ WILEY.COM.

For ordering and customer service, call 1(800) CALL-WILEY.

Library of Congress Cataloging-in-Publication Data:

Shi, Yun-Bo.
   Amphibian metamorphosis : from morphology to molecular biology / Yun-Bo Shi,
      p. cm.
   Includes bibliographical references.
   ISBN 0-471-24475-9 (cloth : alk. paper)
   1. Ampibians--Metamorphosis. 2. Amphibians--Molecular aspects.
  I. Title.
QL669.2.S48   1999
571.8'76178--dc21   99-21920
                                                                     CIP

Printed in the United States of America

10 9 8 7 6 5 4 3 2 1

Contents

Preface   *xi*

Abbreviations   *xiii*

1 An Overview of Amphibian Metamorphosis / 1

    1.1 Introduction / 1
    1.2 Classification of Amphibians / 2
    1.3 Metamorphosis of Different Amphibians / 4

        1.3.1 Metamorphosis in Urodeles / 4
        1.3.2 Metamorphosis in Caecilians / 5
        1.3.3 Metamorphosis in Anurans / 5

    1.4 Direct Development / 6
    1.5 Metamorphosis and Thyroid Gland / 9
    1.6 Environmental Effects on Metamorphosis / 10
    1.7 Metamorphosis is a Unique Model for Postembryonic Vertebrate Organ Development / 13

2 Hormonal Regulation / 15

    2.1 Introduction / 15
    2.2 Thyroid Hormone / 15

        2.2.1 Evidence for the Causative Effect of Thyroid Hormone on Amphibian Metamorphosis / 15
        2.2.2 Thyroid Gland Development / 19

    2.3 Neuroendocrine Control / 21

2.3.1 Pituitary Regulation of Thyroid Gland Function / 22
2.3.2 Regulation of Pituitary Secretion by the Hypothalamus / 23
2.4 Influence of Steroid Hormones on Metamorphosis / 25
2.4.1 Corticoids / 25
2.4.2 Gonadal Steroids / 27
2.5 Inhibition of Anuran Metamorphosis by Prolactin / 27
2.5.1 Dual Functions of Prolactin During Anuran Development / 27
2.5.2 Regulation of Prolactin Levels During Development / 28
2.5.3 Function and Mechanism of PRL Action During Metamorphosis / 31
2.6 Other Hormones / 35

3 Morphological Changes During Anuran Metamorphosis / 36

3.1 Introduction / 36
3.2 Resorption of Tadplole-specific Organs / 37
3.2.1 The Tail / 37
3.2.2 The Gills / 39
3.3 De Novo Development of Frog Limbs / 41
3.4 Remodeling of Existing Organs for Frog Use / 41
3.4.1 The Liver / 42
3.4.2 The Nervous System / 42
3.4.3 The Intestine / 43

4 Cellular and Biochemical Changes / 50

4.1 Introduction / 50
4.2 Programmed Cell Death or Apoptosis During Tissue Resorption / 50
4.2.1 Cell Death and Its Executioners / 50
4.2.2 Apoptosis During Intestinal Remodeling / 55
4.2.3 Cell Death During Tail Resorption / 61
4.2.4 Apoptosis in Other Organs and Tissues / 64
4.3 Biochemical Changes Associated with Tissue Resorption and Remodeling / 65
4.3.1 Upregulation of Degradative Enzymes During Tail Resorption / 65
4.3.2 Biochemical Changes Associated with Intestinal Remodeling / 67
4.3.3 Alterations in the Blood / 68
4.3.4 Regulation of Enzyme Activities in the Liver / 71

5 Mechanism of Thyroid Hormone Action / 74

- 5.1 Introduction / 74
- 5.2 Thyroid Hormone and Its Binding Proteins / 74
- 5.3 Thyroid Hormone Receptors / 78
  - 5.3.1 Thyroid Hormone Binding by Thyroid Hormone Receptors / 82
  - 5.3.2 DNA binding by Thyroid Hormone Receptors / 84
- 5.4 Mechanism of Transcriptional Regulation / 87
  - 5.4.1 Activation versus Repression / 88
  - 5.4.2 Coactivators and Corepressors / 90
  - 5.4.3 Role of Chromatin / 91
  - 5.4.4 Participation of Histone Acetyltransferases and Deacetylases / 96
  - 5.4.5 A Putative Model of Thyroid Hormone Receptor Action / 100

6 Gene Regulation by Thyroid Hormone / 102

- 6.1 Introduction / 102
- 6.2 Methods for Isolation of Thyroid Hormone-response Genes / 104
  - 6.2.1 Differential Hybridization / 104
  - 6.2.2 Subtractive Differential Screen / 105
  - 6.2.3 Differential Display / 107
  - 6.2.4 Other Methods / 107
- 6.3 Early and Late Thyroid Hormone Response Genes / 108
  - 6.3.1 Early Thyroid Hormone Response Gene / 108
  - 6.3.2 Late Thyroid Hormone Response Genes / 116
- 6.4 Tissue-dependent Variation in the Gene Expression Program Induced by Thyroid Hormone / 119

7 Functional Implication of Thyroid Hormone Response Genes: Transcription Factors in the Thyroid Hormone Signal Transduction Cascade / 123

- 7.1 Introduction / 123
- 7.2 Expression of Genes Encoding Transcription Factors During Metamorphosis / 124
- 7.3 Biochemical and Molecular Characterization of Thyroid Hormone-induced Transcription Factors / 128
- 7.4 Target Genes of the Thyroid Hormone-induced Transcription Factors / 130

8 Competence and Tissue-specific Temporal Regulation of Amphibian Metamorphosis / 135

- 8.1 Introduction / 135

8.2 Expression and Function of Thyroid Hormone and 9-cis Retinoic Acid Receptors in Pre- and Metamorphosing Tadpoles / 136
    8.2.1 Thyroid Hormone and 9-cis Retinoic Acid Receptor Expression / 136
    8.2.2 Functions of Thyroid Hormone Receptors in Frog Development / 140
8.3 Regulation of Cellular Thyroid Hormone Levels / 145
    8.3.1 Circulating Thyroid Hormone / 145
    8.3.2 Cytosolic Thyroid Hormone Binding Proteins / 146
    8.3.3 Deiodinases / 147
8.4 Molecular Basis for Competence and Timing of Tissue-Specific Transformation / 150

9 Thyroid Hormone Regulation and Functional Implication of Cell–Cell and Cell–Extracellular Matrix Interactions during Tissue Remodeling / 154

9.1 Introduction / 154
9.2 Extracellular Matrix Remodeling and Cell–Cell Interaction During Intestinal Metamorphosis / 155
9.3 Regulation of Cell–Cell Interactions Through Secreted Molecules in the Intestine / 159
    9.3.1 Sonic Hedgehog as a Signaling Molecule During Intestinal Remodeling / 159
    9.3.2 Mechanism of Hedgehog Signaling / 162
9.4 Matrix Metalloproteinases as Regulators of Extracellular Matrix / 163
    9.4.1 Correlation of *Xenopus* Stromelysin-3 Expression with Cell Death During Metamorphosis / 166
    9.4.2 Differential Regulation of Different Matrix Metalloproteinases During Metamorphosis / 168
9.5 Multiple Factors Contributing to the Regulation of Cell–Cell and Cell-Extracellular Matrix Interactions in Different Organs / 175
9.6 Extracellular Matrix as a Regulator of Cell Fate / 176
9.7 Summary / 177

10 Model Systems and Approaches to Study the Molecular Mechanisms of Amphibian Metamorphosis / 180

10.1 Introduction / 180
10.2 Thyroid Hormone Regulation of Cell Fate in Cell Cultures / 180
10.3 Organ Culture Systems / 183

10.4 Frog Embryos as Models for Studying Gene Regulation and Function / 185
10.5 Transgenic *Xenopus Laevis* Animals / 187

11 Comparison with Insect Metamorphosis / 190

11.1 Introduction / 190
11.2 Hormonal Regulation of Insect Metamorphosis / 191
11.3 Ecdysone Receptors / 194
11.4 Ecdysone-Induced Gene Regulation Cascade / 196
  11.4.1 The Ashburner Model / 196
  11.4.2 Genes Regulated by Ecdysone / 198
  11.4.3 A Gene Regulation Cascade for Tissue Resorption / 200
11.5 Similarities and Differences Between Amphibian and Insect Metamorphosis / 202

12 Roles of Thyroid Hormone and Its Receptors in Mammalian Development and Human Diseases / 205

12.1 Introduction / 205
12.2 Thyroid Hormone in Human Development and Diseases / 206
12.3 Roles of Thyroid Hormone Receptors in Thyroid Hormone Resistance Syndromes / 209
  12.3.1 Thyroid Hormone Resistance Syndromes / 209
  12.3.2 Human Thyroid Hormone Receptors / 211
  12.3.3 Thyroid Hormone Receptor Mutations in Human Diseases / 216
12.4 TR Knockout Studies in Mice / 218
  12.4.1 TRα Gene / 218
  12.4.2 TRβ Gene / 220
  12.4.3 Double Knockout / 222
12.5 Conclusions / 222

References / 224

Index / 279

Preface

The century-old phenomenon of thyroid hormone-dependent amphibian metamorphosis has fascinated biologists ever since its discovery. It has served as an excellent model to study postembryonic organogenesis and mechanisms of hormonal regulation during vertebrate development. It will continue to do so even in this age when advanced molecular and genetic tools, especially gene knockout and transgenic technologies in vertebrates, are allowing developmental studies in various model systems.

Amphibian metamorphosis, in particular, anuran metamorphosis, is developmentally similar to the postembryonic organogenesis in mammals. This process occurs externally and can be easily controlled by regulating the availability of a single hormone, the thyroid hormone. This has made it popular not only for studies of organ development in vertebrates but also for investigations on endocrine regulation of animal development. Early work on metamorphosis focused mainly on the morphological changes associated with the tadpole-to-frog transformation. Subsequently, advances in cell biological and biochemical tools allowed extensive studies to determine the cellular and biochemical bases of this process. However, studies on molecular aspects of metamorphosis were largely limited to the characterization of genes that are expressed in terminally differentiated cells. The cloning of avian and mammalian thyroid hormone receptors helped push forward the current molecular undertaking to dissect the pathways of tissue transformations induced by the hormone. The primary goal of this book is to integrate the findings from recent studies with earlier observations and to provide some molecular and mechanistic insights into the signal transduction pathways underlying tissue-specific transformations during metamorphosis. For the ease of discussion, this book deals exclusively on anuran metamorphosis, focusing mainly on the South African clawed toad *Xenopus laevis*. Some references are made to another anuran species *Rana catesbeiana*.

The first two chapters provide an overview of metamorphosis in different classes of amphibians and various factors that influence this process. In particular, the role of thyroid hormone and other endocrine factors are discussed. A review of earlier work on the morphological, cellular, and biochemical changes during metamorphosis follows, focusing mainly on those organs/tissues that have been studied extensively at molecular levels.

The bulk of the book is on the more recent molecular analyses. This includes a discussion on thyroid hormone signal transduction pathway, especially transcriptional regulation mechanisms by thyroid hormone receptors. A summary and comparison is provided for the gene regulation programs induced by thyroid hormone in several organs that undergo distinct metamorphic transformations. Then several chapters are devoted to discussions on functional and mechanistic implications of the molecular findings on the thyroid hormone response genes in tissue transformations.

Metamorphosis is not unique to amphibians; invertebrate insects also undergo varying degrees of metamorphic changes during their life cycle. In particular, *Drosophila melanogaster* metamorphosis bears strong similarities to that in anurans. Thus a chapter is included to describe insect metamorphosis and compare the molecular pathways underlying *Drosophila* and *Xenopus* metamorphosis.

Thyroid hormone is clearly essential for amphibian metamorphosis, it also plays critical roles in mammalian development and organ function. A final chapter discusses the similarities between amphibian and mammalian development, and the conservation of the role of thyroid hormone and its underlying signal transduction mechanisms in such diverse animal species.

I am indebted to Dr. Donald Brown for introducing this interesting subject to me as my postdoctoral advisor and to Dr. Alan Wolffe for his encouragement, insightful discussions, and collaborations. I am grateful to current and former members of my laboratory for their contributions. I thank my long-term collaborator Dr. Atsuko Ishizuya-Oka as well as other current and former collaborators for many fruitful discussions and successful collaborations. I also appreciate the critical reading of the manuscript by Drs. Tosikazu Amano, Trevor Collingwood, Sashko Damjanovski, Robert Denver, Tyrone Hayes, Peter Jones, and Laurent Sachs. I owe my thanks to my former editor Robert Harington and current editor Luna Han for their help in bringing this book to publication, to Ms. Thuy Vo for preparing the manuscript, and to my managing editor Liz Adams for guiding the manuscript through production. Finally, I thank my wife Theresa Ng for her patience and continuous support.

YUN-BO SHI

Abbreviations

ACTH	Adrenocorticotropin
ATA	Tricarboxylic acid
BMP	Bone morphogenetic protein
CAD	Caspase-activated DNase
Ci	Cubitus interruptus
CNS	Central nervous system
Col1	Collagenase-1
Col3	Collagenase-3
Col4	Collagenase-4
COS2	Costal-2
CPP32	Caspase-3
CPS	Carbamyl phosphate synthetase
CRF	Corticotropin-releasing factor
CsA	Cyclosporin A
CTHBP	Cytosolic thyroid hormone binding protein
DBD	DNA binding domain
dUTP	deoxy-uridine triphosphate
Ec	Ecdysone or 20 hydroxyecdysone
ECM	Extracellular matrix
EcR	Ecdysone receptor
FAK	Focal adhesion tyrosine kinase
FU	Fused
GelA	Gelatinase A
GR	Glucocorticoid receptor

GRTH	General resistance to thyroid hormone
IAP	Inhibitor of apoptosis
ICAD	Inhibitor of caspase-activated DNase
ICE	Interleukin 1β converting enzyme
IFABP	Intestinal fatty acid binding protein
JAK	Janus kinase
LBD	Ligand binding domain
MMP	Matrix metalloproteinase
NFI	Nuclear factor I
OTC	Ornithine transcarbamylase
PCR	Polymerase chain reaction
PRL	Prolactin
PRL-R	Prolactin receptor
PRTH	Pituitary resistance to thyroid hormone
rT_3	Reverse T_3
RTH	Resistance to thyroid hormone
RXR	9-*cis* retinoic acid receptor or retinoid x receptor
ST3	Stromelysin-3
ST1	Stromelysin-1
Stat	Signal transducer and activator of transcription
T_2	Diiodothyronine
T_3	3, 3′, 5-triiodothyronine
T_4	3, 3′, 5, 5′-tetraiodothyronine or thyroxine
TF	Transcription factor
TH	Thyroid hormone
TH/bZip	TH-induced basic leucine-zipper containing transcription factor gene
TIMP	Tissue inhibitor of MMP
TR	Thyroid hormone receptor
TRE	Thyroid hormone response element
TRH	Thyrotropin-releasing hormone
TSA	Trichostatin A
TSH	Thyroid stimulating hormone or thyrotropin
TUNEL	Terminal deoxynucleotidyl transferase-mediated dUTP–biotin nick-end labeling
UPS	Ultraspiracle
Z-VAD	Z-Val-Ala-Asp-fluoromethalketone

1

An Overview of Amphibian Metamorphosis

1.1 INTRODUCTION

The central theme of developmental biology has been to understand how the fertilization of a single cell, the egg, gives rise to a complex organism. For centuries, biologists have experimented with many different animal systems to address this problem. Perhaps the oldest and most widely used model system to study development has been the frog (Duellmann and Trueb, 1986).

There are many advantages to the use of frogs in developmental biology. First, the frog embryo can be easily manipulated because it is 10 to 20 times larger than other vertebrate embryos such as the mouse or human. For example, the egg of the South African clawed toad *Xenopus laevis* is about 1 mm in diameter. This large size makes the fertilized frog egg very useful for studying events in early development. One can actually observe early cell divisions with the naked eye. As development progresses the different morphological changes can also be easily observed. The second advantage is that the entire developmental process, including fertilization, occurs externally. This makes the embryo easily accessible to observation and experimental manipulation.

By contrast with other vertebrates, the development of the frog can be divided into two stages, a larval stage and an adult stage. The process by which the larval tadpole transforms into the adult frog is referred to as metamorphosis. The most commonly known form of amphibian metamorphosis, anuran metamorphosis, brings about drastic changes in essentially every organ/tissue of the tadpole. The consequence of such a complex but systematic transformation is the conversion of an aquatic herbivorous tadpole into a terrestrial carnivorous frog. Other amphibians undergo less dramatic changes as their juveniles transform to adults. In still other amphibians, metamorphosis takes place only in the presence of exogenous inducing hormone or can not take place at all throughout their life-span.

Although the extent of organ transformation during metamorphosis varies depending on the species, such changes lead to the formation of adult organs that are structurally and functionally similar to those in adult mammals. Thus amphibian metamorphosis can be used as a model system to study postembryonic organ development in vertebrate animals. It is also an extreme example showing the ability of organisms to adapt to their environment, a key factor in the evolution of higher vertebrates.

1.2 CLASSIFICATION OF AMPHIBIANS

Amphibians are a class of vertebrates that generally make a change in habitat at some point of their lives (Just et al., 1981). Such a change normally involves the transformation from an aquatic larva to a terrestrial adult. However, some modern toads spend most of their lives on land, whereas other amphibians remain in water throughout life. Most amphibians remain near ponds and streams even in adulthood.

Modern amphibians can be grouped into three orders. These are Anura (Salientia), Urodela (Caudata), and Caecilia (Apoda and Gymnophiona) (Just et al., 1981; Fox, 1983; Duellman and Trueb, 1986). Anurans include tailless frogs and toads. Toads can be considered as a subclass of frogs. Similarly, urodeles include newts and salamanders, and newts are aquatic members of the family Salamandridae and are salamanders. Anurans represent the vast majority of the existing amphibians species. Of the over 3000 living amphibian species, 85% are anurans whereas only 10% are urodeles and 5% are caecilians (Carroll, 1977).

Being the largest group of amphibians, anurans include over 2500 species and have the broadest geographical distribution. Anurans are present on all continents with possible exception of Antarctica (Just et al., 1981; Duellman and Trueb, 1986). Anurans have a characteristic shape of the head grading into the body with no neck movement (Fig. 1.1). The adults are tail-less but have elongated hind legs, which provide an efficient means for leaping. The foot is lengthened by the elongated proximal tarsal elements, which are fused at least proximally and distally. Anurans have a diversity of reproductive modes. On the other hand, with some exceptions, fertilization is external. The eggs and larvae are aquatic but adults of most species are terrestrial.

There are about 300 species of known urodeles. Most of them are restricted to the northern hemisphere with the exception of a radiation into Central and South America, which accounts for about 40% of the existing urodele species (Wake, 1966; Savage, 1973; Just et al., 1981; Duellman and Trueb, 1986). In contrast to anurans, urodeles maintain their long tails throughout their adult, terrestrial life (Fig. 1.1). Furthermore, their four legs are of equal size, although

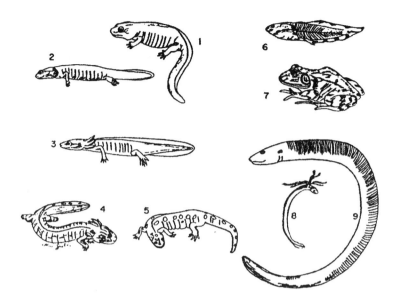

Figure 1.1 Representatives of the three classes of amphibians. Larval urodeles are shown in #2 (stream type) and #3 (pond type). Adult urodeles are shown in #1 and #5. #4 is a neotenic urodele. A larval and adult anuran are shown in #6 and #7, respectively. Finally, a very young and older caecilians are depicted in #8 and #9, respectively. Adapted from Just et al., (1981). With permission from Plenum Press.

some have tiny limbs or no hind limbs (Fox, 1983: Duellman and Trueb, 1986). An otic notch and middle ear are absent. Their aquatic larvae have teeth on both jaws, gill slits, and external gills. Finally, unlike anurans, fertilization is internal for the vast majority of urodeles (Duellman and Trueb, 1986). Copulation is not required for fertilization. Instead, males have specialized cloacal glands that produce gelatinous pyramidal structures, spermatophores, that are capped with sperm. Females pick up the sperm cap in their cloaca, whose roof is modified into a spermatheca to store sperm. As eggs pass through the cloaca, they are fertilized and then deposited singly or in clumps or strings in the water.

The least known group of amphibians are the caecilians. They are elongate, legless, wormlike animals that live primarily in tropics (Just et al., 1981; Duellman and Trueb, 1986). Their body is greatly elongated and segmented by annular grooves. They have short and pointed tails and small eyes covered with skin or bone (Fig. 1.1). Their aquatic larvae have gill slits but no external gills. The males have a single, median, protrusible copulatory organ. All caecilians employs internal fertilization for reproduction. Primative caecilians are oviparous and have aquatic larvae. Some advanced ones are oviparous whereas most advanced ones are viviparous.

1.3 METAMORPHOSIS OF DIFFERENT AMPHIBIANS

Metamorphosis is a postembryonic period of profound morphological changes by which the animal alters its mode of living. Metamorphosis is known to occur in all major living chordate groups except amniotes (Dent, 1968; Just et al., 1981). To provide a common terminology for different chordate groups, Just et al (1981) suggested the following three criteria as a standard approach for metamorphosis:

1. There must be some change in form of nonreproductive structures between the time the embryo hatches from the egg and sexual maturity. With this criterion, all structural changes having to do with embryonic development, sexual maturation, and aging are eliminated.
2. The form of the larva enables it to occupy an ecological niche different from that of the adult, clearly setting it apart from the embryo or adult. This criterion ensures that the change in form does not represent late embryonic changes or developmental changes in the young organism, but aids species survival by permitting the exploitation of new ecological niches.
3. The morphological changes that take place at the end of larval life (climax) depend on some environmental cue, either external (e.g., light, salinity, temperature, food supply) or internal (e.g., hormonal changes, changes in yolk reserves).

Just et al. (1981) further suggested that metamorphosis in any one taxonomic group may not meet all three criteria; however, at least two criteria are met before a given process can be termed "metamorphosis." Based on such a definition, all three classes of amphibians undergo metamorphosis, although not all species within each do so. There are three major types of changes associated with metamorphosis. These are the resorption or regression of organs/tissues that have significant functions only in the larval form, the remodeling of larval organs to their adult forms suitable for function in the adult, and the de novo development of organs/tissues that function in adult but are not required for the larva. These changes are most dramatic and comprehensive in anurans but quite limited in urodeles and caecilians.

1.3.1 Metamorphosis in Urodeles

Urodeles undergo fairly subtle morphological changes during the larva-to-adult transition. The animal body remains largely unchanged because both the larva and adult have the limbs and the tail is retained in the adult (Just et al., 1981; Fig. 1.1). However, some changes in the limbs do occur during metamorphosis. The earliest changes are the growth of the balancers, and forelimb growth and differentiation. The balancers degenerate in the midlarval stages when the hind limbs grow and differentiate. The most noticeable morphological changes are the resorption of the three sets of external gills and the tail fin at the final stages of metamorphosis (Fig. 1.1, compare #1 and #5 for adults to #2 and #3 for larvae). Other changes also take place during urodele metamorphosis. These other changes, for example, pigmentary development in the skin and alterations in the skeleton and musculature, are generally less obvious (Etkin, 1964; Fox, 1983; Dent, 1968).

Urodele metamorphosis is controlled by thyroid hormone (TH). Treatment of premetamorphic larvae of urodeles can lead to precocious metamorphic changes even in facultative neotenes or pedomorphic salamanders such as axolotl that do not undergo natural metamorphosis (Prahlad and DeLanney, 1965; Shaffer and Voss, 1996; Rosenkilde and Ussing, 1996; Kuhn and Jocobs, 1989; Brown, 1997). However, the obligatory neotenes such as *Necturus maculosus* do not metamorphose either in nature or when treated with exogenous TH (Dent, 1968). This group of neotenic urodeles have lost the "adult" terrestrial forms and instead reproduce as larvae.

1.3.2 Metamorphosis in Caecilians

Caecilian metamorphosis is the least studied among the three classes of amphibians. Larval and adult caecilians are similar in morphology with the notable exception of the presence of three pairs of external gills in the larvae (Just et al., 1981; Fig. 1.1). Developing caecilians are limbless and have three pair of gills that are not covered by an operculum (Fig. 1.1, #8). These external gills degenerate during late embryonic, larval, or fetal life, although it is unclear whether the gills are resorbed or broken off. In addition, there are other morphological changes in the transition from larvae to adults, particularly in the eyes and teeth. Toward the final stages of metamorphosis, the skin thickens and skin glands develop as in the other classes of amphibians (Just et al., 1981).

Although it has been well established that anurans and urodeles undergo thyroid hormone-dependent metamorphosis, the role of TH in caecilian metamorphosis is less certain. Treatment of premetamorphic anuran and urodele larvae induces precocious metamorphic transformations. However, no such experimental manipulations have been performed in caecilians. On the other hand, studies of thyroid histology suggest that there is increased gland activity at the time of transformation. Furthermore, TH receptor genes are conserved and expressed in diverse amphibian species from anurans such as *Xenopus laevis* to obligatory neotenic urodeles such as *Necturus maculosus* (Safi et al., 1997a,b). These findings, together with the demonstration of a role for TH in the development of several fish species (Inui and Miwa, 1985; Brown, 1997), argue that TH is likely to be the causative agent of caecilian metamorphosis.

1.3.3 Metamorphosis in Anurans

Anuran metamorphosis is both the most studied and the most dramatic metamorphosis (Fig. 1.1, compare #6 and #7). The vast majority of the

Figure 1.2 Phylogenetic relationships between Xenopus and Rana. Based on Ford et al. (1993). For simplicity, additional branches not directly related to Xenopus or Rana at various evolutionary stages are indicated by the second arm of the brackets.

experiments on amphibian metamorphosis are on only three species of anurans: *Xenopus laevis*, *Rana catesbeiana*, and *Rana pipiens*. Although *Xenopus* and *Rana* are evolutionarily far apart in the anuran phylogenetic tree (Fig. 1.2), metamorphosis in the two genera is conserved at both morphological and molecular levels.

Many different methods have been developed to stage anurans during development, especially during metamorphosis (Just et al., 1981). A comparison of three widely used staging methods is shown in Table 1.1. The most commonly used staging method for *Rana pipiens* and *Rana catesbeiana* is that of Taylor and Kollros (1946) (Table 1.1) whereas that for *Xenopus laevis* is that of Nieuwkoop and Faber (1956) (Table 1.1, Fig. 1.3). This book deals exclusively with anurans, with emphasis on one species, *Xenopus laevis*, and the Neiuwkoop and Faber method is used throughout the book. Cross-species comparisons can be easily obtained based on the stage alignment in Table 1.1.

Anuran metamorphosis is separated into three specific periods: premetamorphosis, prometamorphosis, and metamorphic climax (Etkin, 1964, 1968; Dodd and Dodd, 1976; Table 1.1). Premetamorphosis refers to a period when embryogenesis and early tadpole growth and development take place in the absence of thyroid hormone. However, some morphological changes such as limited development of the hind limbs do occur. During prometamorphosis, hind limbs undergo morphogenesis as exemplified by the differentiation of the toes and rapid and extensive growth of the hind limbs. This period is characterized by rising concentrations of endogenous thyroid hormone. Metamorphic climax is the period when endogeneous TH is at highest levels and when rapid morphological changes take place. Most noticeable is the complete resorption of the tadpole's tail. In addition to the external changes, drastic internal transformations also take place during anuran metamorphosis. These changes are covered in more detail in Chapter 3.

1.4 DIRECT DEVELOPMENT

It should be pointed out that although most anurans develop through two stages, the tadpole and frog, some can undergo direct development. Direct development involves the suppression of the free-living larval (tadpole) stage

TABLE 1.1 Comparison of Larval Anuran Stages[a]

Species	Limb Bud Growth							Toe Differentiation							Rapid Hind Limb Growth				Tail Resorption						
Bufo bufo[b]	23	24	25	I	II	III	IV	V	VI	VII	VIII	VIII	VIII	IX	IX	X	X	XI	XI	XII	XIII	XIV	XIV	XIV	XV
Rana pipiens[c]	I	II	III	IV	V	VI	VII	VIII	IX	X	XI	XII	XIII	XIV	XV	XVI	XVII	XVIII	XIX	XX	XXI	XXII	XXIII	XXIV	XXV
Xenopus laevis[d]	46	47/48	49/50	51	52	53	53	53	54	55	55	55	56	57	57	58	59	60	60	61	62	63	64	65	66

Premetamorphosis[e] — Prometamorphosis[e] — Climax[e]

[a] Just et al., (1981).
[b] Rossi (1959).
[c] Taylor and Kollros (1946).
[d] Nieuwkoop and Faber (1956).
[e] Dodd and Dodd (1976).

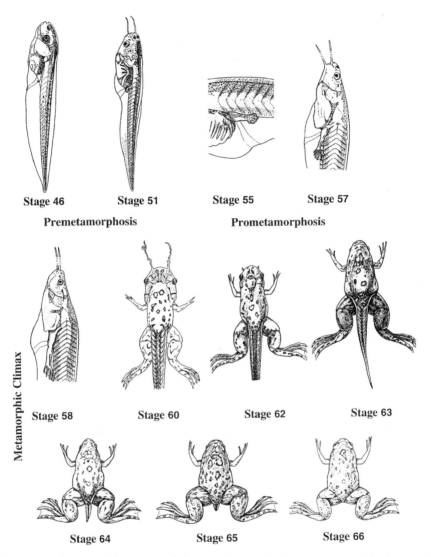

Figure 1.3 *Representative stages of* Xenopus laevis *tadpoles according to Niewkoop and Faber (1956).*

so that the young animal hatches directly as a miniature adult (Etkin, 1964; Dent, 1968). This strategy is found in species that have adapted to essentially complete terrestrial living. The terrestrial habitat has several advantages because more predators exist for aquatic larvae, which have no parental protection. In addition, pond drying is not a threat for direct developing larvae and the animal can avoid the traumatic metamorphic process (Dent, 1968). Examination of various anuran species shows that this adaptation accompanies various degrees of withdrawal from water (Dent, 1968). Species such as *Rana pipiens*, *Rana catesbeiana*, and *Xenopus laevis* have the most complete aquatic larval period, which extends from eggs laid in the water to the tadpole growth and the entire metamorphic process.

At the other extreme, anurans such as *Cornufer hazelae* and *Eleutherodactylus nubicola* develop directly and do not have any larval period. In the case of *Eleutherodactylus nubicola*, the eggs are laid in damp cavities under stones. The developing embryos and larvae are encased within a gelatinous membrane during the entire developmental period up to the stage equivalent to stage 66 in *Xenophus laevis* (Table 1.1), that is, the end of metamorphosis. These direct developing larvae have no gill slits because respiration in water is not required. Instead they specialize other organs/tissues for respiration. For example, *Cornufer hazelae* larvae use abdominal respiratory folds and *Elentherodactylus nubicolad* larvae employ a highly vascularized tail that expands into a balloon-like structure for respiration (Dent, 1968).

Between these extremes are anurans having variable lengths of aquatic larval periods (Dent, 1968). The larvae of these species use the aquatic period to adapt to the varying availability of water. In this regard, it is interesting to note that even fully metamorphosing tadpoles such as *Rana* tadpoles can respond to pond drying by reducing the lengths of their larval period (see Section 1.6).

The direct development in anurans is quite similar to the oviparous development typical of birds and other reptiles and allows the animal to withdraw from water. Many anurans capable of direct or near direct development also provide varying degrees of parental care and handling ranging up to true ovoviviparity (Dent, 1968). Despite these interesting developing patterns, relatively few studies have been carried out to determine the hormonal requirements during the developmental process that leads to the formation of the miniature adult. On the other hand, recent studies on the direct developing frog *Eleutherodacylus coqui* (*Leptodactylidae*) have revealed that the embryonic thyroid activity peaks at stages that are comparable to metamorphic stages in frogs that undergo biphasic development, that is, embryogenesis to produce tadpoles and metamorphosis to produce the frog (Jennings and Hanken, 1998). Thus thyroid hormone may play a similar role in direct developing anurans as in metamorphosing anurans.

1.5 METAMORPHOSIS AND THYROID GLAND

Anuran metamorphosis changes a larval tadpole into an adult frog. During metamorphosis, essentially every organ/tissue is remodeled to achieve this transformation. However, individual organs undergo vastly different changes (Dodd and Dodd, 1976, also see Chapter 3). For example, in the transition from tadpole to adult frog many new structures must be formed. One of these is the limb that develops from undifferentiated blastema cells during the early stages of metamorphosis (around stage 55 for hind limbs in *X. laevis*, Table 1.1). In contrast, other organs must be removed. An example of this is the total resorption of the tadpole tail, which begins around the climax of metamorphosis (stage 62 in *X. laevis*, Table 1.1). The completion of tail resorption marks the end of the metamorphic period.

Unlike the limb and tail, which are unique to the adult frog and larval tadpole, respectively, most other organs are present in both phases of development. These organs also undergo extensive remodeling during metamorphosis (Chapter 3). Thus metamorphosis involves systematic, coordinate transformations of different organs. The question then is how this process is regulated.

As early as 1912, a German scientist named J. F. Gudernatsch (1912) performed a landmark experiment. He fed various horse tissues to young tadpoles and looked for morphological changes. To his surprise, he found that the thyroid gland and only this gland could speed up the transition from a tadpole to a frog. A few years later, B. M. Allen of the University of Kansas showed that a simple thyroidectomy could prevent metamorphosis and consequently produced giant tadpoles (Allen, 1916, 1918).

In 1915, shortly after Gudernatsch showed the ability of thyroid gland extract to promote metamorphosis, E. C. Kendell (1915) was able to show that the active ingredient in the thyroid gland was thyroid hormone. These early studies led to extensive investigations on the nature of the hormone as well as its mechanisms of action, which are discussed in detail in the later chapters.

1.6 ENVIRONMENTAL EFFECTS ON METAMORPHOSIS

Like all other processes in animal development, environmental factors can regulate amphibian metamorphosis. These factors include temperature, water availability, crowding, light, diet, and environmental iodine levels (Dodd and Dodd, 1976). Amphibian larvae respond to alterations in these factors through high levels of plasticity in the developmental phenotypes (Stearns, 1989). Such plasticity may involve changes in the rate of metamorphic transition, adopting alternative morphologies, and so on (Newman, 1992; McCollum and Van Buskirk, 1996). Some of these factors that may inhibit growth when present during premetamorphic stages can stimulate metamorphosis when present during prometamorphosis. This class of factors includes crowding, resource limitation, habitat desiccation, and predation (Denver, 1997a; Denver, 1998).

Temperature is perhaps the best studied environmental factor that can regulate metamorphosis (Dodd and Dodd, 1976; Denver, 1997a). Higher temperatures stimulate both tadpole growth and the rate of metamorphosis (Dodd and Dodd, 1976; Hayes et al., 1993). In *Xenopus laevis*, the growth period from hatching to eruption of the forelimbs (stage 58) takes about 8–9 months at 15°C but only 35–42 days at 22°C (Dodd and Dodd, 1976). Similarly, the period of metamorphic climax (from stage 58 to 66) occupies 12,

16, and 36 days at 22°C, 18°C, and 15°C, respectively (Fig. 1.4; Dodd and Dodd, 1976). Lower temperatures also slow down TH-induced metamorphosis. Below 20°C, the response of tadpoles to exogenous TH is markedly reduced compared to that at 25–30°C in terms of both tail shrinkage and the shift from ammonium excretion to urea excretion (Frieden et al., 1965; Ashley et al., 1968). In addition, at extremely low temperatures, that is, 5°C, natural or TH-induced metamorphosis is completely blocked (Huxley, 1929; Lynn and Wachowski, 1951; Frieden et al., 1965; Ashley et al., 1968). These effects of temperature may be due to the reduction in the ability of tadpole tissues to bind TH, alterations in the neuroendocrine system that inhibit the synthesis and/or release of endogenous TH, reduced uptake of exogenous TH, perhaps more generalized effects on metabolism, and so on (Tata, 1972; Yamamoto et al. 1966; Dodd and Dodd, 1976).

The diet, water level, and iodine content in the water can also affect metamorphosis (Dent, 1968; Dodd and Dodd, 1976; Kaltenbach, 1968). Iodine is essential for the synthesis of thyroid hormone; thus, not surprisingly, sufficient amounts of iodine must be present in the diet and/or water. On the other hand, evaporation of the pond leads to the concentration of nonvolatile compounds and metabolic wastes, which may affect animal development. It has been shown that tadpole growth rate decreases and development rate increases as the pond dries (Newman, 1989; Denver, 1997a, b). For example, gradually reducing the water level leads to accelerated development of spadefoot toad (*Scaphiopus hammondii*) tadpoles (Fig. 1.5; Denver, 1998; Denver et al., 1998).

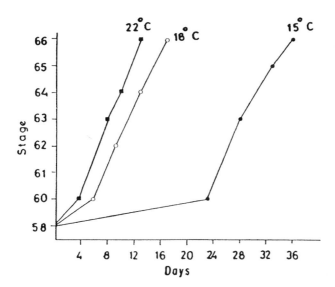

Figure 1.4 Effect of temperature on the duration of metamorphic climax in a batch of X. laevis larvae reared at different temperatures from stage 56. Adapted from Dodd and Dodd (1976) with permission from Academic Press.

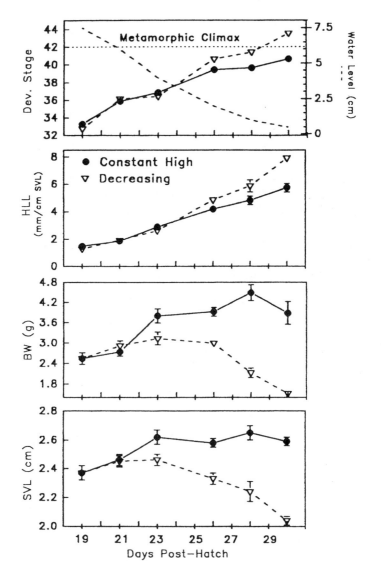

Figure 1.5 Acceleration of development of S. hammondii *tadpoles exposed to water volume reduction. Tadpoles [starting with Gosner Stage 32 (Gosner, 1960), equivalent to* Xenopus laevis *stage 53] were raised in rat cages (12 animals/cage) with either a constant volume of water (10 L; constant high) or a decreasing volume [decreasing; the dashed line (right axis) in the top graph shows the decline in the water level over time]. Measurements are developmental (dev.) stage (Gosner, 1960), body weight (BW), snout–vent length (SVL), and hind limb length (HLL). Points show the mean for each measurement (n = 12/treatment/time) at the designated day post-hatch, and vertical lines are the SEM. The horizontal dotted line in the top graph designates metamorphic climax (Gosner Stage 42; forelimb emergence, equivalent to* Xenopus laevis *stage 62). Adapted from Denver (1998) with permission from Academic Press.*

Because of these changes, the tadpoles are smaller when they metamorphose compared with those raised in unchanging, high water levels. It is thought that such plasticity is adaptive for those amphibians living in unpredictable environments.

The mechanisms governing the environmental regulation of metamorphosis are largely unknown. Interestingly, by using different regimes to reduce the aqueous space available to the tadpoles of *Scaphiopus hammondii*, Denver et al. (1998) found that the developmental acceleration due to water level reduction appeared to be related to the reduced swimming volume and perhaps the proximity to the water surface but not to the concentration of compounds in the water or interactions among the conspecifics. One potential mediator of such environmental change may be stress neuropeptide corticotropin-releasing hormone (Denver, 1997a, b), which can directly stimulate the release of thyrotropin (thyroid-stimulating hormone or TSH) by the tadpole pituitary (Denver, 1998; Denver and Licht, 1989; Malagon et al., 1989, 1991). Thus the effect of water level reduction is to increase the levels of thyroid stimulating hormone through corticotropin-releasing hormone. The TSH in turn acts on the thyroid gland to increase the synthesis and release of TH, consequently activating/stimulating metamorphosis. Therefore, at least some environmental factors can influence metamorphosis through the neurodocrine pathway.

Another factor that is likely to function through a neuroendocrine pathway is light (Dodd and Dodd, 1976; Wright et al., 1990). For example, it has been reported that tadpoles of *R. esculenta* and *R. temporaria* do not metamorphose in the dark and continued illumination stimulates metamorphosis of *R. temporaria* and *Alytes obstetricans* tadpoles. Such effects of light may be mediated through melatonin to influence thyroid activity (Wright et al., 1996b, 1997).

Although there are also other environmental factors that can influence tadpole development and metamorphosis, only very limited studies on selected species have been carried out. The neuroendocrine axis undoubtedly plays an important role in mediating the effects by many of these factors. Currently, it is unclear which aspects of the axis are affected by most of these factors. However, the role of the neuroendocrine system in metamorphosis is discussed in more detail in Chapter 2.

1.7 METAMORPHOSIS IS A UNIQUE MODEL FOR POSTEMBRYONIC VERTEBRATE ORGAN DEVELOPMENT

Embryogenesis has for a long time been a favorite subject of study by developmental biologists. This has led to impressive progress in identifying key factors and genes critical for different embryonic processes, such as germ layer specification and pattern formation, especially in recent years. Compared with embryonic development, postembryonic processes are poorly understood. This lack of understanding is in part due to the lack of appropriate systems for separating maternal and zygotic signals, as is the case in mammals.

Amphibian metamorphosis is developmentally equivalent to postembryonic organogenesis in mammals (Tata, 1993). Both processes involve changes of larval/fetal organs to their adult forms, even though many of the changes in amphibians are much more dramatic and may include the destruction of existing, fully functional tadpole organs, followed by the development of adult organs. Such similarities are particularly well reflected at the biochemical and molecular levels. For example, the switch from fetal to adult hemoglobin in red blood cells during late mammalian development also occurs during TH-dependent amphibian metamorphosis. Similarly, skin keratinization and urea cycle enzyme induction bear strong resemblance in the two systems. In addition, thyroid hormone is not only essential for brain development during amphibian metamorphosis but also critical for the neural development in postembryonic mammals (Kandel and Schwarts, 1985; Tata, 1991; McEwen et al., 1991; also see Chapter 12).

These similarities argue that amphibian metamorphosis can serve as a model for vertebrate postembryonic development. However, unlike mammals, amphibian tadpoles undergo metamorphosis as free-living animals without any maternal input. Furthermore, as is reviewed in more detail below, this process can be easily manipulated by controlling the levels of thyroid hormone, making it very user-friendly for developmental studies. The recent cloning of the nuclear receptors for TH, that is, the thyroid hormone receptors (TRs), and the demonstration that TRs are transcription factors have further stimulated the research on the molecular mechanisms, especially in terms of gene regulation (Tata, 1993; Atkinson, 1994; Shi, 1994). A major objective of this book is to cover recent advances in this area by focusing on anuran metamorphosis.

2

Hormonal Regulation

2.1 INTRODUCTION

Nearly nine decades ago, J. F. Gudernatsch (1912) found that a substance(s) in the thyroid gland could induce metamorphosis. Shortly after, E. C. Kendall (1915, 1919) showed that the active ingredient in the thyroid gland is thyroid hormone (TH). These studies led to the subsequent isolation and structural characterization of two natural thyroid hormones and demonstration that TH is the causative agent of amphibian metamorphosis.

Like any other developmental processes, metamorphosis is under the influence of other hormones. Some of the hormones from the pituitary and adrenal glands can either accelerate or inhibit metamorphosis depending upon the hormones, the tissue, and the developmental stage of the animal. In addition, a complex neuroendocrine cascade appears to be involved in the regulation and fine tuning of amphibian metamorphosis.

2.2 THYDROID HORMONE

2.2.1 Evidence for the Causative Effect of Thyroid Hormone on Amphibian Metamorphosis

There are two naturally occurring thyroid hormones. These are 3,5,3',5'-tetraiodothyronine (T_4), commonly known as thyroxine, and 3,5,3'-triiodothyronine (T_3) (Fig. 2.1). Three independent lines of evidence have firmly established that TH is the causative factor governing anuran metamorphosis.

First, elevations in the circulating plasma concentrations of the thyroid hormone T_3 and T_4 correlate with metamorphosis (Fig. 2.2 for *Xenopus laevis*, Leloup and Buscaglia, 1977; White and Nicoll, 1981). In *Xenopus laevis*, there is little TH before stage 54, when tadpoles grow rapidly but exhibit little

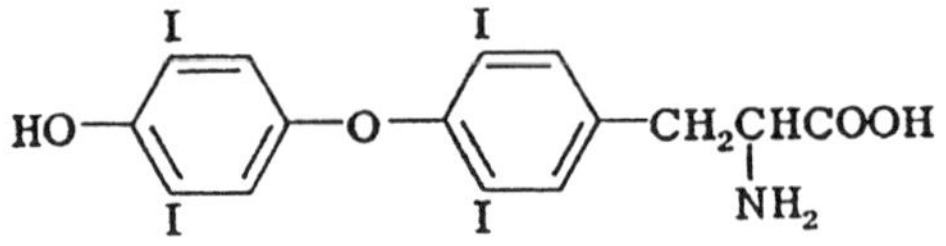

T4: thyroxine =
3, 5, 3', 5'-tetraiodothyronine

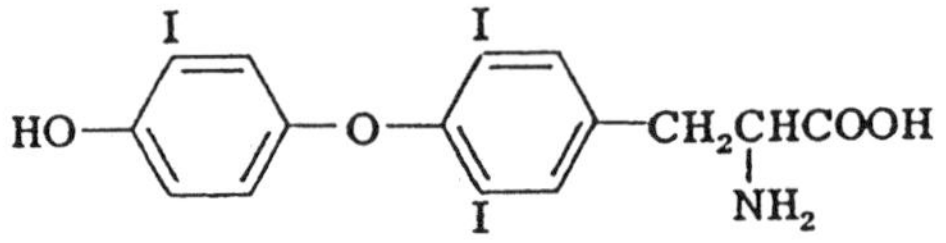

T3: 3, 5, 3'-triiodothyronine

Figure 2.1 *Structure of two natural thyroid hormones. Both T_4 and T_3 are synthesized in the thyroid gland. T_4 is the precursor for T_3 and can be converted to T_3 in many different tissues. T_3 is biologically more active.*

morphological change (Nieuwkoup and Faber, 1956). During prometamorphosis (stages 54–58), the synthesis of endogenous TH allows for the accumulation of increasing levels of T_3 and T_4 in the plasma (Fig. 2.2). Accompanying this, the tadpole undergoes both growth and morphological transformations, most noticeable of which is the development of the hind limbs. Finally, at the climax of metamorphosis (stages 58–66), the TH is at peak levels and the tadpole stops feeding and undergoes a rapid metamorphic transition. Upon the completion of metamorphosis at stage 66, the plasma TH levels are also reduced.

The second line of evidence supporting the role of thyroid hormone in metamorphosis comes from the ability of TH to induce precocious metamorphosis. The first such experiment was the landmark study by Gudernatsch (1912). It has since been reproduced for many different anuran species with pure T_3 and T_4 at concentrations comparable to endogenous plasma TH levels (Tata,1968; Dodd and Dodd, 1976; White and Nicoll, 1981). For example, the addition of T_3 to the rearing water ($5nM$) of *Xenopus laevis* tadpoles at stage 54 leads to drastic changes within 5 days (Fig. 2.3). Most notable are the external changes, including hind limb morphogenesis and cranial restructuring. In addition, interior organs also undergo transformations. The best example is the reduction of the length of the intestine (Fig. 2.4; Shi and Hayes, 1994), which mimics the up to 10-fold length reduction of the small intestine during natural metamorphosis (Marshall and Dixon, 1978; Ishizuya-Oka and Shimozawa 1987a; Shi and Hayes, 1994).

Noticeably difficult to observe in this kind of experiment is the resorption of the tadpole tail, although some shortening and other changes can occur during TH treatment of whole tadpoles (Tata, 1968). On the other hand, it is

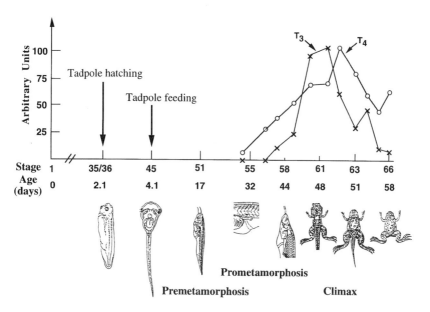

Figure 2.2 Correlation of plasma thyroid hormone concentration with Xenopus laevis metamorphosis. Development stages are based on Nieuwkoop and Faber (1956) and plasma T_3 and T_4 levels on Leloup and Buscaglia (1977). The peak concentrations in the plasma are about 10 nM and 8 nM for T_4 and T_3, respectively, at the climax of metamorphosis.

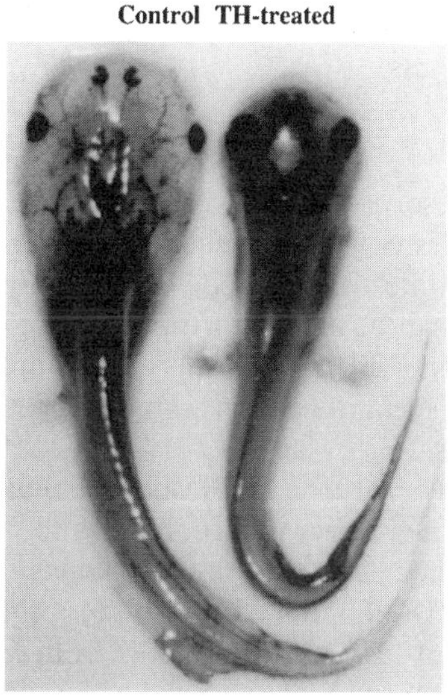

Figure 2.3 TH induces precocious metamorphosis when added to tadpole rearing water. On the left is a stage 55/56 Xenopus laevis tadpole maintained in control solution for 5 days and the one the right was treated with 10 nM T_3 with the solution changed every other day.

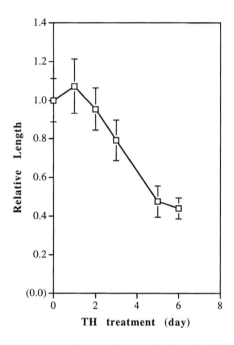

Figure 2.4 TH treatment of premetamorphic (stage 52–54) tadpoles leads to drastic reduction of the intestinal length. The tadpoles were treated for a total of 6 days with or without 5 nM T_3 and the solution was changed daily. The time of TH treatment refers the number of days when TH was present immediately prior to the end of the treatment when all tadpoles were sacrificed. The length of the entire intestine was measured and expressed as a ratio to the length of the intestine in the control tadpoles (0 day TH). (Reprinted from Shi and Hayes (1994 with permission from Academic Press).

easy to induce tail resorption in organ cultures of dissected tails from premetamorphic tadpoles (Fig. 2.5; Weber, 1967; Tata et al., 1991). Similarly, limb development and intestinal remodeling (Ishizuya-Oka and Shimozawa, 1991; Tata et al., 1991) can be induced with physiological concentrations of T_3 in vitro. Thus the effects of TH on metamorphosis are organ autonomous.

Finally, amphibian metamorphosis can be prevented by blocking the synthesis of endogenous TH. In fact, shortly after the landmark experiment of Gudernatsch (1912), Allen (1916, 1929) showed that thyroid gland removal resulted in the formation of larger than normal tadpoles that were not capable of metamorphosis. Figure 2.6a shows a *Xenopus laevis* tadpole that was congenitally athyroid (Dodd and Dodd, 1976). Although normal *Xenopus laevis* larval development lasts only 2 months from fertilization to the end of metamorphosis, the mutant tadpole retained a tadpole morphology even after 2 years. Although it was much larger than normal tadpoles at any stages, its hind limb bud morphologically resembled that of a tadpole at stage 51 (Fig. 2.6c; Nieuwkoop and Faber, 1956), which is just prior to the synthesis of

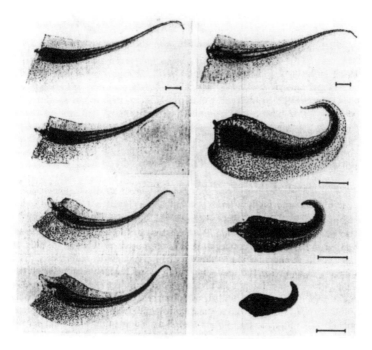

Figure 2.5 Resorption of isolated tail tips under the influence of thyroxine. Left, from top to bottom: control tips of Xenopus larvae maintained in Holtfreter solution for 6, 8, 10, and 12 days after amputation. Right, from top to bottom: tail tips of the same age as the controls but treated for 3, 5, 7, and 9 days with DL-thyroxine (1:5 million). Scale = mm. Reprinted from Weber (1967) with permission from Academic Press.

detectable levels of endogenous TH (Leloup and Buscaglia, 1977). However, such giant tadpoles can resume metamorphosis when exogenous TH is added to their rearing water.

In addition to thyroidectomy, chemical inhibitors (goitrogens) can be used to inhibit the synthesis of endogenous TH (Dodd and Dodd, 1976). These inhibitors include thiocarbamide-containing compounds such as thiourea and thiouracil, sulfonamides, thiocyanate, potassium perchlorate, and methimazole (Dodd and Dodd, 1976; Brown et al., 1997). Although the mechanisms of their action vary, all inhibit the synthesis of TH by the thyroid gland and block metamorphosis. Their action can again be reversed by exogenous TH. Thus all of the above evidence together demonstrates unambiguously that TH is the causative agent of anuran metamorphosis.

2.2.2 Thyroid Gland Development

Thyroid glands develops around late embryogenesis (Dodd and Dodd, 1976; Regard, 1978). In *Xenopus laevis*, the thyroid glands develops initially as a

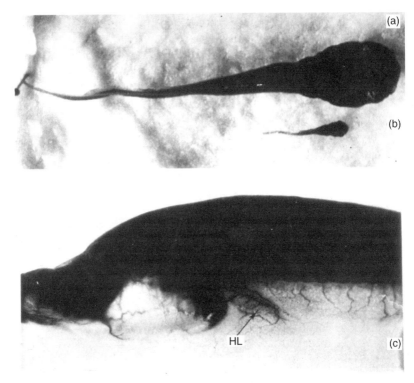

Figure 2.6 Thyroid hormone is required for metamorphosis. A 2-year-old congenitally thyroid-deficient tadpole of Xenopus laevis (a) was morphologically similar, although much larger (~ 13 cm long, 12 g) than a normal 26-day old stage 54 tadpole (b). The morphology of the hind limb (HL) of the mutant tadpole (c) resembled that of a normal 17-day-old stage 51 tadpole. Reprinted from Dodd and Dodd (1976) with permission from Academic Press.

median thickening of the pharyngeal epithelium around tadpole hatching (stages 35/36, Nieuwkoop and Faber, 1956). This rudiment then divides into two at stages 40–44. By stage 44, follicular structure is already present and follicles increase in number subsequently. A functional thyroid is developed by stage 53, with each gland containing about 20 follicles (Nieuwkoop and Faber, 1956; Saxen et al., 1957a,b). This process is quite similar to that in other anurans (Dodd and Dodd, 1976).

As the tadpole develops further, the thyroid glands increase in size owing to proliferation of follicles and an increase in their volume (Dodd and Dodd, 1976). Concurrently, TH synthesis and secretion into the plasma increases, leading to the accumulation of increasing levels of plasma TH to trigger metamorphosis. Toward the end of metamorphosis, the glands regress somewhat, which may be responsible for the reduced levels of plasma TH (Fig. 2.2).

2.3 NEUROENDOCRINE CONTROL

Although TH is the controlling agent of metamorphosis, its synthesis in the thyroid gland is under complex neuroendocrine control. Thyroid hormone in turn can influence neuroendocrine function during metamorphosis. These interactions are manifested in the hypothalamus–pituitary–thyroid axis through the actions of several hormones (Fig. 2.7).

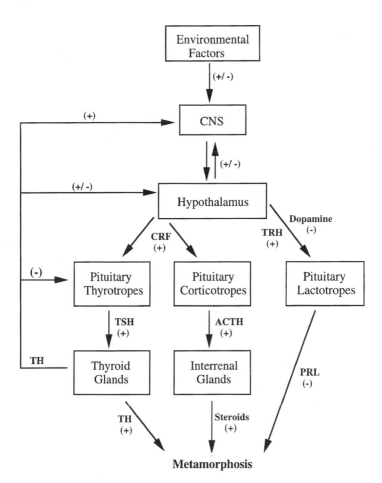

Figure 2.7 *Hormonal pathways that affect amphibian metamorphosis. A plus sign indicates a positive or stimulatory action and a minus sign indicates a negative or inhibitory action. TH and steroids (corticoids) exert positive effects on metamorphosis whereas prolactin inhibits metamorphosis. Their synthesis and secretion are under complex neuroendocrine regulation involving both positive and negative feedbacks. CNS, central nervous system; CRF, corticotropin releasing factor; TRH, thyrotropin-releasing hormone; TSH, thyrotropin; ACTH, adrenocorticotropin; TH, thyroid hormone; PRL, prolactin. Based on Denver (1996) and Kaltenbach (1996).*

2.3.1 Pituitary Regulation of Thyroid Gland Function

It has long been established that the thyroid gland is under the control of the pituitary gland or hypophysis. As early as 1914, Adler showed that the removal of the hypophysis led to a retardation in the growth of the thyroid gland and prevented it from storing colloid. Conversely, thyroid removal caused the pars distalis of the hypophysis to increase in size (reviewed in Allen, 1929). Thus the pituitary gland positively regulates the thyroid gland whereas the thyroid gland negatively feeds back to regulate the pituitary gland secretion (Fig. 2.7). The consequence of this interaction is that hypophysectomy inhibits frog metamorphosis (Allen, 1916, 1929; Smith, 1916).

The effect of the pituitary gland on the thyroid gland is mediated by the pituitary hormone TSH (thyroid-stimulating hormone or thyrotropin) from the TSH cells (thyrotropes) in the pars distalis (Dodd and Dodd, 1976; White and Nicoll, 1981; Kikuyama et al., 1993; Kaltenbach, 1996; Denver, 1996). The thyroid hormone in turn feeds back on the pituitary to reduce TSH release (Fig. 2.7).

TSH is a glycoprotein consisting of two polypeptides, the α and β subunits. The α subunit is common among several glycoprotein hormones including follicle-stimulating hormone and luteinizing hormone, whereas the β subunit is unique and provides for hormonal specificity (Pierce and Parsons, 1981; Kaltenbach, 1996; Denver, 1996). Using antibodies against mammalian TSHβ subunit, it was possible to detect amphibian TSHβ during the ontogenic differentiation of pituitary TSH cells (reviewed in Kikuyama et al., 1993). However, it has been impossible to determine the plasma levels of TSH in metamorphosing tadpoles owing to the lack of a sensitive assay.

Using a bioassay that measured the ability to stimulate the release of T_4 from cultured tadpole thyroid glands, Sakai et al. (1991) showed that purified frog TSH functioned as expected, similar to the bovine TSH, although the frog TSH was more potent. Using a different bioassay, which measured the pituitary content of TSH based on radioiodine uptake by hypophysectomized *Xenopus laevis* tadpoles, Dodd and Dodd (1976) determined TSH levels in crude pituitary extracts during development. They found that TSH was present in prometamorphic *Xenopus laevis* tadpoles (stage 56) and its level rose to high levels at early metamorphic climax (stages 59/60). There was a subsequent fall in TSH content in the pituitary at stage 61 followed by a rise to a peak value at stage 62. TSH levels dropped to below prometamorphic levels by the end of metamorphosis. Thus high levels of TSH are present during metamorphosis when it is needed to stimulate T_4 release. The drop in TSH level in the pituitary at stage 61 coincides with the peak of plasma T_4 (Leloup and Buscaglia, 1977; Fig. 2.2). Thus this drop is likely due to increased secretion of TSH from the pituitary to the plasma to accelerate TH release, so that less TSH remains in the pituitary (Dodd and Dodd, 1976). However, no systematic qualification of TSH levels in the plasma is available.

The cDNAs encoding the *Xenopus laevis* TSH subunits have been cloned (Buckbinder and Brown, 1993). In agreement with the findings that mammalian TSH could work in frogs in bioassays and that mammalian TSH antibodies could recognize frog TSH (Dodd and Dodd, 1976; Kikuyama et al., 1993), the frog TSHα and β cDNA share high degrees of homology with their mammalian counterparts (60–70% identities at the amino acid sequence level). Analysis of their mRNA levels during development revealed that the genes were activated around stages 52/53, just prior to the stage when T_4 is detectable in the plasma (stage 54, Leloup and Buscaglia, 1977; Fig. 2.2). Subsequently, the mRNA levels rise to peak levels by stages 58/59 and then drop to lower levels toward the end of metamorphosis (Fig. 2.8; Buckbinder and Brown, 1993). Thus the mRNA levels are in general agreement with the bioassay studies by Dodd and Dodd (1976). The repression of the genes after stage 59 coincides with high levels of plasma TH, consistent with the negative-feedback on the pituitary by TH (Fig. 2.7).

2.3.2 Regulation of Pituitary Secretion by the Hypothalamus

The involvement of the hypothalamus in regulating amphibian metamorphosis has long been established (for reviews see Dodd and Dodd, 1976; White and Nicoll, 1981; Kikuyama et al., 1993; Kaltenbach, 1996; Denver, 1996). These experiments involved blocking the influence of the hypothalamus on the

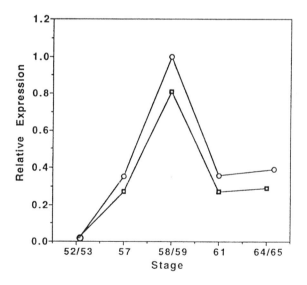

Figure 2.8 Developmental expression of TSH mRNAs in Xenopus laevis. *Total RNA from pituitaries of tadpoles at indicated stages or frogs (F) was analyzed by Northern blot hybridization with ^{32}P-labeled TSHα and TSHβ cDNA probes. Each lane represents one pituitary RNA equivalent. Autoradiograms were quantified by scanning laser densitometry. Results for TSHα (□) and TSHβ (○) were given relative to the highest value, arbitrarily defined as 1.0. Reprinted from Buckbinder and Brown (1993) with permission from the National Academy of Sciences, U.S.A.*

pituitary, thus leading to metamorphic stasis. The approaches used included hypothalectomy, transplanting the pituitary gland to the tail, and placing an impermeable barrier between the hypothalamus and the pituitary.

The stimulatory effect of the hypothalamus on metamorphosis is mainly through its ability to stimulate the synthesis and secretion of TSH from the thyrotropes of the pituitary (Fig. 2.7). In mammals, this stimulation is mediated by the tripeptide amide TRH (thyrotropin-releasing hormone, pyro-Glu-His-Pro-NH_2) (Morley, 1981; Kikuyama et al., 1993). TRH is present in amphibians (Kikuyama et al., 1993; Kaltenbach, 1996). For example, high concentrations of TRH were found in the hypothalamus and other parts of the brain as well as in the skin of the frog *Rana pipiens* (Jackson and Reichlin, 1977). In *Xenopus laevis*, TRH concentrations in the brain increase during prometamorphosis and climax (King and Miller, 1981; Bray and Sicard, 1982; Millar et al., 1983; Balls et al., 1985). Immunohistochemical studies on *Rana catesbeiana* yielded similar conclusions (Mimnagh et al., 1987; Taniguchi et al., 1990). Despite its presence during amphibian metamorphosis, the role of TRH in metamorphosis has been controversial (Dodd and Dodd, 1976; White and Nicoll, 1981; Kikuyama et al., 1993; Kaltenbach, 1996; Denver, 1996). Although experiments involving injecting TRH into adult frogs showed that TRH appears to increase TSH release (Denver, 1988, 1993; Denver and Licht, 1989), most experiments have failed to show an acceleration of metamorphosis or stimulation of TSH release in tadpoles by TRH.

Interestingly, Denver and colleagues (Denver, 1988; Denver and Licht, 1989) reported that mammalian corticotropin-releasing factor (CRF) stimulates the release of a thyroid-stimulating substance from the pituitary. CRF is a 41 amino acid polypeptide that was first isolated from ovine hypothalamus and shown to be a potent stimulator of pituitary adrenocorticotropin (ACTH) secretion in mammals (Vale et al., 1981). Mammalian CRF can also stimulate the release of ACTH from the distal lobe of frog pituitaries (Tonon et al., 1986) or dispersed frog pituitary cells (Gracia-Navarro et al., 1992). However, ACTH has no thyroid-stimulating activity (Sakai et al., 1991). Thus CRF may enhance TH release by acting through TSH. Indeed, CRF acts directly on the pituitary to stimulate the release of TSH (Denver, 1988; Denver and Licht, 1989; Jacobs and Kuhn, 1992).

Unlike TRH, mammalian CRF injected into tadpoles can raise the levels of TH and accelerate metamorphosis of *Rana perezi* tadpoles (Gancedo et al., 1992). Similar observations have been obtained for the anurans *Scaphiopus hammondii* and *Rana catesbeiana* (Denver, 1993). In addition, anti-CRF antibodies and the CRF receptor antagonist α-helical CRF (9-41) can slow down metamorphosis (Rivier et al., 1984; Denver, 1993; Denver, 1997b). Thus it is most likely that CRF functions as the larval amphibian equivalent of mammalian TRH and mediates the regulatory role of the hypothalamus on the pituitary.

Two genes encoding CRF have been cloned in *Xenopus laevis*. The deduced amino acid sequences share more than 93% similarity with mammalian CRFs

(Stenzel-Poore et al., 1992), consistent with the ability of mammalian CRFs and their antibodies to function in tadpoles. Immunohistochemical studies using heterologous antibodies or *in situ* hybridization with the *Xenopus* gene probe have demonstrated the presence of CRF-expressing cells in the hypothalamus (Verhaert et al., 1984; Olivereau et al., 1987; Gonzalez and Lederis, 1988; Carr and Norris, 1990; Stenzel-Poore et al., 1992). In addition, the expression of CRF gene in *Xenopus laevis* is TH dependent. CRF gene is upregulated by a 4-hr TH treatment of premetamorphic tadpoles, and downregulated after a 24-hr TH treatment. These results support the existence of a feedback mechanism by the thyroid gland (Denver et al., 1997b). More importantly, CRF immunoreactivity and its mRNA level increase significantly during prometamorphosis and peak at the climax (Carr and Norris, 1990; Denver et al., 1997b), consistent with the likely role of CRF as a hypophysiotropin during metamorphosis (Denver, 1996).

2.4 INFLUENCE OF STEROID HORMONES ON METAMORPHOSIS

2.4.1 Corticoids

It has long been known that corticoids can accelerate TH-induced metamorphosis (Fig. 2.7; Kaltenbach, 1968; Kikuyama et al., 1993; Hayes, 1997). Corticosterone and aldosterone are the major corticoids secreted by the interrenal gland of amphibians (Carstensen et al., 1961; Macchi and Phillips, 1966; Kikuyama et al., 1993). The concentrations of these hormones rise in synchrony with the plasma TH levels (Jaffe, 1981; Krug et al., 1983; Jolivet-Jaudet and Leloup-Hatey, 1984; Kikuyama et al., 1986, 1993; Hayes, 1997). Furthermore, corticoid binding sites are also present in the tail of developing tadpoles (Woody and Jaffe, 1984; Yamamoto and Kikuyama, 1993). These data are supportive of a role for these hormones in regulating metamorphosis.

Direct evidence for the role of corticoids in metamorphosis comes from several experiments. Exogenous corticoids accelerate tail reduction when added to tadpoles rearing water (Kaltenbach, 1958, 1985; Kikuyama et al., 1983). Corticoids also increases the activity of a hepatic enzyme (carbamoyl-phosphate synthase) (Galton, 1990). The effects of corticoids have been shown even in in vitro organ cultures of tails and for different tadpole species including *Rana japonica* and *Bufo boreas* (Kobayashi, 1958; Kikuyama et al., 1983; Hayes et al., 1993; Hayes and Wu, 1995a,b; Hayes, 1997). Among the steroid hormones tested, aldosterone, corticosterone, and deoxycorticosterone, which are from the interrenal gland, are very effective in potentiating the action of TH (Kikuyama et al., 1993). However, the gonadal steroids estradiol and progesterone are not (Kikuyama et al., 1993; also see below).

Interestingly, exogenous corticoids have dual effects on anuran metamorphosis (Hayes, 1997). During TH-induced metamorphosis, Wright et al. (1994) found that exogenous corticoids inhibit limb development and antagonize the effects of low doses of exogenous TH, but synergize with higher doses of TH. This seems to agree with the observation that corticoids inhibit development when added to tadpoles at early stages (TH levels are low) but accelerate development when applied to older tadpoles (TH levels are high) (Hayes et al., 1993; Hayes and Wu, 1995b; Hayes, 1997). The exact mechanisms underlying such opposite effects of corticoids are unknown. However, corticoids can directly act on metamorphosing tissues, for example, in the in vitro cultures of the tail, epidermal, or red blood cells (Galton, 1990; Nishikawa et al., 1992; Shimizu-Nishikawa and Miller, 1992; Schneider and Galton, 1995; Tata,1997). In addition, it has been speculated that corticoids have a negative feedback on the pituitary and the hypothalamus (Denver and Licht, 1989; Gancedo et al., 1992; Denver, 1993; Hayes, 1997). Thus, depending on the tissues and stages of the tadpoles, the effect of corticoids may vary (Kikuyama et al., 1993; Hayes, 1997).

Complementary to the above experiments with exogenously added corticoids, blocking the synthesis of endogenous corticoids can inhibit metamorphosis. Kikuyama et al. (1982) blocked the synthesis of endogenous TH by keeping tadpoles in thiourea and induced metamorphosis with exogenous T_4. To test the role of endogenous corticoids, they added Amphenone B to inhibit corticoid synthesis and found that the inhibitor drastically retarded T_4-induced tail resorption. Addition of exogenous corticoids was able to prevent this retardation, indicating the effects were due to the inhibition of the corticoid synthesis.

The studies of the effects of corticoids on metamorphosis are mostly based on the shrinkage of the tail. However, similar synergies between corticoids and TH have been observed on other organs/cells such as the limb and erythrocytes (Galton, 1990; Kikuyama et al., 1993; Hayes, 1997). At the molecular level, Shimizu-Nishikawa and Miller (1992) showed that corticoids can synergize with TH to induce the expression of adult type keratin gene expression in the epidermis of *Xenopus laevis*. However, the molecular mechanism governing such synergies remains to be defined. Corticoids presumably function through their nuclear receptors, the glucocorticoid receptor (GR). The GR belongs to the same superfamily of nuclear receptors that includes TH receptors (Evans,1988; Green and Chambon, 1988; Mangelsdorf et al., 1995; also see Chapter 5). Thus the effects of corticoids on different tissues/organs are likely mediated through transcriptional regulation by the GR (see Chapter 5 for more details on transcriptional regulation by TRs; also, see Tsai and O'Malley, 1994 for a review on GR action mechanisms).

The synthesis and secretion of endogenous corticoids are under the direct or indirect control of TH, adrenocorticotropin (ACTH), and CRF (Fig. 2.7). As in mammals, ACTH is believed to stimulate the synthesis and release of corticoids from the interrenal glands (Kikuyama et al., 1993). Hypophysectomy in *Xenopus laevis* results in degenerative changes in the interrenal cells and loss of 3β-hydroxysteroid dehydrogenase, an enzyme involved in the biosynthesis of corticoids (Dodd and Dodd, 1976). On the other hand, ACTH injections into

Xenopus laevis produce the same changes in lipid, carbohydrate, and protein metabolism as do injections of corticoids, supporting a positive regulatory role of ACTH on corticoid synthesis and secretion (Dodd and Dodd, 1976).

ACTH is in turn under the control of CRF from the hypothalamus (Fig. 2.7). As described above, immunohistochemical staining suggests that CRF levels increase at metamorphic climax; thus CRF is available during metamorphosis to stimulate ACTH synthesis and secretion. Furthermore, bovine CRF stimulates ACTH secretion from frog pituitary (Tonon et al., 1986). Thus CRF appears to have dual functions, stimulating the release of both TSH and ACTH from two different regions of the pituitary, that is, thyrotropes and corticotropes, respectively (Denver and Licht, 1989).

2.4.2 Gonadal Steroids

Unlike corticoids, the roles of gonadal steroids on metamorphosis are less clear. Although there is one report showing that estrone enhances T_4 action in *Bufo bufo bufo* (Frieden and Naile, 1955), gonadal steroids in general appear to inhibit metamorphosis. As early as the 1940s, it was shown that testosterone antagonized the effects of T_4 in *Rana temporaria* (Roth, 1941). Similarly, estradiol can inhibit the action of T_4 in *Rana temporaria* (Roth, 1948). Subsequently, Richards and Nace (1978) demonstrated that sex steroids inhibit larval development in *Rana pipiens*. A systematic study in *Xenopus laevis* by Gray and Janssens (1990) showed that testosterone and estradiol inhibit-metamorphosis (based on head length, body weight, etc.) in vivo but not tail resorption in vitro in organ cultures. These results suggest that sex steroids may not act directly on the metamorphosing organ. Instead, they may function by influencing the hypothalamic–pituitary–thyroid axis ultimately to downregulate the levels of circulating TH levels or upregulate metamorphic inhibitors such as prolactin (see below) (Gray and Janssens, 1990; Hayes, 1997). Direct evidence for such possibilities is lacking in amphibians, although these or similar mechanisms are possible based on studies in other animal species (Hayes, 1997).

2.5 INHIBITION OF ANURAN METAMORPHOSIS BY PROLACTIN

2.5.1 Dual Functions of Prolactin During Anuran Development

Prolactin (PRL) was first known as a growth hormone for amphibian larvae and later found to have an antimetamorphic activity (Fig. 2.7; Etkin and Lehrer, 1960; Dodd and Dodd, 1976; White and Nicoll, 1981; Kikuyama et al., 1993; Kaltenbach, 1996; Denver, 1996). As early as 1964; Berman et al. demonstrated that mammalian prolactin had growth-promoting effects in the larvae of *Rana catesbeiana*. This effect was confirmed by others in different anuran species (Dodd and Dodd, 1976).

Interestingly, these earlier studies found that mammalian prolactin could also inhibit normal metamorphosis when given to prometamorphic tadpoles (Nicoll et al., 1965; Dodd and Dodd, 1976; also the references above). In addition, organ cultures of the tail and limb can be induced by TH to resorb and develop, respectively, and this TH-induced metamorphosis can be blocked by adding prolactin to the medium (Fig. 2.9; Dodd and Dodd, 1976; Tata et al., 1991; Iwamuro and Tata, 1995). In support of this inhibitory effect of PRL, injecting antibodies against prolactin can accelerate spontaneous metamorphosis (Eddy and Lipner, 1975; Clemons and Nicoll, 1977b).

Most of these studies used mammalian prolactins owing to the lack of amphibian prolactins. However, the conclusions are the same with amphibian prolactins. First, Yoshizato et al. (1972) demonstrated the presence of growth-promoting and metamorphosis-inhibiting activity in the bullfrog anterior pituitary gland by injecting its homogenate into tadpoles. In addition, Nicoll and Nichols (1971) reported the presence of prolactin-like hormone in the amphibian pituitary gland. Finally, using prolactin partially purified from adult and larval bullfrog pituitary, Kikuyama et al. (1980) demonstrated that amphibian prolactin could similarly inhibit T_4-induced shrinkage of tail culture in vitro.

2.5.2 Regulation of Prolactin Levels During Development

Anuran prolactin was first purified and cloned from bullfrog and found to be highly homologous to its mammalian counterpart (Yamamoto and Kikuyama, 1981; Yasuda et al., 1991; Takahashi et al., 1990). Later the *Xenopus laevis* prolactin gene was cloned and again found to be conserved at both cDNA and amino acid sequence levels, sharing more than 60% identical amino acid sequences with mammalian prolactin (Buckbinder and Brown, 1993).

The availability of anuran prolactin and its cDNAs has allowed the analysis of prolactin expression during metamorphosis. Immunohistochemical analyses indicate that prolactin is produced in the distal lobe of the pituitary gland (Doerr-Schott, 1980; Yamamoto et al., 1986; Tanaka et al., 1991). Once synthesized, the prolactin is secreted and transported to target tissues through the plasma. Using radioimmunoassays, several groups determined the circulating prolactin levels (Clemons and Nicoll, 1977a; Yamamoto and Kikuyama, 1982; Yamamoto et al., 1986). Contrary to the expectation that as an inhibitor of metamorphosis, prolactin levels should be low during metamorphosis but high in premetamorphic tadpoles, the plasma prolactin levels were found to be relatively low during pre- and prometamorphosis but to rise to peak levels late in climax. These observation were later confirmed at the prolactin mRNA levels after the cloning of the bullfrog and *Xenopus laevis* prolactin cDNAs (Fig. 2.10; Takahashi et al., 1990; Buckbinder and Brown, 1993).

Prolactin secretion is under both inhibitory and stimulatory regulation by the hypothalamus (Fig. 2.7; Kikuyema et al., 1993; Kaltenbach, 1996). Although TRH is not a stimulatory hormone for TSH as in mammals (see

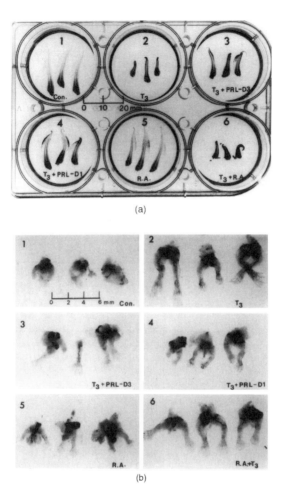

Figure 2.9 (A) Effects of retinoic acid (R.A.) and ovine prolactin (PRL) on T_3 induced resorption of Xenopus tadpole tails in organ culture. Batches of 12 stage 54/55 tadpole tails were cultured in multiwell dishes for 24 hr before T_3 or RA was added and three tail explants, representative of 12, were photographed 7 days later. Additions: well 1, control; well 2, 2×10^{-9} M T_3; well 3, 0.2 U PRL added 3 days after T_3; well 4, PRL added 1 day after T_3; well 5, 10^{-7} M RA; well 6, T_3 and RA added together 1 day after setting up cultures. (B) Inhibition by PRL of T_3-induced development of Xenopus tadpole hind limb buds in organ culture. Twelve pairs of limb buds were excised from the same stage 54/55 tadpoles whose tails were removed for culture in A. Three pairs of limb buds or limbs, representing the full range of variation within each group of 12, were photographed at the end of the culture period with different regimes of treatment. The dose and duration of treatment with T_3, PRL, and RA, as shown in panels 1–6, are the same as for the tails in wells 1–6 in (A). The dark patches are due to concentration of pigment. The single limb in panel 3 was separated during preparation for photography at the end of the culture period. Note that PRL inhibited T_3 action in both tail resorption and limb development. However, RA had no effect even though RA functions similarly as T_3 by regulating gene transcription through its nuclear receptors belonging to the same family as T_3 receptors (see Chapter 5). Thus the effect of PRL is specific. Reprinted from Tata et al. (1991) with permission from Academic Press.

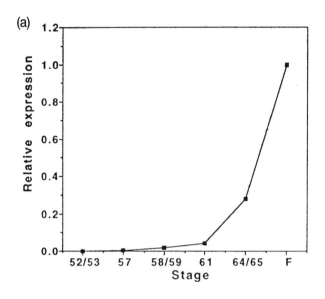
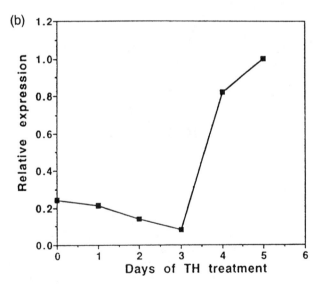

Figure 2.10 Developmental and T_3-induced expression of PRL mRNA. (A) Pituitaries were isolated from animals of the stages indicated (F, frog). Total RNA was prepared for RNA blot analysis, and each lane represents one pituitary RNA equivalent. X. laevis ^{32}P PRL cDNA was the probe. Multiple exposures of the autoradiograms were quantitated by scanning laser densitometry. Results are given relative to the highest value, arbitrarily defined as 1.0. (B) Effect of TH of PRL expressin in premetamorphic tadpoles. Stages -52 to -53 tadpoles were chosen to examine the response of pituitary PRL to exogenous TH treatment (5 nM T_3). Pituitaries were collected each day for 5 days. Total RNA was prepared for RNA blot analysis as described in (A). Reprinted from Buckbinder and Brown (1993) with permission from the National Academy of Sciences, U.S.A.

Section 2.3.2), it is a major prolactin-releasing neuropeptide. On the other hand, dopamine serves as a major inhibitor of prolactin release. The exact mechanisms underlying their effects remain to be investigated. More recently, a novel mammalian prolactin-releasing peptide was identified and cloned from bovine hypothalamus based on its ability to bind to a pituitary receptor with seven transmembrane domains and to stimulate prolactin release from pituitary (Hinuma et al., 1998). Its homolog in amphibians, if any, remains to be demonstrated.

None of the above hypothalamic factors appear likely to affect prolactin gene expression during metamorphosis. The upregulation of the prolactin gene expression appears to be controlled by TH. Blocking the synthesis of the endogenous TH in *Xenopus laevis* with methimazole prevents the prolactin gene expression whereas treatment of premetamorphic tadpoles of *Xenopus laevis* (stages 52/53) with exogenous T_3 leads to precocious upregulation of the PRL mRNA levels (Fig. 2.10, Buckbinder and Brown, 1993). However, this regulation by TH may be indirect, for it requires more than 3 days of continuous TH treatment. Regardless of the mechanism, this TH-dependent regulation may be responsible for the upregulation of pituitary PRL mRNA levels, which in turn leads to higher levels of PRL synthesis and secretion to give rise to the high levels of plasma PRL at metamorphic climax.

2.5.3 Function and Mechanism of PRL Action During Metamorphosis

It is clear that prolactin can antagonize the effects of TH. Thus prolactin was initially considered a "juvenile" hormone by analogy to the juvenile hormone in insects, which inhibits metamorphosis (Riddiford, 1996; Chapter 11). On the other hand, the expression profile of prolactin both at the mRNA and protein levels argues against its role as a classical juvenile hormone. The upregulation of PRL during both natural and TH-induced metamorphosis argues instead for an alternative function for PRL in metamorphosis. One possible role for prolactin is to counteract high concentrations of TH at the climax of metamorphosis to coordinate sequential transformations of different organs/tissues. Such a role may be important for the tadpole–frog transition because different tissues/organs may require different levels of TH action for their metamorphic transformations. For example, the hind limb, tail, and intestine metamorphose at very different stages when the plasma TH levels are quite different (Fig. 2.2; Niewkoop and Faber, 1956; Leloup and Buscaglia, 1977; for more detail, see Chapter 8).

For the above alternative model to be valid, PRL has to act at levels of individual tissues/organs and influence TH action early in the TH signal transduction cascade. As indicated above, PRL can block TH action in organ cultures (Fig. 2.9). Furthermore, by analyzing the expression of thyroid hormone receptor TRα and TRβ genes in *Xenopus laevis* tail organ cultures,

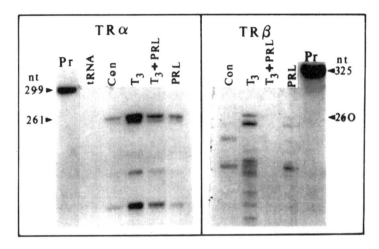

Figure 2.11 Inhibition by PRL of autoinduction of TRα and β mRNAs in organ cultures of stage 54 Xenopus tadpole tails. The tails were treated with control solution (Con), 2 nM T_3 (T_3), or 0.2 IU/ml ovine PRL along with T_3. After 5 days of treatment, total RNA was isolated and TRα and β transcripts were determined in total tail RNA by RNase protection assay. The probe (Pr) was 299 or 325 nucleotides (nt) long for TRα or TRβ, respectively, and the expected protected band was 261 or 260 nt, respectively. The lane labeled tRNA was control RNase protection reaction with tRNA instead of tail RNA. Reprinted from Tata (1997) with permission from Plenum Press.

Tata and colleagues demonstrated that PRL can prevent the autoinduction of the TRβ genes by TH (Fig. 2.11; Baker and Tata, 1992; Tata, 1997). Because the activation of at least the TRβ genes is a direct or immediate early response to TH (Kanamori and Brown, 1992; Ranjan et al., 1994; Machuca et al., 1995; Wong et al., 1998b), this observation indicates that PRL directly blocks the action of TH.

TH functions by regulating gene transcription through its nuclear receptors or TRs, which act as heterodimers with the 9-*cis*-retinoic acid receptors (RXRs) (Fig. 2.12; for more details see Chapter 5). On the other hand, PRL acts through its cell surface receptors, the prolactin receptors (PRL-Rs) (Fig. 2.12). Although PRL-Rs have not been reported for amphibians, at least two PRL-R mRNA forms have been characterized in mammals and birds (Boutin et al., 1988; Edery et al., 1989; Scott et al., 1992; Shirota et al., 1990; Ali et al., 1991; Tanaka et al., 1992; Clarke and Linzer, 1993; Moore and Oka, 1993; Bignon et al., 1997; Schuler et al., 1997). These receptors are transmembrane proteins lacking intrinsic protein kinase activity, in contrast to many other membrane receptors such as growth factor receptors (Boutin et al., 1988; Kishimoto et al., 1994). Instead, PRL-R can regulate the activities of the Janus kinases (JAKs Fig. 2.12; Darnell et al., 1994; Darnell, 1997).

PRL-Rs function through JAK2 (Fig. 2.12; Campbell et al., 1994; David et al., 1994; Lebrun et al., 1994; Rui et al., 1994; Han et al., 1997). Prolactin

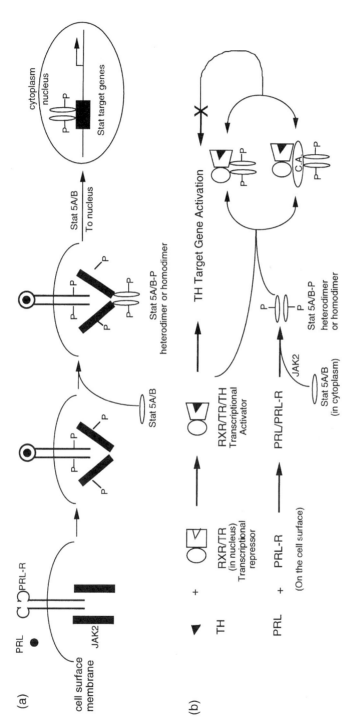

Figure 2.12 Proposed models for prolactin action. (A) Prolactin (PRL) function through a cell surface receptor (PRL-R). The binding of PRL to PRL-R leads to autophosphorylation of PRL-R and JAK2 kinase, which in turn recruits and phosphorylates Stat 5A/B homodimer or heterodimer. The phosphorylated Stat 5A/B then migrates into the nucleus, where they can bind to their recognition sequences to regulate target gene expression or interact with other transcription factors such as GR and possibly TR (see text and below). (B) A potential mechanism for PRL to inhibit T_3 action during metamorphosis. T_3 function by binding to its receptor TR in the heterodimeric complex RXR/TR. The resulting complex then regulate gene transcription (see Chapter 5 for more details). PRL binds to PRL-R and the ligand/receptor complex then phosphorylates Stat 5A/B through JAK2 kinase. The phosphorylated Stat 5A/B subsequently migrates into the nucleus, where it may bind to RXR/TR/TH complex either directly, or indirectly through TR-coactivators (C.A.) or other factors to inhibit TR function.

binding to the receptors leads to rapid but transient tyrosine phosphorylation of several intracellular proteins, including JAK2 and the transcription factors Stats (signal transducer and activators of transcription) (Campbell, et al., 1994; David et al., 1994; Dusanter-Fourt et al., 1994; Gilmour and Reich, 1994; Gouilleux et al., 1994; Lebrun et al., 1994; Rui et al., 1994; Wakao et al., 1994; Han et al., 1997). The phosphorylated Stats then migrate into the nucleus to regulate gene transcription (Fu, 1992; Darnell et al., 1994; Fu, 1995; Darnell, 1997).

There are multiple members of the JAK and Stat families that are involved in regulating a variety of cellular processes including growth induction by growth factors and interferons and cell death activation (Darnell et al., 1994; Fu, 1995; Chin et al., 1997; Darnell, 1997). Prolactin appears to act through the JAK2–Stat5 pathway to regulate gene transcription (Fig. 2.12; Wakao et al., 1994; Han 1997). It is possible that one or more target genes of PRL may inhibit the function of TH–TR complex, thus blocking metamorphosis. However, a more direct mechanism can be envisioned based on the recent finding of the direct interaction between Stats and the glucocorticoid receptor (GR) (Stoecklin et al., 1996; 1997; Zhang et al., 1997). GR and Stat3 or Stat5 can form a complex that synergizes the effects of glucocorticoid and interleukin-6 or prolactin, respectively, in gene transcription. Furthermore, this synergism can occur through DNA element recognized by either GR or Stat.

GR and TR belong to the same superfamily of nuclear receptor transcription factors. They have nearly identical structural organization, especially in the DNA and hormone binding domains (see Chapter 5 for more details). Thus it is possible that TR may interact with Stat5 or a related Stat that is regulated by PRL. Such an interaction could possibly lead to the inhibition of TR function, thereby blocking the effect of TH during amphibian metamorphosis (Fig. 2.12). Such an effect is, however, different from the synergistic effect on gene expression by the GR–Stat interaction. On the other hand, the mechanism is consistent with the ability of PRL to block one of the earliest events induced by TH, that is, the activation of TRβ genes (Fig. 2.11), which is independent of new protein synthesis (Kanamori and Brown, 1992). According to this model, the effects of PRL on TH action vary depending upon the tissue/cell types because the receptor levels (and JAK/Stat levels in the case of PRL) likely differ in different cell types. Such variations may provide a mechanism to coordinate systematic transformations of different tissues during metamorphosis. The test of such a model will require the cloning of amphibian PRL-R, JAKs, and Stats, and studying their expression, function, and effects on transcriptional regulation by TRs.

2.6 OTHER HORMONES

There are a number of other hormones that may influence anuran metamorphosis (Dodd and Dodd,1976; Kaltenbach, 1996; Denver, 1996). For example, melatonin from the pineal gland appears to retard metamorphosis. Similarly, somatostatin is known to inhibit TSH secretion and thus may retard metamorphosis. On the other hand, gonadotropin-releasing hormone can elevate circulating TH levels in axolotls and frogs. Thus it may have a positive role in metamorphosis. Unfortunately, the studies on these and other hormones are fragmentary or inconclusive. Their roles during metamorphosis, if any, remain to be established.

3

Morphological Changes During Anuran Metamorphosis

3.1 INTRODUCTION

Metamorphosis in anurans involves the most comprehensive and most dramatic transformations of all major living chordate groups. This postembryonic process systematically transforms most, if not all, organs of a tadpole to their adult forms. In addition, it also brings about the development of organs that only function in the adult frog. There are three major types of changes that take place during metamorphosis. The first is the complete resorption of tadpole-specific organs such as the tail (Fig. 3.1). On the other extreme, frog-specific organs like the hind limb develop de novo from undifferentiated stem cells in a process that involves first the proliferation of stem cells and subsequent cell differentiation and tissue morphogenesis. The last major type of transformation is the partial but profound remodeling of the existing organs like the liver and intestine into their adult forms. Not only do different organs undergo different changes, but they also occur at distinct developmental stages to coordinate the effective transition of a tadpole to a frog. Furthermore, even within a single organ, different tissues undergo temporally regulated, specific transformations. There are many excellent reviews on the morphological changes of different organs and tissues during anuran metamorphosis (Dodd and Dodd, 1976; Hourdry and Dauca, 1977; Gilbert and Frieden, 1981; Fox, 1983; Balls and Bownes, 1985; Yoshizato, 1989). Thus only a few organs/tissues are described here to illustrate the contrasting changes during metamorphosis.

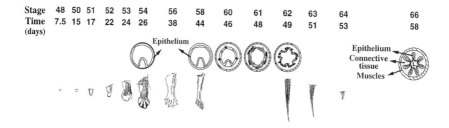

Figure 3.1 *Stage-dependent transformation of the hind limb, intestine, and tail of* Xenopous laevis *tadpoles. The developmental stages and ages of tadpoles are from Nieuwkoop and Faber (1956). The tails at stages 62–66 are drawn to the same scale to show the resorption (no tail is left by stage 66), but the tadpoles' intestinal cross-sections and the hind limbs at different stages are not drawn to the same scales in order to show the morphological diffences at different stages. Tadpole small intestine has a single epithelial fold, the typhlosole, where connective tissue (C) is abundant, whereas a frog has multiply folded intestinal epithelium (E), with elaborate connective tissue and muscle (M). Dots, proliferating adult intestinal epithelial cells; open circles, apoptotic primary intestinal epithelial cells; L, intestinal lumen.*

3.2 RESORPTION OF TADPOLE-SPECIFIC ORGANS

By definition, tadpole-specific organs are removed during metamorphosis. Two tadpole organs degenerate completely during this transition. These are the tail and gills. Of the two, tail resorption has been extensively studied but gill degeneration has received relatively little attention (Dodd and Dodd, 1976; Atkinson, 1981; Yoshizato, 1989).

3.2.1 The Tail

The tail represents the largest organ of the tadpole and undergoes the most dramatic changes during metamorphosis. Structurally, the tail consists of several major tissues. These include the dorsal and ventral tail fins, epidermis, connective tissue, blood vessels, notochord, and muscles. The connective tissue surrounds the centrally located notochord and nerve cord and also supports the structure of the tail fins, but skeletal muscles occupy the largest volume in the tail. Despite such diversity in tissue types, all tail tissues are resorbed by the end of metamorphosis.

Generally, tail resorption occurs during late metamorphosis (Nieuwkoop and Faber, 1956; Dodd and Dodd, 1976). The completion of tail resorption also marks the end of metamorphosis in anurans (Fig. 3.1). However, some structural changes can be detected around stage 60 in *Xenopus laevis*, shortly after the onset of metamorphic climax (stage 58) (Weber, 1964) but before any measurable reduction in tail length (after stage 62; Fig. 3.2; Nieuwkoop and Faber, 1956). These early changes include the loss of cross-striations of the myofibrilla and disintegration of the cristae in the mitochondria (Weber, 1964;

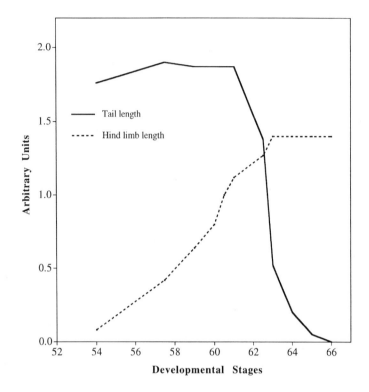

Figure 3.2 *Stage-dependent resorption of the tadpole tail and growth of the hind limb. The lengths of the tail and hind limb are plotted as the ratio of the length of the respective organ to that of the tadpole body. Based on Atkinson (1981) and normalized to* Xenopus *stages according to Nieuwkoop and Faber (1956) (also see Table 1.1).*

Dodd and Dodd, 1976). On the other hand, the most noticeable early change in the tail is tail fin degeneration, which occurs around stage 60 in *Xenopus laevis* (Nieuwkoop and Faber, 1956).

Massive tail resorption begins around stage 62 in *Xenopus laevis* when its length undergoes rapid reduction (Fig. 3.2; Nieuwkoop and Faber, 1956). Within 10 days, the tail is completely resorbed. This takes place in conjunction with the changes of the shapes of the body and particularly the head. Tail resorption in other anuran species such as *Rana catesbeiana* (Atkinson, 1981) occurs during similar stages of development. Such a rapid resorption appears to be facilitated by the concurrent changes throughout the entire tail, although the rate of resorption (length reduction) is faster (about two-fold) at the distal end (tail tip) compared to that at the proximal end (near the body) (Dmytrenko and Kirby, 1981).

Two different processes appear to contribute to tail resorption (Yoshizato, 1989). These are condensation and histolysis. Condensation is an important factor contributing to the length reduction and is caused by water loss, which results in the compaction of the cells and the extracellular matrix (ECM) (Frieden, 1961; Lapiere and Gross, 1963; Yoshizato, 1989, 1996). Although the mechanism of condensation is unclear, extensive studies have been carried out on histolysis. The tail cells undergo programmed cell death or apoptosis with distinct sequential morphological changes whereas the ECM are degraded by various enzymes (see Chapter 4).

Histolysis in the tail has been best studied in the skin and fin. Electron microscopic examination has shown that the earliest changes in the tail epidermis are the formation of vacuoles with acid phosphatase activity and the breakdown of the membranes in the outermost cells (Kerr et al., 1974; Kinoshita et al., 1985; Yoshizato, 1989). These changes then propagate toward the inner layers. Following these changes the epidermal cells undergo nuclear and cellular fragmentation, resulting in the formation of membrane-bound vesicles (apoptotic bodies) containing cellular and/or nuclear fragments, including condensed chromatin fragments. These cellular changes are typical of programmed cell death or apoptosis in vertebrates (Kerr et al., 1974; see Chapter 4).

Underneath the epidermis is the dermis (mesenchyma), which contains the basement lamella made of collagen fibers (Yoshizato, 1989). During tail resorption, the basement lamella first swells when its collagen fibers are fragmented and frayed. Following this breakdown, the collagen layers are removed through phagocytosis by the underlying mesenchymal cells (Gross, 1964; Usuku and Gross, 1965; Gona, 1969). This degeneration process proceeds proximally from the tip of the tail. Accompanying this is the death of the epidermal cells, occurring inward from the outermost cells. The dead cells are subsequently sloughed and/or phagocytosed (Dodd and Dodd, 1976; Yoshizato, 1989).

As described in Chapter 2, tail resorption is under the direct control of TH. Its response to TH is organ autonomous because isolated tails cultured in vitro also undergo resorption. Furthermore, this resorption is highly specific, being unaffected by other neighboring tissues or organs. For example, a tadpole tail transplanted to the body of another tadpole undergoes simultaneous resorption as the host tail, whereas the host body continues its normal development (Fig. 3.3; Weber, 1967). Similarly, placing a tadpole eye cup onto the tail does not affect the resorption of the tail and conversely, the tail resorption process does not lead to the degeneration at the eye cup (Fig. 3.3; Weber, 1967). Thus the tadpole tail is genetically predetermined toward destruction, waiting only for sufficiently high levels of TH to trigger the process.

3.2.2 The Gills

Gills are internal in most anurans, for example, *Rana catesbeina*, but external in urodeles and caecillians (Dodd and Dodd, 1976; Fox, 1983). They are the main respiratory organs prior to the development of the frog lungs. Their

40 MORPHOLOGICAL CHANGES DURING ANURAN METAMORPHOSIS

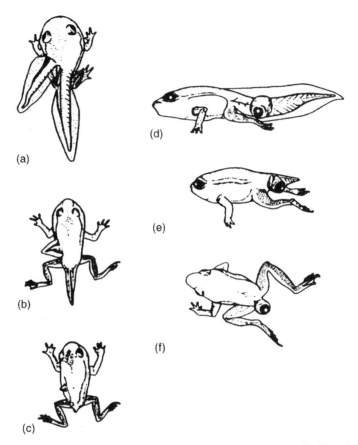

Figure 3.3 Organ specificity of metamorphic responses in tadpoles. (A–C) Tail tip transplanted to the trunk region undergoes resorption simultaneously with the host's tail. (D–F) Eye cup transplanted to the tail remains unaffected by the resorbing tail tissue. After Weber (1967) with permission from Academic Press.

function is taken over by the lungs during metamorphosis. Thus it is not surprising that total resorption of the gills (based on gill weight, Atkinson, 1981) occurs relatively late, around the same stages or slightly earlier than tail resorption.

Gill resorption begins around stages XX–XXI in *Rana catesbeiana* (equivalent to stages 61/62 in *Xenopus laevis*). As this process takes place, gill morphology changes extensively (Atkinson and Just, 1975; Atkinson, 1981). The epithelium becomes separated from the vascular network as it undergoes histolysis. The gills turn black in appearance by stage XXIII (equivalent to stage 64 in *Xenopus laevis*) and concurrently are rapidly degenerating. Their spaces are occupied by melanocytes in later stages as the gills are completely resorbed.

3.3 DE NOVO DEVELOPMENT OF FROG LIMBS

The limbs are frog-specific organs that develop during metamorphosis. Even though the limbs are made of similar tissue/cell types as the tail, including skin, muscle, connective tissue, and cartilage, they develop in a TH-dependent process whereas the tail resorbs when TH is present.

Hind limb development is one of the earliest changes during anuran metamorphosis (Fig. 3.1). In *Xenopus laevis*, the hind limb buds are first visible at stage 46, right after the onset of tadpole feeding at stage 45 (Nieuwkoop and Faber, 1956). The limb buds subsequently increase in size as the tadpoles grow. However, little morphological change in the hind limb buds takes place before stage 54 (Fig. 3.1), when the thyroid glands become functional (Dodd and Dodd, 1976). Thus hind limb bud formation and growth do not require TH. Between stages 54 and 58, the hind limb buds begin to differentiate and undergo morphogenesis to form the toes as endogenous TH levels rise (Fig. 2.2; Leloup and Buscagia, 1977). After stage 58, the hind limbs increase in size but change little morphologically (Fig. 3.2).

The requirement for TH in hind limb morphogenesis has been demonstrated by experiments that manipulate the levels of TH. Treating premetamorphic tadpoles as early as stage 45 in *Xenopus laevis* can stimulate first growth, then morphogenesis of the hind limb (Fig. 2.3; Kaltenbach, 1953; Dodd and Dodd, 1976). On the other hand, blocking the synthesis of endogenous TH through-thyroidectomy or with inhibitors can prevent the morphogenesis of the hind limb but not the development of the hind limb bud (Fig. 2.6; Allen, 1929; Dodd and Dodd, 1976).

The forelimbs develop somewhat later. In *Xenopus*, the forelimb buds are first visible at stage 48 (Nieuwkoop and Faber, 1956). Again, the initial bud development occurs in enclosed opercular or peribranchial sacs, and growth is independent of TH. Like the hind limbs, the forelimbs undergo morphogenesis after stage 53 when endogenous TH becomes available. By stage 58, the forelimbs break through the skin and forelimb morphogenesis is essentially complete. Subsequently, the forelimbs merely grow in size, similar to the hind limbs.

Forelimb development can also be induced by treating premetamorphic tadpoles with TH (Kaltenbach, 1953), and blocking the synthesis of endogenous TH inhibits forelimb development.

3.4 REMODELING OF EXISTING ORGANS FOR FROG USE

The majority of the organs are present in both tadpoles and frogs. During metamorphosis, they undergo partial but profound transformations. Several excellent reviews have described these processes (Dodd and Dodd, 1976; Gilbert and Frieden, 1981; Fox, 1983; Balls and Bownes, 1985). This Section covers only three organs representing different types of remodeling.

3.4.1 The Liver

The liver is best known for its biochemical changes associated with the transition from ammonotelism to ureotelism during metamorphosis (see Chapter 4). These biochemical changes in the liver are thought to be due to reprogramming of the gene expression profiles of existing hepatocytes, which account for the majority of the liver cells, during both natural and TH-induced metamorphosis (Atkinson et al., 1994; 1996; Chen et al., 1994). Larval hepatocytes are capable of expressing adult type genes when tadpoles are treated with TH (Chen et al., 1994). There are no dramatic morphological changes that accompany the transformation of the tadpole liver to the frog one (Atkinson et al., 1994).

On the other hand, fine structural changes have been observed in liver cells during metamorphosis (Dodd and Dodd, 1976; Fox, 1983). As summarized by Dodd and Dodd (1976), these changes include (1) the nuclei becoming more heterochromatic and more irregular with higher number of nucleoli at late stages of metamorphosis, (2) the mitochondria increasing in size and their cristae changing from a lamellar to a more tubular appearance, (3) the rough endoplasmic reticulum increasing in mass and its cisternae becoming dilated, and (4) the Golgi complexes increasing in size. Toward the end of metamorphosis, the sizes of the Golgi complexes and rough endoplasmic reticulum are reduced again, implying that at least some of the fine structural changes are associated with the increased biosynthetic activity during metamorphosis but are not required in the frog hepatocytes.

3.4.2 The Nervous System

The brain undergoes extensive reorganization during metamorphosis (Dodd and Dodd, 1976; Kollros, 1981; Fox, 1983; Gona et al., 1988; Tata, 1993). Regions associated with new organs/tissues, such as the limbs that are formed during metamorphosis, develop de novo, whereas those connected to larval organs/tissues, such as the tail, degenerate. At the gross morphological level, some of the changes include the widening and shortening of the diencephalon, and the narrowing and shortening of the fossa rhomboidalis of the medulla (Kollros, 1981). In the cerebellum, a major metamorphic event is the appearance of a thick bed of the external granular layer cells, many of which migrate past the Purkinje cells and differentiate into neurons of the definitive internal granule layer (Gona et al., 1988).

Many studies on the metamorphic transitions of the nervous system have focused on individual neuron types. A detailed review on these studies can be found in Kollros (1981) and Fox (1983). One of the most dramatic changes among the findings is the death of the Mauthner cells. These cells are a pair of giant neurons involved in tail function and degenerate as the tail is resorbed. Likewise, the Rohon–Beard neurons also die during metamorphosis, although some of them degenerate in premetamorphic stages. The Purkinje cells, on the other hand, undergo maturation such that they acquire elaborate dendritic trees (Gona et al., 1988). Several other types of neurons, including lateral motor column neurons, mesencephalic V nucleus, and dorsal root ganglia neurons, first proliferate during late premetamorphosis and early prometamorphosis. Some of those neurons then mature/differentiate while others die. Such processes are associated with the development of the adult organ/tissues. For example, the development of the lateral motor column neurons in the lumbosacral region of the nerve cord accompanies limb development (Dodd and Dodd 1976; Kollros, 1981; Fox, 1983).

Like other processes during metamorphosis, the transformation of the brain also depends upon TH. Thus thyroidectomy or hypophysectomy results in the larger size of the ventricles and thinner walls of the cerebellum and medulla oblongata compared to those in metamorphosed control animals (Allen, 1918; Kollros, 1981). On the other hand, TH treatment of premetamorphic tadpoles leads to precocious brain metamorphosis, including the establishment of the thick bed of external granule layer cells and alternations in the visual system (Kollros, 1981; Fox, 1983; Gona et al., 1988; Hoskins, 1990).

3.4.3 The Intestine

The intestine is another organ that is remodeled during metamorphosis. The morphological changes that take place during intestinal remodeling are more drastic than those in the liver and brain. The intestinal epithelium is a complex structure that provides an enormous luminal surface area for efficient food processing and absorption, the primary function of the organ (Glass, 1968; Segal and Petras, 1992). In mammals and birds, the large luminal surface of the epithelium is achieved by first forming multiple circular epithelial folds (Fig. 3.4; Glass, 1968). Then numerous fingerlike villi and crypts are formed along these folds. Finally, each villus/crypt consists of a densely packed monolayer of columnar epithelial cells, which themselves have a large number of microvilli (the brush border) to further amplify the absorptive surface. The epithelium is constantly renewed through cell proliferation exclusively in the crypts. As cells migrate up along the crypt–villus axis, they gradually differentiate. Eventually at the tip of the villus, the fully differentiated epithelial cells undergo cell death after a finite period of time and are sloughed into the lumen.

The intestine in adult amphibians resembles that in higher vertebrates (Fig. 3.4, Reeder, 1964; McAvoy and Dixon, 1977; Dauca and Hourdry, 1985; Ishizuya-Oka and Shimozawa, 1987a; Shi and Ishizuya-Oka, 1996). It has elaborate connective tissue and muscles. The epithelium also forms multiple circular folds; however, the villi and crypts are absent (the amphibian epithelial folds have been referred to as villi/crypts in some publications). Instead, the epithelial cells with numerous microvilli line the luminal surface of the folds with the proliferative cells confined toward the trough and differentiated cells toward the crest, thus generating a cell renewal system along the trough–crest axis similar to that in higher vertebrates (McAvoy and Dixon, 1977; Marshall and Dixon, 1978; Shi and Ishizuya-Oka, 1996).

44 MORPHOLOGICAL CHANGES DURING ANURAN METAMORPHOSIS

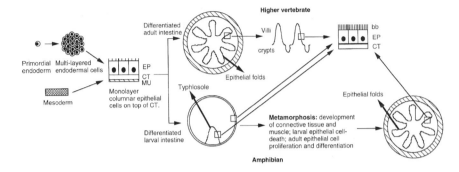

Figure 3.4 *Comparison of intestinal development in amphibians and higher vertebrates. The primordial endodermal cells first form a multilayered cell mass. The endodermal cells are then converted into a monolayer of columnar epithelial cells (EP) tightly associated with the connective tissue (CT), which is derived from the mesoderm, through a basement membrane. Further development in amphibians diverges from that in higher vertebrates. In the latter, the columnar cells develop into multiply folded epithelium surrounded by elaborate connective tissue (stippled area) and muscles (derived from the mesoderm) (hatched area, MU). In amphibians, the epithelium remains as a simple tubular structure with only a single fold, the typhlosole. The differentiated epithelial cells in both cases have numerous microvilli in the brush border (bb) on the luminar surface for efficient nutrient processing and absorption. Unlike in higher vertebrates, the amphibian intestine then undergoes a second phase of development that results in the replacement of larval epithelium with adult epithelium as well as extensive development of connective tissues and muscles.*

The tadpole intestine, on the other hand, has a much longer but simpler structure. It consists of a single layer of columnar epithelium surrounded by thin layers of muscles with little intervening connective tissue (Fig. 3.4, McAvoy and Dixon, 1977; Kordylewski, 1983; Ishizuya-Oka and Shimozawa, 1987a; Shi and Ishizuya-Oka, 1996). There is only a single epithelial fold, the typhlosole, present in the anterior part of the intestine where larval connective tissue is abundant. These structural differences between larval and adult intestines presumably reflect changes in the physiological functions between herbivorous tadpoles and carnivorous frogs.

This contrasting morphology of the intestine in tadpoles and frogs together with the relatively simple spatial organization of its different tissues, for example, the epithelium, connective tissue, and muscles, has encouraged extensive investigation of the morphological changes of the organ during metamorphosis. At the gross anatomic level, the long larval small intestine suddenly begins to shorten around the onset of metamorphic climax (stage 58 in *Xenopus laevis*), and this process continues until the end of metamorphosis (Fig. 3.5A; Marshall and Dixon, 1978; Ishizuya-Oka and Shimozawa, 1987a). The region of the small intestine containing the typhlosole remains relatively constant, about one-third of the small intestine, in tadpoles and during early metamorphic climax (Fig. 3.5B). After stage 61, the typhlosole is no longer recognizable as the morphogenesis of intestinal folds takes place (Marshall and Dixon, 1978; Ishizuya-Oka and Shimozawa, 1987a). These intestinal folds appear as several circular folds that run longitudinally and are straight along the gut axis, gradually increasing in number and height, and finally being modified into longitudinally zigzagged folds (Shi and Ishizuya-Oka, 1996). The zigzag folds then remain throughout adulthood (Fig. 3.4; McAvoy and Dixon, 1978a,b).

3.4.3.1 Epithelial Transformation The epithelial transition from larval to adult form in the amphibian intestine can be divided into two processes, degeneration of the primary epithelium and development of the adult (secondary) epithelium. The degeneration of the larval epithelium begins around the onset of metamorphic climax. For example, the microvilli composing the brush border decrease in number and height, whereas lysosomes increase in number and in hydrolytic activity (Bonneville, 1963; Bonneville and Weinstock, 1970; Hourdry and Dauca, 1977). Electron microscopic studies show that cell death of the primary epithelium in the amphibian intestine occurs by apoptosis (programmed cell death), producing membrane-encircled cellular and nuclear, chromatin-containing fragments, that is, the apoptotic bodies (Ishizuya-Oka and Shimozawa, 1992b; see Chapter 4 for more details). The apoptotic bodies are at least partially phagocytosed by macrophages localized in the degenerating primary epithelium (Ishizuya-Oka and Shimozawa, 1992b). The macrophages are eventually extruded into the lumen while still retaining the apoptotic bodies.

Just after the beginning of the primary epithelial degeneration, primordia of the secondary epithelium are detected at the epithelial–connective tissue interface as small islets consisting of undifferentiated epithelial cells (Fig. 3.6). It is still unclear whether these secondary epithelial cells are derived from a pool of undifferentiated cells in the primary epithelium (Bonneville, 1963) or transformed from differentiated primary cells (Marshall and Dixon, 1978; Shi and Ishizuya-Oka, 1996). In any case, the primordia rapidly grow into the connective tissue through active cell proliferation and differentiate to form the secondary epithelium, replacing the degenerating primary epithelium (Fig. 3.6; McAvoy and Dixon, 1977; Ishizuya-Oka and Shimozawa, 1987a). With the progression of fold formation, proliferative cells of the secondary epithelium become localized in the trough of the folds like those in the mammalian crypts (Cheng and Leblond, 1974), in contrast to the primary epithelium. As the secondary epithelial cells differentiate, they migrate from the trough to the crest of the intestinal fold, similar to the processes in mammalian and avian intestine (McAvoy and Dixon, 1977, 1978b). The differentiated adult epithelial cells have shorter microvilli in the brush border than the larval ones (Bonneville, 1963; Bonneville and Weinstock, 1970; Shi and Ishizuya-Oka, 1996). Thus, during metamorphosis, amphibian intestinal folds seem to acquire a basic structure analogous to the cell renewal system in the mammalian small intestine (Bjerknes and Cheng, 1981).

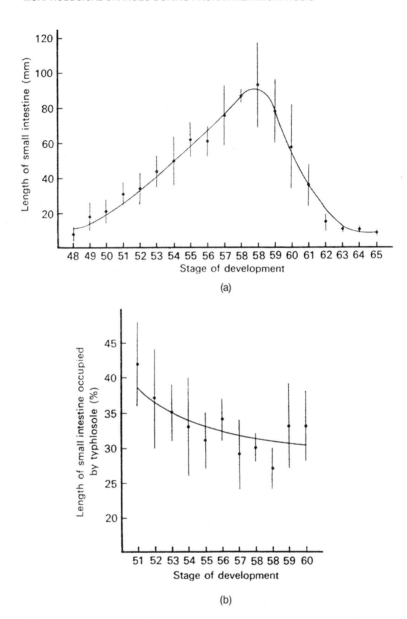

Figure 3.5 Changes in the lengths of the intestine and typholosole during Xenopus laevis development. (A) Growth of the small intestine of larval Xenopus laevis, based on measurements of the outstretched gut from bile duct to junction of small and large intestines. Each point represents the mean of at least five separate measurements. Bars indicate standard deviations. (B) Proportion of the small intestine occupied by the typhlosole. Each point represents the mean of at least five separate measurements. Bar indicate standard deviations. The typhlosole is no longer recognizable after stage 60. Reprinted from Marshall and Dixon (1978) with permission from Cambridge University Press.

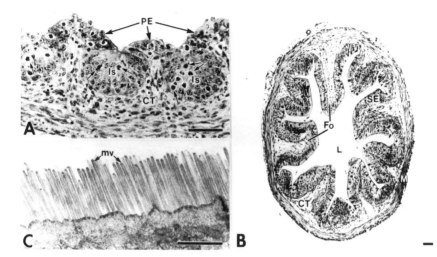

Figure 3.6 Development of the secondary epithelium (SE) or adult epithelium of the Xenopus small intestine during metamorphosis. (A) Islets (Is) growing into the connective tissue (CT) at the onset of metamorphic climax. Mitotic cells (arrows) are numerous in the islets. The connective tissue forms a thick layer. PE, primary epithelium. (B) Cross-section of the small intestine at the late stage of metamorphic climax. The secondary epithelium is differentiated and covers well-developed intestinal folds (Fo). L, lumen. M, muscle. (C) Higher magnification of the apical surface of the secondary epithelium at the end of metamorphosis. The brush border consists of microvilli (mv) shorter than those of the primary epithelium. Scale bars, 50 μm (A and B) and 1 μm (C). Reprinted from Shi and Ishizuya-Oka (1996) with permission from Academic Press.

3.4.3.2 Development of the Connective Tissue

When the epithelial transition from the larval to adult form begins, the connective tissue suddenly increases in mitotic activity, cell number, and thickness (Fig. 3.1). The connective tissue at this time consists of various types of cells such as immature mesenchymal cells, fibroblasts, macrophages, and mast cells (Ishizuya-Oka and Shimozawa, 1987b, 1992b; Yoshizato, 1989).

Accompanying these changes in the connective tissue is the profound remodeling of the interface between the connective tissue and the epithelium. When the primary epithelium begins to degenerate, the basal lamina, which is thin throughout the larval period, becomes thick in the entire region beneath the epithelium and remains thick until the primary epithelium disappears (Ishizuya-Oka and Shimozawa, 1987a,b; Shi and Ishizuya-Oka, 1996). In addition, throughout the thick basal lamina, fibroblasts that possess well-developed rough endoplasmic reticulum often contact the epithelial cells. These cell–cell contacts are most frequently observed around the primordia of the secondary epithelium when the epithelial cells most actively proliferate. These observations suggest that the thickening of the basal lamina and the cell contacts are related to the primary epithelial cell death and the secondary epithelial cell proliferation.

At later stages, the basal lamina becomes thin beneath the secondary epithelium, and the cell-cell contacts and all cell types of the connective tissue, except fibroblasts, decrease in number (Ishizuya-Oka and Shimozawa, 1987a,b; Shi and Ishizuya-Oka, 1996). By the end of metamorphosis, almost all of the connective tissue cells are ordinary fibroblasts. In the trough of the epithelial folds, these fibroblasts are close to the epithelium and aligned parallel to the curvature of the epithelial basal surface. This structure is similar to the subepithelial fibroblastic sheath reported to be present in the crypt of the mammalian small intestine. This sheath has been thought to play important roles in epithelial cell proliferation and/or differentiation (Marsh and Trier, 1974).

3.4.3.3 Other Tissues Intestinal metamorphosis also changes other tissue types constituting the organ. One such tissue, the muscle, becomes considerably thicker during metamorphosis primarily due to the increase of the inner circular muscle layer. In contrast, after metamorphosis, the thickening is mainly due to that of the outer longitudinal muscle layer. Consequently, in adult *Xenopus*, the thickness of each layer is almost the same (Kordylewski, 1983). Thus the longitudinal muscles are delayed in development compared to the other major tissues of the intestine.

There has also been evidence for the changes in other cell types in the intestine. For example, Torihashi (1990) reported that the neurons in the myenteric plexus of the bullfrog intestine are replaced from larval to adult type during metamorphosis. In addition, a sharp transition of major histocompatability complexes during metamorphosis has also been observed (Flajnik et al., 1987; Du Pasquier and Flajnik, 1990), implicating immune cell participation in intestinal remodeling.

3.4.3.4 Thyroid Hormone Regulation of Intestinal Remodeling Like other organs, the intestine can also be induced to undergo precocious remodeling by treating premetamorphic tadpoles with TH. In addition, this regulation is organ autonomous, for intestinal fragments cultured in vitro can be induced to metamorphose by including TH in the culturing medium.

The changes induced by TH treatment of premetamorphic tadpoles or intestine organ cultures mimics those in vivo, that is, intestinal length reduction, degeneration of the larval epithelium, and development of the adult epithelium (Figs. 2.4, 3.7; Ishizuya-Oka and Shimozawa, 1991; Shi and Hayes, 1994), although different regions of the small intestine respond to TH differently in organ cultures (Fig. 3.7B vs. C) (see Chapter 9, section 9.2 for more details). Interestingly, in the organ cultures, the development of the adult epithelium requires the addition of glucocorticoid and insulin in addition to TH whereas the larval epithelial degeneration can occur in the presence of TH alone (Ishizuya-Oka and Shimozawa, 1991). Thus, even though TH acts directly on the intestine, other nonintestinal factors also influence intestinal remodeling in vivo.

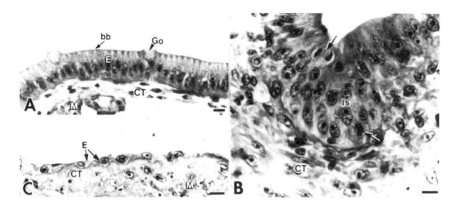

Figure 3.7 Organ culture of the larval small intestine of Xenopus tadpoles. (A) Control explant of the anterior intestine cultured in the medium deprived of hormones. The epithelium (E) is maintained in good conditions, and metamorphic changes do not occur. Both the connective tissue (CT) and the muscle (M) remain undeveloped. Go, globlet cell; bb, brush border. (B) Explant of the anterior intestine cultured with T_3, insulin, and cortisol. A typical islet (Is) grows through rapid cell proliferation (arrows). The connective tissue also develops. (C) Explant of the posterior intestine cultured in the presence of TH. No islets are formed. The number of epithelial cells is small. Connective tissue cells are few. Scale bars, 10 μm. Reprinted from Shi and Ishizuya-Oka (1996) with permission from Academic Press.

4

Cellular and Biochemical Changes

4.1 INTRODUCTION

The preceding chapter describes the morphological changes in various organs during anuran metamorphosis. Those changes are accomplished in part through the proliferation and differentiation of adult cells and development of the extracellular matrix, which provides the structural support for the organs. Another major contributor is the degeneration of larval cell types. Finally, some of the larval cells are reprogrammed to become adult types. Accompanying such cell replacement and differentiation are the alterations in the biochemical properties of the cells and organs. This chapter discusses some of the cell replacements and biochemical changes during metamorphosis. (Additional information on these subjects can be found in Dodd and Dodd, 1976; Gilbert and Frieden, 1981; Fox, 1983). A major focus is on degeneration of larval cells through programmed cell death as it occurs in diverse tissues/organs, such as tail, intestine, stomach, and brain, and affects large fractions of the cells in many tissues/organs (Fig. 4.1; Kerr et al., 1974; Ishizuya-Oka and Ueda,1996; Ishizuya-Oka et al., 1998; Yoshizata, 1996; Tata, 1997).

4.2 PROGRAMMED CELL DEATH OR APOPTOSIS DURING TISSUE RESORPTION

4.2.1 Cell Death and Its Executioners

Normal organ function and development require proper balance among different cell types. This is achieved through well controlled cell death and cell proliferation/differentiation (Wyllie et al., 1980; Schwartzman and Cidlowski,

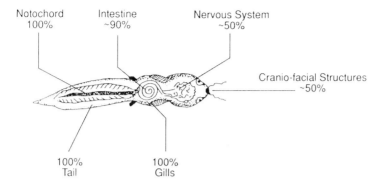

Figure 4.1 Schematic representation of the major tissues of an anuran tadpole undergoing degeneration during metamorphosis. The numbers indicate the extent of programmed cell death expressed as percent of larval cells lost upon completion of metamorphosis. Reprinted from Tata (1997) with permission from Plenum Press.

1993; Thompson, 1995; Jacobson et al., 1997; Shi et al., 1997). For a long time, it had been difficult to study cell death owing to the lack of proper detection methods for dying or dead cells in vivo because these cells are promptly removed through processes such as phagocytosis by neighboring cells or macrophages. Complicating the matter further is the existence of two types of cell death, that is, physiological and pathological cell death (now referred to as programmed cell death or apoptosis) and traumatic or accidental cell death known as necrosis.

Through a series of microscopic examinations, Kerr et al. (1972) found that cell death under both physiological and pathological conditions occurs with distinct morphological changes, including membrane blebbing, chromatin condensation, and cytoplasmic and nuclear fragmentation, to form membrane-enclosed vesicles containing chromatin fragments and/or cellular organelles. Such morphological changes have led to the term "apoptosis," from a Greek word to describe falling or dropping off of petals from flowers or leaves from trees. The vesicles containing cellular organelles and/or chromatin fragments are referred to as "apoptotic" bodies.

Apoptosis has distinct characteristics compared to necrosis (Table 4.1). Apoptotic cells are normally scattered among healthy cells whereas necrosis affects contiguous tracts of cells. The apoptotic cells are quickly removed in tissues by macrophages or other nearby cells, thus preventing inflammation normally associated with necrosis. Ultrastructurally, necrosis occurs through swelling and rupture of cellular organelles and plasma membrane. In contrast, the organelles remain largely intact during early stages of apoptosis and are later enclosed into apoptotic bodies for removal through phagocytosis or other processes. Another major characteristic of apoptosis in vertebrates is the

TABLE 4.1 Comparison of Apoptosis and Necrosis[a]

Feature	Apoptosis	Necrosis
Cause and Control	Physiological and pathological triggers. Determined by gene expression programs within the dying cells	Traumatic injury. Not controlled by cellular gene expression programs
Histological appearances	Scattered individual cells. No exudative inflammation	Tracts of contiguous cells. Exudative inflammation present
Ulstrastructural changes		
Plasma membrane	Membrane blebbing at early stages, then breaking up to enclose apoptotic bodies	Rupture
Cellular organelles	Condensation of cytosol; mitochondria and other organelles are structurally intact but are compacted to form apoptotic bodies eventually	Swelling followed by rupture of the membranes and destruction of the organelles
Nucleus and chromatin	Nuclear membrane progressively convolutes, eventually becomes discontinuous as the chromatin condenses, fragments, and then lies among cytoplasmic organelles. Finally, the chromatin fragments are enclosed in apoptotic bodies	The chromatin marginates in small loosely textured aggregates as in apoptosis. The nuclear membrane initially retains pore structures. As cytoplasmic degradation advances, the membrane and chromatin are destroyed together with other organelles

[a]Based on Wyllie et al. (1980) and Schwartzman and Cidlowski (1993).

fragmentation of the chromatin to produce a nucleosomal sized DNA ladder (multiples of about 180 bp in lengths) (Wyllie et al., 1980; Schwartzman and Cidlowski, 1993).

Many genes involved in cell death have been isolated and characterized. These studies were made possible largely because of the genetic studies in the nematode *Caenorhaditis elegans* that identified genes involved in cell death (Horvitz et al., 1982; Ellis and Horvitz, 1986). During normal development of *C. elegans*, 131 out of the 1090 somatic cells in the developing animal undergo programmed cell death (Sulston, 1988). Although the signals for the death of these 131 cells are still under investigation, 11 genes have been found to play critical roles at the death execution or subsequent steps (Fig. 4.2; Table 4.2; Driscoll, 1992; White, 1993). Subsequently, the homologs of some of these genes were found in mammals and shown to be able to function similarly

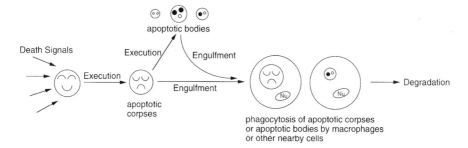

Figure 4.2 Pathway of programmed cell death. Upon receiving various cell death signals, cells activate their own apoptosis execution apparatus to kill themselves and often produce apoptotic bodies, which are membrane-bound vesicles containing cytoplasmic organelles (open circles) and/or fragmented, condensed chromatin (filled objects). The cell corpses and apoptotic bodies are subsequently phagocytosed by macrophages or other neighboring cells and degraded. Nu: nucleus.

TABLE 4.2 C. Elegans Genes Involved in Apoptosis and Their Likely Homologs in Mammals[a]

C. Elegans	Mammals
Execution	
Ced-9 (+)	Bcl-2 (−), Bax (+)
Ced-3 (+)	Caspases (+)
Ced-4 (+)	Apaf-1 (+)
	IAP (−)
	Apaf-3 (cytochrome c) (+)
	Cyclophilins A-C (chromatin fragmentation) (+)
	CAD: Caspase-activated DNase (chromatin fragmentation) (+)
	ICAD: inhibitor of CAD (DFF45, or DNA fragmentation factor 45) (−)
Engulfment	
Ced-1	
Ced-2	
Ced-5	DOCK 180
Ced-6	
Ced-7	ABC1
Ced-8	
Ced-10	
	Scavenger receptors
	Integrins
	CD14: glycosylphosphatidyl-inositol-(GPI) anchored protein
Degradation	
Nuc-1	

[a] See text for references. A (+) sign indicates a factor that promotes apoptosis and a (−) sign indicates a factor that inhibits apoptosis.

(Table 4.2; Yuan et al., 1993; Hengartner and Horvitz, 1994a,b; Zou et al., 1997).

At the execution step, three *C. elegans* genes are involved (Table 4.2). The ced-3 and ced-4 are cell death promoters, whereas ced-9 inhibits cell death. The dead cells are subsequently removed through phagocytosis. Seven *C. elegans* genes (ced-1, ced-2, ced-5, ced-6, ced-7, ced-8, and ced-10, Table 4.2; Horvitz et al., 1982; Driscoll, 1992) have been shown to be involved in the engulfment of the dead cells. Only one gene (nuc-1) has been identified in the next and the final step of cell death, that is, the degradation of dead corpses, and nuc-1 gene product controls the activity of a nuclease required for the degradation of the DNA in the dead corpses (Table 4.2; Horvitz et al., 1982; Ellis and Horvitz, 1986; Driscoll, 1992).

The mammalian homologs of some *C. elegans* death execution genes have been cloned and their gene products have been characterized biochemically. The most extensively studied are the ced-3 homologs, which encode cysteine proteases capable of cleaving after an aspartic acid residue within a substrate (Yuan et al., 1993; Alnemri et al., 1996; Cryns and Yuan, 1998). At least 11 such proteases, now referred to as caspases, exist in mammals (Table 4.2; Alnemri et al., 1996; Cryns and Yuan, 1998). Upregulation of these caspases through increased expression and/or activation of the proenzymes leads to the degradation of various cellular substrates and cell death (Salvesen and Dixit, 1997; Nicholson and Thornberry, 1997; Rao and White, 1997; Cryns and Yuan, 1998). The exact roles of the various caspases in apoptosis are largely unclear. However, evidence exists to support that at least in some circumstances, protease cascades involving multiple caspases participate in the sequential activation of different steps of the cell death program (Orth et al., 1996; Thornberry and Lazebnik, 1998; Wang et al., 1998). Furthermore, gene knockout analyses in mouse indicate that different caspases have distinct roles in regulating apoptosis during development (Kuida et al., 1995, 1996, 1998; Wang et al., 1998; Hakem et al., 1998; Bergeron et al., 1998).

The homologs of ced-9 are the bcl-2 family members (Hengartner and Horvitz, 1994a; White, 1996; Rao and White, 1997). Interestingly, there are two subfamilies within the bcl-2 superfamily. The bcl-2 subfamily members protect cells from apoptosis whereas the bax subfamily members promote cell death. They can form homodimers as well as heterodimers and a critical balance of the two types of bcl-2 family members appears to be important for cell fate determination (Adams and Cory, 1998; Minn et al., 1999). The exact mechanisms by which bcl-2 subfamily members inhibits cell death are yet unknown. A possible direct inhibition mechanism was suggested more recently by the cloning and characterization of ced-4 and its mammalian homolog Apaf-1 (Table 4.2; Zou et al., 1997; Li et al., 1997b; Xue and Horvitz, 1997; Spector et al., 1997; Chinnaiyan et al., 1997; Wu et al., 1997; Cecconi et al., 1998; Yang et al., 1998; Yoshida et al., 1998). Apaf-1 (ced-4) can interact directly with caspases (ced-3), thus participating directly in caspase activation. Interestingly, ced-4 also interacts with ced-9 directly, suggesting that ced-9 (bcl-2) may directly inhibit caspase activation through this interaction (Vaux, 1997). In mammals, in addition to Apaf-1, at least another factor, Apaf-3 or cytochrome c, is also required for caspase activation (Liu et al., 1996; Kluck et al., 1997; Yang et al., 1997). Cytochrome c resides in the space between the inner and outer membranes of mitochondria. Thus Bcl-2 family members may regulate caspases activation by influencing cytochrome c release through their potential roles as ion channels or adaptors, or through other indirect mechanisms (docking protein) (Vander Heiden et al., 1997; Vaux, 1997; Green and Reed, 1998; Li et al., 1998a; Luo et al., 1998; Narita et al., 1998; Wang and Reed, 1998; Heiden et al., 1999).

There are at least two other classes of protein involved in apoptotic execution in mammals. These are the inhibitors of apoptosis (IAPs), which blocks caspase function (Roy et al., 1995; Duckett et al., 1996; Uren et al., 1996; Thornberry and Lazebnik, 1998), and proteins likely involved in the degradation of nuclear DNAs, that is, cyclophilins, caspase-activated DNase (CAD), and inhibitor of CAD (ICAD or DFF) (Table 4.2; Montague et al., 1994, 1997; Liu et al., 1997, 1998; Su et al., 1997a; Enari et al., 1998; Mukae et al., 1998; Sakahira et al., 1998).

Relatively little is known about the subsequent steps, that is, the engulfment and degradation, in mammals. Recent studies have identified that the human protein Dock 180 is similar to ced-5 and that it may participate in cell surface extension to engulf dead cell corpse or apoptotic bodies (Wu and Horvitz, 1998). In addition, a number of other mammalian proteins have also been implicated in the engulfment step (Table 4.2; Duvall et al., 1985; Savill et al., 1993; Platt et al., 1998; Savill,1998). How these proteins participate in the engulfment is yet unclear. They are likely involved in the recognition of cell surface changes such as the externalization of phosphatidyl serines on cell membrane. Interestingly, recent evidence suggests that this cell surface recognition can be dissociated from other steps of apoptosis such as caspase activation and nuclear DNA fragmentation (Zhuang et al., 1998).

The genes participating in execution of apoptosis and subsequent steps are likely to be common in many systems where cell death takes place. On the other hand, those participating in the earlier steps that lead to apoptosis are presumed to vary depending upon the nature of the signals that trigger cell death in developmental and pathological processes, for example, TH during apoptotic metamorphic tissue remodeling. Understanding a given process involving cell death will thus require the characterization of its signal transduction pathway and identification of the genes involved.

4.2.2 Apoptosis During Intestinal Remodeling

The larval epithelium is the predominant tissue of the intestine in premetamorphic tadpoles. It degenerates through programmed cell death and is replaced by the adult epithelium during metamorphosis (Smith-Gill and Carver, 1981; Dauca and Hourdry, 1985; Yoshizato, 1989). Like apoptosis in mammals, the intestinal cell death process involves the condensation and margination of chromatin against the nuclear envelope followed by the fragmentation of the chromatin and cell itself to form the membrane-bound apoptotic bodies containing condensed chromatin fragments and/or cellular organelles (Ishizuya-Oka, 1996). The larval epithelium begins to degenerate around the onset of the metamorphic climax. Electron microscopic examination of the *Xenopus laevis* small intestine during metamorphosis showed that as early as stage 59, apoptotic bodies can be detected among and within intact larval epithelial cells, and within macrophage-like cells in the epithelium (Fig. 4.3; Ishizuya-Oka and Shimozawa, 1992a). The presence of apoptotic bodies in intact epithelial cells indicates that larval epithelial cells can participate in the removal of their apoptotic neighbors even though they themselves are destined to die eventually.

The macrophages appear to play an important role in removing the apoptotic cells. They are from the connective tissue underlying the degenerating epithelium. During metamorphosis, these cells migrate across the remodeled basal lamina separating the connective tissue and the epithelium (Ishizuya-Oka and Shimozawa, 1992a), although how they are triggered to migrate is unclear. As the intestinal epithelial cell death begins, the number of

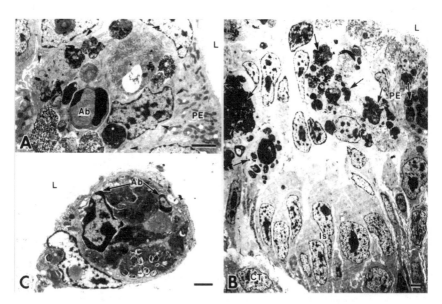

Figure 4.3 Degeneration of the primary epithelium (PE) of the Xenopus small intestine during metamorphosis. (A) An intraepithelial macrophage (arrowheads), including an apoptotic body (Ab) characteristic of condensed chromatin close to its nuclear membrane. L, lumen. (B) Many macrophages (arrows) are localized in the primary epithelium, but not in the islet (Is) of undifferentiated epithelial cells. CT, connective tissue. (C) An intraluminal macrophage including apoptotic bodies. Scale bars, 2 μm. Reprinted from Shi and Ishizuya-Oka (1996) with permission from Academic Press.

macrophages (relative to the epithelial cells) begins to increase, reaching the maximal levels at stage 61, the peak of larval epithelial degeneration (Fig. 4.4A; Ishizuya-Oka and Shimozawa, 1992a). After stage 61 (in *Xenopus laevis*), the macrophages in the epithelium decrease in number as the degeneration is completed and adult epithelial cells proliferate and differentiate. Likewise, during TH-induced intestinal remodeling, the number of macrophages in the epithelium undergoes similar changes (Fig.4.4B; Ishizuya-Oka and Shimozawa, 1992a).

Interestingly, the macrophages seem to recognize specifically the dying cells or apoptotic bodies, but not the intact larval epithelial cells or the proliferating adult epithelial cells in the epithelium (Fig. 4.3; Ishizuya-Oka and Shimozawa, 1992a; Ishizuya-Oka, 1996). The precise mechanism for this recognition is yet unclear. However, apoptosis is accompanied by changes in the cell surface, including the flipping of the phosphatidyl serines from inner to outer surface of the plasma membranes and the alteration of glycoconjugate compositions on the cell surface (Ishizuya-Oka and Shimozawa, 1990; Savill, 1998). Such changes are likely important for the recognition of dying cells by macrophages

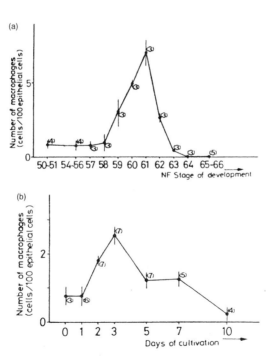

Figure 4.4 Developmental changes in the number of macrophage-like cells in the intestinal epithelium during spontaneous metamorphosis (A) and during TH-induced metamorphosis in vitro (B). The number of specimens examined is shown in parentheses. Filled circles and vertical lines represent the means SD. From Ishizuya-Oka and Shimozawa (1992a).

through their surface receptors, such as integrins, scavenger receptors, and so on (Table 4.2; Duvall et al., 1985; Savill et al., 1993; Savill, 1998). The macrophages together with the engulfed apoptotic bodies are removed at least in part through extrusion into the intestinal lumen (Fig. 4.3; Ishizuya-Oka and Shimozawa, 1992a; Ishizuya-Oka, 1996).

A hallmark of apoptosis is the fragmentation of the chromatin, which can be detected by the so-called TUNEL method (terminal deoxynucleotidyl transferase-mediated dUTP–biotin nick-end labeling) even before the fragmentation of the nucleus and cytoplasm (Gavrieli et al., 1992). This method allows the in situ incorporation of biotin–dUTP into the DNA ends generated during early stages of apoptosis through the action of the transferase and its subsequent color detection through peroxidase conjugated to the biotin-binding protein streptavidin. Using this method, cell death can be detected throughout the larval epithelium at stage 60/61 in the small intestine of *Xenopus laevis* but not in pre- and prometamorphic tadpoles or even at the onset of metamorphic climax (stage 59; Fig. 4.5; Ishizuya-Oka and Ueda, 1996; Ishizuya-Oka et al., 1997c; Shi and Ishizuya-Oka, 1997b). Expectedly, little cell death is detected in other tissues, including the adult epithelium. In postmetamorphic frog, the method can detect cell death only at the tip of the epithelial fold (Fig. 4.5), equivalent to the villus in mammals and birds. This reflects the constant self-renewal of the epithelial cells as the fully differentiated cells at the tip die off and are replaced by the newly arrived differentiated epithelial cells generated from cell proliferation at the trough of the fold (the crypt in mammals and birds) (Shi and Ishizuya-Oka, 1996).

The epithelial cell death is organ autonomous and TH-dependent, for it can occur even when intestinal fragments are cultured in vitro in the presence of physiological levels of TH (Ishizuya-Oka and Shimozawa, 1992; Ishizuya-Oka et al., 1997c). Furthermore, when the larval epithelial cells are cultured in vitro, they too are induced to die by TH (Fig. 4.6; Su et al., 1997a,b). On the other hand, the fibroblasts of the intestine are refractory to the TH-induced apoptosis; instead, they are stimulated to proliferate (Fig. 4.6). Interestingly, the epithelial cells can proliferate in vitro and this proliferation is also stimulated by TH (based on DNA synthesis) (Su et al., 1997b). As TH simultaneously induce the cells to die, the net effect is a gradual decrease in epithelial cell number (Fig. 4.6; Su et al., 1997b). These results indicate that the primary cell cultures reproduce at least some of the TH-dependent cell transformations during intestinal metamorphosis and that the TH-dependent epithelial cell death is cell autonomous.

The in vitro cell cultures have allowed the characterization of the TH-dependent cell death in more detail. The intestinal epithelial cell death induced by TH is typical of apoptosis in mammals. The cells have distinct apoptotic morphology (Fig. 4.7*A*; Su et al., 1997a,b) and produce a population of cells with subdiploid levels of DNA detectable by flow cytometry (Fig. 4.7*C*). In addition, a ladder of DNA fragments with sizes of multiples of nucleosomes is also present (Fig. 4.7*D*). Dose-dependence analysis

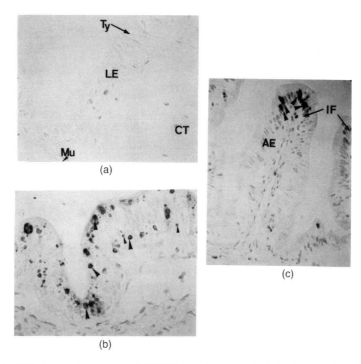

Figure 4.5 DNA fragmentation assay (TUNEL) detects apoptosis during intestinal remodeling. Intestinal cross-sections from Xenopus laevis tadpoles at different stages were analyzed by the TUNEL assay (Gavrieli et al., 1992), by which cells undergoing apoptosis are labeled. (A) No cell death was observed in premetamorphic tadpole intestine at stage 59 (X630). (B) Most of the larval epithelial cells were labeled at the metamorphic climax (stage 60). Note the apoptotic morphology of the dying cells. Small and large arrowheads refer to labeled nuclear fragments of apoptotic bodies and intact nuclei of the larval cells, respectively (X630). (C) Cell death (arrowheads) in postmetamorphic frog intestine (stage 66) was limited to the crests of intestinal folds (IF), where the fully differentiated epithelial cells degenerate after functioning for a finite period of time (X420). Ty; typholosole; LE; larval epithelium; AE; adult epithelium; Mu; muscle; CT, connective tissue. Reprinted from Shi and Ishizuya-Oka (1997) with permission from Plenum Press.

shows that physiological concentrations of T_3 (5–10 nM) added to the culture medium are sufficient to induce apoptosis, although higher concentrations have greater effects (Fig. 4.7B). Interestingly, by sorting cells and cell fragments based on their DNA contents and cellular granularity (cell death increases granularity), it has been shown that populations of cells at all different stages of the cell cycle (with DNA contents varying from that in a diploid cell in G1 phase, to twice as much in M and G2 phases, and those in between the two in the S phase) are present in the primary cell cultures in both the presence and absence of T_3. However, the treatment with T_3 leads to increases in granularity of cells at these different stages,

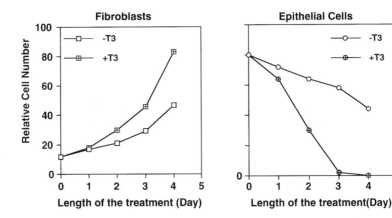

Figure 4.6 Contrasting effects of thyroid hormone on tadpole intestinal epithelial and fibroblastic cells. The fibroblasts and epithelial cells were isolated from stage 57/58 of tadpole small intestine and then cultured on a six-well plastic dish in 60% L-15 medium containing 10% T_3-depleted fetal bovine serum at 25°C in the presence or absence of 100 nM T_3. The live cells were counted daily by trypan blue staining. Note that the epithelial cell number decreased even when no exogenous T_3 was present. This could be because of the residual T_3 in the treated serum.

showing that all these cells are capable of T_3-induced apoptosis, independent of their cell cycle positions (Su et al., 1997b).

The T_3-induced epithelial cell death can be blocked by known inhibitors of mammalian apoptosis. For example, using a quantitative assay for DNA fragmentation, it has been shown that ATA and Z-VAD, which inhibit nucleases and caspases, respectively, can inhibit the epithelial apoptosis (Fig. 4.8; Su et al., 1997b), suggesting the involvement of nuclease(s) and caspases(s), just like that during apoptosis in other systems (Table 4.2). Similarly, the immunosuppressant drug cyclosporin A (CsA), which can inhibit activation-induced T-cell death in mammals (Shi et al., 1989), also blocks T_3-induced intestinal epithelial cell death (Fig. 4.8). On the other hand, another immunosuppressant FK506, which inhibits activation-induced T-cell death (Birer et al., 1990), fails to have any effect on the intestinal cells (Fig. 4.8). Cyclosporin A and FK506 are believed to inhibit activation-induced T-cell death by inhibiting the calmodulin-dependent phosphatase calcinurin through at least some of their binding proteins, cyclophilins and FK506-binding proteins, respectively (Shi et al., 1989; Liu et al., 1991). In addition, cyclophilins have nuclease activity (Table 4.2; Montague et al., 1994; 1997; Su et al., 1997a). Thus, although the cause for the different effects of FK506 in the two cell death processes is unclear, CsA may inhibit T_3-induced cell death by blocking signal transduction pathways involving phosphorylation and/or inhibiting the nuclease activity of its binding proteins. These inhibitor studies further show that cell death execution during amphibian metamorphosis employs genes similar to those in other species (Table 4.2).

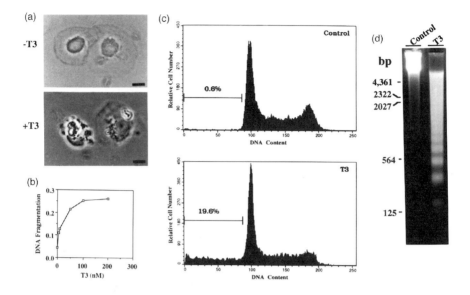

Figure 4.7 Thyroid hormone (T_3) induces apoptosis in primary culture of tadpole intestinal epithelial cells. (A) T_3 treatment results in apoptotic morphology of the cultured Xenopus laevis tadpole intestinal epithelial cells. Intestinal epithelial cells were cultured on coverslips with or without 100 nM T_3 for 2 days. The coverslips were fixed with 0.5% of paraformaldehyde on ice for 20 min and mounted on slides. Note that the control cells showed epithelial morphology. The bars represent 7 μm. (B) T_3 dose response of apoptosis induction in cultured cells. Tadpole epithelial cells were cultured on a 96-well plastic culture plate in the presence of different concentrations of T_3 for 2 days and lysed for the DNA fragmentation ELISA assay. (C) DNA content analysis by flow cytometry shows a subdiploid population, representing apoptotic cells, after 1 day of T_3 treatment. (D) T_3 induces the formation of nucleosome-size DNA ladder in epithelial cells. The primary epithelial cells were treated with or without 100 nM T_3 for 1 day. The genomic DNA was isolated, analyzed on an agarose gel, stained with ethidium bromide, and visualized under ultraviolet light. Reprinted with permission from Su et al. (1997a).

4.2.3 Cell Death During Tail Resorption

The tail is a larva-specific organ in anurans and is completely resorbed during metamorphosis. The epidermis and muscles are two major tissues in the tail. Electron microscopic study of the tail resorption in tadpoles of the dwarf tree frog, *Litoria glauerti*, demonstrated that apoptosis occurs in both of these two tissues during tail resorption (Kerr et al., 1974). The earliest changes in the epithelial cells are the aggregation of condensed chromatin beneath the nuclear envelope and the condensation of the cytoplasm, similar to those observed in mammals (Kerr et al., 1972; Wyllie et al., 1980). The nucleus and condensing cells then fragment to form apoptotic bodies of various sizes.

In the muscle, the first sign of degeneration is the occurrence of longitudinal clefts between myofibrils. Under an electron microscope, the nuclei of striated

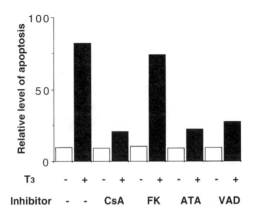

Figure 4.8 T_3-induced intestinal epithelial cell death can be inhibited by some but not all known inhibitors of mammalian apoptosis. The epithelial cells were cultured on plastic dishes for 3 days in the presence or absence of 100 nM T_3 and/or 300 ng/ml CsA, 10 ng/ml FK506 (FK), 100 μM ATA, and 50 μM Z-VAD (VAD). Relative level of cell death was determined by measuring DNA fragmentation using an apoptosis ELISA method. Note that with the exception of FK506, all inhibitors blocked T_3-induced epithelial cell DNA fragmentation. None of the drugs had any effect on DNA fragmentation by themselves.

muscle fibers that display the cytoplasmic changes have the same peripheral aggregation of condensed chromatin as in other types of apoptosis (Kerr et al., 1974). The cytoplasm and myofibrils are also condensed and there is widespread dilatation of the sarcoplasmic reticulum. Eventually, the muscle fibers are fragmented into many so-called sarcolytes with well preserved cross-striations (Fig. 4.9). These sarcolytes or apoptotic bodies are then removed through phagocytosis by macrophages (Fig. 4.9; Kerr et al., 1974).

Similarly findings have also been reported for the tail epidermal cells in *Rana japonica* (Kinoshita et al., 1985) and tail muscle cells in *Xenopus laevis* (Nishikawa and Hayashi, 1995; Nishikawa et al., 1998). As early as stage 59 (early metamorphosic climax) in *Xenopus laevis*, apoptotic bodies (sarcolytes) can be detected. However, the cell death is quite limited (occurring in less than 5% of the muscle area) up to stage 61 and rapidly increases, occupying 50% or more of the muscle area, after stage 62 when massive tail resorption occurs. Accompanying this temporal increase in cell death is a dramatic rise of the number of macrophages per unit muscle area to remove the sarcolytes. Essentially identical observations are also found during TH treatment of isolated tails in cultures in vitro.

Two studies involving skin transplantation and organ cultures have suggested that adult-type non-T leukocytes may participate in the specific elimination of the tail tissues (Izutsu and Yoshizato, 1993; Izutsu et al., 1996). In addition, removing the tadpole epidermis also prevents TH-induced tail resorption in *Rana catesbeiana* organ cultures, suggesting that an inductive factor from the epidermis is necessary for tail resorption (Niki et al., 1982; Niki and Yoshizato,

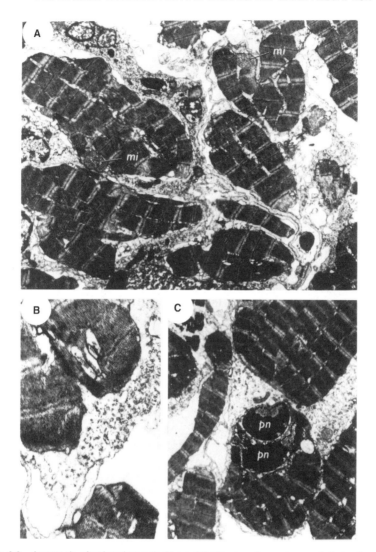

Figure 4.9 Apoptosis of striated muscle fibers. (A) *Cluster of muscle apoptotic bodies. Some lie within macrophages; others are still extracellular. Note the well preserved mitochondria (mi) in two of the bodies. Even at this magnification, it is clear that the muscle fragments are bounded by membranes. X6000.* (B) *The extracellular apoptotic body at the bottom of the micrograph is bounded by a membrane; the ingested body within the macrophage is surrounded by a narrow, membrane-enclosed space, the outer limit of which constitutes the phagosome membrane. X16800.* (C) *Extracellular and phagocytosed muscle apoptotic bodies, one with dense nuclear remnants (pn). X5300. Reprinted from Kerr et al. (1974) with permission from the Company of Biologists LTD.*

1986). A further study has shown that the factor is secreted by cultured tail skin and that the factor is resistance to protease digestion and heat inactivation up to 100°C, indicating that it is nonproteinaceous. Thus the epidermis appears to play a role in tail resorption. On the other hand, the subepidermal fibroblast layer might be removed along with the epidermis in these experiments and the fibroblasts, but not the epidermal cells, express high levels of proteinases. Therefore it has been suggested that it is the subepidermal fibroblasts that are responsible for the tail resorption in these organ cultures (Berry et al., 1998a). Regardless of the exact identity of the tissue, these studies suggest that cell–cell interactions are important for tissue resorption.

In contrast to the organ culture work, primary cell culture studies show that the tadpole tail epidermal cells are induced to die in vitro by TH (Nishikawa and Yoshizato, 1986; Nishikawa et al., 1989), just like the intestinal epithelial cells (Su et al., 1997a,b). This suggests that the tail epidermal cell death is also cell autonomous, at least in vitro. Similarly, a stable cell line derived from muscle cells of *Xenopus laevis* tadpole tails can undergo T_3-dependent apoptosis (Yaoita and Nakajima, 1997).

Cell death in the intestine and tail shares many similar characteristics. Both are typical of vertebrate apoptosis, with chromatin condensation and fragmentation that can be detected by TUNEL assay in situ (Fig. 4.5) and produce a nucleosomal sized DNA ladder (Fig. 4.7; Kerr et al., 1974; Nishikawa and Hayashi, 1995; Ishizuya-Oka and Ueda, 1996; Izutsu et al., 1996; Ishizuya-Oka et al., 1997; Su et al., 1997a,b; Yaoita and Nakajima, 1997). Both can be inhibited by the immunosuppressant CsA and caspase inhibitors (Little and Flores, 1992; Su et al., 1997a,b; Yaoita and Nakajima, 1997), suggesting the involvement of similar signal transduction pathways and cell death executioners as during cell death in mammals. Evidence for the roles of various cell death executioners is quite limited. Two bcl-2 family members have been cloned in *Xenopus laevis*, and they are expressed in tadpole tissues (Cruz-Reyes and Tata, 1995). However, their expression levels do not change during metamorphosis. On the other hand, studies on *Xenopus* tadpole tail muscle degeneration have implicated the involvement of bax in muscle cell death (Sachs et al., 1997a). Likewise, expression analysis has shown that although caspase-1 (ICE, or interleukin 1β converting enzyme) has little expression in metamorphosing *Xenopus laevis* tadpoles, the mRNA levels of caspase-3 (cpp32/apopain/Yama) are upregulated during metamorphosis or by TH treatment of a cell line derived from *Xenopus* tail muscle cells (Yaoita and Nakajma, 1997), implicating the participation of caspase-3 in cell death during metamorphosis. Clearly, further studies are required to elucidate the roles of various potential cell death regulators and executioners in apoptosis during metamorphosis.

4.2.4 Apoptosis in Other Organs and Tissues

Many organs and tissues other than the intestine also undergo extensive cell death as they degenerate or remodel into the adult forms. The extents of tissue degeneration are shown for some of them in Figure 4.1. Among them, the gills and notochord are tadpole specific and degenerate completely during metamorphosis. As described in the preceding chapter, the central nervous system also remodels extensively with the neurons specific for tadpole organs/tissues being degenerated. Even for neurons targeting adult tissues they first proliferate during metamorphosis to generate higher than necessary numbers. Those excess ones, which fail to reach target tissues, are destined to die as well. Similarly, apoptosis occurs during the development of the limbs, whose morphogenesis is manifested in part through the degeneration of the interdigital cells. It is believed all these processes are mediated by programmed cell death and share similar properties as the intestinal and tail apoptosis. However, detailed characterization of these processes remains to be done.

4.3 BIOCHEMICAL CHANGES ASSOCIATED WITH TISSUE RESORPTION AND REMODELING

The biochemistry of amphibian metamorphosis has been studied extensively since the beginning of this century (Frieden, 1961; Weber, 1967; Etkin and Gilbert, 1968; Dodd and Dodd, 1976; Gilbert and Frieden, 1981). At the gross biochemical level, metamorphosis leads to a drastic reduction of all major components of the animal, that is, body water, organic matters, and mineral substances, ranging from 25% to more than 60% relative to those at the stage of maximal growth (Weber, 1967). There is general reduction in most, if not all, components such as nitrogen, lipids, and carbohydrates, in the animal. This is likely due to the resorption of tadpole tissues/organs such as the tail and to removal of intestinal contents as the tadpole stops feeding during metamorphosis. However, different components do reduce to somewhat different levels, thus a frog has different biochemical compositions from those in the tadpole.

The biochemical changes in individual tissues and organs are reflected in the alteration, elimination, and de novo systhesis of various enzymes and other proteins. These changes are likely required to accommodate the adaptation of the animal from being free swimming, aquatic, and herbivorous to being terrestrial, air breathing, and carnivorous in most anuran species. Some of the biochemical changes that contribute to this adaptation include (1) the shift from ammonotelism to ureotelism, (2) increase in serum albumin levels, (3) changes in hemoglobin isotypes, (4) alterations in the digestive system, and (5) changes in the respiration system (Frieden, 1961; Weber, 1967; Broyles, 1981). This section describes the individual enzymes and other proteins that are responsible for some of these changes in different organs/tissues.

4.3.1 Upregulation of Degradative Enzymes During Tail Resorption

As the aquatic tadpole is transformed into a terrestrial frog, the tail is no longer required and is thus completely resorbed. The diverse nature of the various tissues within the tail requires the participation of various enzymes to digest tail components.

Earlier studies suggested that acid hydrolases are important players in tail resorption. An injection of small amounts of lactic acid or butyric acid can cause tail atrophy, which begins at the site of the injection and spreads to the rest of the tail (Helff, 1926). The pH of the tail also drops from about 7.5 in normal tissues to about 6.5 in resorbing tails (Aleschin, 1926). Consistently, many acid hydrolases have been found to change during metamorphosis. These include cathepsin, acid phosphatase, β-glucuronidase, deoxyribonuclease, and ribonuclease (Table 4.3; Weber, 1967; Price and Frieden, 1963; Coleman, 1962, 1963; Kubler and Frieden, 1964; Eeckhout, 1969; Dodd and Dodd, 1976). The activities of all these enzymes increase at the climax of metamorphosis when the tail is being resorbed, although the magnitudes of the increases vary for different enzymes and in different studies, likely due to experimental variations (e.g., difference in staging; Robinson, 1970, 1972) and the use of different animal species (Table 4.3). Furthermore, the changes in the enzymatic activities appear to be regulated temporally and spatially within the tail. For example, the acid phosphatase activity in the tail of *Xenopus laevis* begins to increase dramatically when the tail length is reduced by 20% and reaches higher levels later; and it is upregulated first in the dorsal fin, which is the first to resorb (Robinson, 1970, 1972). Finally, the activities of these enzymes are also

TABLE 4.3 Changes in Enzyme Activity in Tadpole Tails[a]

Enzymes	Activity Ratio
Proteinases (acid)	2–9
Proteinases (alkaline)	1.5
Proteinases	1
Cathepsin	22–50
Dipeptidases	3–4
Tripeptidases	7
Phosphatase (acid)	1.6–20
Phosphatase (alkaline)	1.5–2.8
ATPase (Mg^{2+})	0.3–0.5
Ribonuclease (alkaline)	0.8
Ribonuclease (acid)	1.3–20
Deoxyribonuclease (acid)	3–50
β-Glucuronidase	12–50
Collagenase	10
Amylase	1.6
Lipase	1.1
Catalase	6
Aldolase	0.2
Succinic dehydrogenase	(decrease)
Glutamic dehydrogenase	(decrease)

[a]Based on Weber (1967). The numbers refer to the ratios of the specific activities of the enzymes at the climax to those at premetamorphosis.

upregulated in tail organ cultures treated with TH (Weber, 1967, 1969), suggesting that the increases are of tail origin, not due to infiltration of leukocytes.

In addition to acid hydrolases, a number of other enzymes are also upregulated during tail resorption (Table 4.3). Among them is collagenase, which can degrade collagen (Gross and Lapiere, 1962; Lapiere and Gross, 1963). Collagen is a major component of the connective tissue, which is critical for the structural integrity of the tail, especially the tail fin. Its degradation is likely to be an important step in tail resorption. The upregulation of collagenase activity both during metamorphosis and after TH treatment is probably responsible for the degradation of collagen during natural and TH-induced tail resorption (Gross, 1964).

The upregulation of the enzymatic activities in the tail is selective, for many other enzymes are not altered in their activity or even reduced during metamorphosis (Table 4.3). Thus considerable interest has been shown in understanding the cellular origin of such regulation. Because lysosomal enzymes are acid hydrolases, one possibility is that the increase in acid hydrolase activity is due to the TH-induced release of hydrolytic enzymes from existing lysosomes (Brachet, 1960). Although some evidence supports the existence of preexisting lysosome-like particles containing such hydrolases, they are insufficient to account for the large increase in the activities of these enzymes (Weber, 1967; Kubler and Frieden, 1964; Eeckhout, 1969; Frieden and Just, 1970). Thus, de novo protein synthesis may be responsible for the upregulation. This was first suggested by the fact that the increase in cathepsin activity in *Xenopus* tails requires a latent period after TH treatment, indicating an indirect mechanism (Weber, 1967). Direct support came from later studies using inhibitors of protein and RNA synthesis (Weber, 1965; Tata, 1966; Eeckhout, 1969), demonstrating that de novo transcription and subsequent translation of the genes encoding these enzymes are, to a large extent, responsible for the increase in their activities during tail resorption. This has now been confirmed at least for collagenase by studies on gene regulation (see Chapter 6).

4.3.2 Biochemical Changes Associated with Intestinal Remodeling

The intestine is present in both tadpoles and frogs but has distinct morphological and structural organizations in the two animal forms. Its remodeling during metamorphosis involves the degeneration of the larval epithelium, which is the predominant tissue within the organ. Thus, just as during tail resorption, lysosomal hydrolases appear to be involved in the epithelial degeneration. The activities of lysosomal hydrolases, for example, acid phosphatase, N-acetylglucosaminidase, β-glucuronidase, aryl sulfatase, and cathepsin, are all upregulated significantly up to the beginning of metamorphic climax (Botte and Buonanno, 1962; Hourdry, 1974; Hourdry and Dauca, 1977; Dauca and Hourdry, 1985), when larval intestinal epithelial cell death takes place. By the end of metamorphosis, their activities are downregulated. Similarly the acid phosphatase activity is also upregulated during TH-induced intestinal remodeling (Pouyet and Hourdry, 1977; Dauca and Hourdry, 1985).

The major function of the intestinal epithelium is food digestion and nutrient absorption, which are performed at the apical surface or the brush border of the epithelial cells. As the larval epithelium degenerates and adult epithelium develops, the activities of brush border enzymes also change considerably. These enzymes, for example, alkaline phosphatase, γ-glutamyl-transpeptidase, glucoamylase, and maltase, are first downregulated as the larval epithelial cell death takes place and then upregulated when adult intestinal epithelial cells differentiate to form the frog epithelium (Kaltenbach et al., 1977; Hourdry et al., 1979; Dauca et al., 1981; Dauca and Hourdry, 1985). By the end of metamorphosis, these enzymes reach higher levels of activities than those in premetamorphic tadpoles (Hourdry et al., 1979). Similarly, when premetamorphic tadpoles are treated with TH, the activities of these enzymes are also first downregulated and then upregulated, reflecting the process of epithelial replacement (Dauca et al., 1980; Dauca and Hourdry, 1985).

4.3.3 Alterations in the Blood

4.3.3.1 Increase in Serum Protein Contents

There are striking quantitative and qualitative changes in the proteins in the blood during metamorphosis. Quantifications of total serum proteins in a number of anuran species reveal that serum protein concentration doubles during metamorphosis (Table 4.4) and that this change also occurs during TH-induced metamorphosis (Frieden, 1961). Electrophoretic analyses of the serum proteins have shown that this increase involves differential changes of different serum proteins. In particular, the serum albumin levels increase most dramatically during metamorphosis, from being essentially nondetectable to being the highest among all serum proteins (Fig. 4.10; Frieden, 1961). This

TABLE 4.4 Changes in Serum Proteins during Metamorphosis[a]

Species	Stage	Serum Protein (g/100 ml)	Ratio of Albumin to Globulin
Rano grylio	Prometamorphosis	1.42 ± 0.30	0.12 ± 0.03
	Postmetamorphosis	2.19 ± 0.17	0.81 ± 0.11
Rana catesbeiana	Prometamorphosis	1.16 ± 0.19	0.12 ± 0.02
	Postmetamorphosis	2.56 ± 0.50	0.64 ± 0.09
Rana heckscheri	Prometamorphosis	0.35 ± 0.04	0.02 ± 0.02
	Postmetamorphosis	2.03 ± 0.19	0.48 ± 0.05
Xenopus laevis	Prometamorphosis	0.20	0.2
	Postmetamorphosis	2.3	0.52

[a] Based on Frieden (1961).

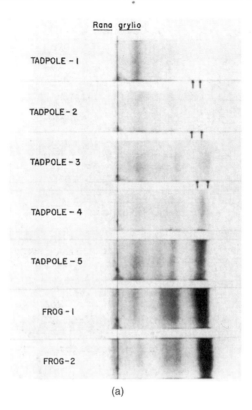

Figure 4.10 *(A) Typical paper strips showing the electrophoretically separated serum proteins stained with bromophenol blue for R. grylio. The gradual increase in the intensity of the albumin fraction during metamorphosis is striking between the two arrows and the redistribution of the globulins can be followed.*

upregulation in albumin levels is also reflected in the ratio of serum albumin to globulin. During both natural and TH-induced metamorphosis, the serum albumin to globulin ratio increases gradually but dramatically, paralleling the increase in the ratio of hind limb length to tail length (Table 4.4; Frieden, 1961).

In addition to albumin, several other serum proteins have been found to increase extensively during metamorphosis. These include the plasma oxidase ceruloplasmin, transferrin, and carbonic anhydrase (Inaba and Frieden, 1967; Frieden and Just, 1970; Wise, 1970).

The increase in serum proteins is believed to play an adaptive role as the tadpole is transformed into the frog. This is especially likely to be the case for albumin. The biochemical and biophysical properties of albumin make it uniquely possible to fulfill not only the osmotic but also the manifold transport needs in connection with the terrestrial living habitat (Frieden, 1961, 1968; Weber, 1967; Broyles, 1981).

70 CELLULAR AND BIOCHEMICAL CHANGES

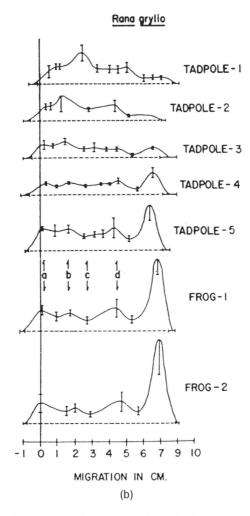

Figure 4.10 (B) *Composite electrophoretic patterns for the serum proteins of R. grylio at various stages of development shown in (A). Spontaneous metamorphosis: Tadpole-1 (early metamorphosis), Tadpole-5 (intermediate), Frog-1 (postmetamorphosis), Frog-2 (adult). T_3 induced metamorphosis: Tadpoles-2, -3, and -4 (3, 5, and 6 days, respectively, after T_3 injection). Reprinted from Frieden (1961) with permission of American Zoology.*

4.3.3.2 Changes in the Hemoglobins and Red Blood Cells

In mammals and birds, multiple changes in hemoglobin synthesis occur during development. Similarly, frog hemoglobins are different from tadpoles. However, in frogs, only a single switch takes place during development. This occurs during metamorphosis and involves the complete replacement of globin chains (Weber, 1996).

It was demonstrated as early as 1936 that hemoglobin isolated from tadpole red blood cells has a greater oxygen affinity and is independent of pH, whereas the oxygen-binding affinity of frog hemoglobin is lower and is lost with decreasing pH (Bohr effect) (McCutcheon, 1936; Riggs, 1951). This reduced affinity for oxygen of frog hemoglobin may be due to the presence of a greater number of available SH groups in the globin part of the molecule (Riggs, 1960). The differences in chemical compositions cause the tadpole hemoglobin to have higher electrophoretic mobility than the frog hemoglobin, although both have similar molecular weights of about 68,000 D (Riggs, 1951; Frieden, 1961).

Electrophoretic and chromatographic analyses indicate that there are actually multiple hemoglobins in both tadpoles and frogs; and amino acid composition and peptide mapping show that these hemoglobins comprise different subunits (Broyles, 1981; Weber, 1996). Like in mammals and birds, the tadpole and frog hemoglobins are derived from distinct genes (Weber, 1996). This switch in gene expression appears to be due to the replacement of larval erythrocytes, which die off during metamorphosis, by adult erythrocytes (Weber et al., 1989, 1991; Weber, 1996).

The switch in hemoglobin occurs at the climax of metamorphosis, although larval hemoglobin persists for some time even in postmetamorphic frogs in some cases (Just and Atkinson, 1972; Dodd and Dodd, 1976; Weber et al., 1991). This switch is believed to facilitate the adaptation of the animal for terrestrial living habitat (Frieden, 1961; Bennett and Frieden, 1962). This is because the frog needs hemoglobins with lower oxygen affinity compared to the tadpole for the latter is limited to the dissolved oxygen in water, and/or the reduced oxygen affinity of the frog homoglobin may facilitate the frog's terrestrial locomotion, which has rapid and prodigious oxygen needs.

4.3.4 Regulation of Enzyme Activities in the Liver

Liver metabolism has been extensively studied at the biochemical level (Frieden, 1961, 1968; Weber, 1967; Cohen, 1970; Frieden and Just, 1970; Dodd and Dodd, 1976). During metamorphosis there are drastic increases in the biosynthesis of nucleic acids and proteins in the liver, which account for, among other things, the dramatic increase in serum albumin levels in the blood (Frieden, 1961; 1968; Tata, 1965; Nakagawa et al., 1967; Smith-Gill and Carver, 1981). In addition to albumin, a number of liver enzymes are also upregulated during metamorphosis. These include catalase, uricase, several phosphatases, and the urea cycle enzymes (Table 4.5; Weber, 1967). The most extensively studied are the urea cycle enzymes, which are responsible for the synthesis of urea for excretion of nitrogenous waste in frogs.

Anuran tadpoles excrete 90% of their nitrogen as ammonia during premetamorphic stages (Munro, 1939, 1953). Beginning at the onset of metamorphic climax, ammonia excretion starts to decrease and urea excretion increases correspondingly. In postmetamorphic frogs, urea represents the predominant (78%) nitrogenous waste (Munro, 1939, 1953; Brown et al., 1959).

TABLE 4.5 Increases in Activity of Enzymes in the Liver of Metamorphosing Tadpoles[a]

Enzymes	Increase in Activity (fold)
Urea cycle enzymes	
Arginase	3–5
Carbamyl phosphate synthetase	29
Ornithine transcarbamylase	5
Argininosuccinate synthetase	3
Phosphatases	
ATPase	4
Glucose-6-phosphatase	4
Acid phosphatase	3
Alkaline phosphatase	2
Catalase	3
Uricase	5

[a] No changes or decreases were observed for esterase, cathepsin, trypsin, succinoxidase, and tryptophan pyrrolase. Based on Weber (1967).

An exception to this is *Xenopus laevis*, which maintains aquatic living habitat and excretes predominantly ammonia (Munro, 1953). However, a transient increase in urea excretion occurs during prometamorphosis in *Xenopus laevis* (Underhay and Baldwin, 1953). Furthermore, restricting water supply can lead to drastic increase in urea excretion in *Xenopus laevis* (Balinsky et al., 1961), indicating that similar transition from ammonotelism to ureotelism can occur in this species as well.

The synthesis of urea is catalyzed by the urea cycle enzymes (Fig. 4.11; Cohen, 1970; Atkinson et al., 1994). There are five enzymes in this cycle. Among them, carbamyl phosphate synthetase I and ornithine transcarbamylase are mitochondrial enzymes, whereas argininosuccinate synthetase, argininosuccinate lyase, and arginase are cytosolic ones (the arginosuccinate synthetase and argininosuccinate lyase are together referred as arginine synthetase). Not surprisingly, the activities of all these enzymes are upregulated during metamorphosis (Fig. 4.12; Table 4.5; Brown and Cohen, 1958; Brown et al., 1959). This upregulation is also observed during TH-induced metamorphosis and shown to be due to de novo protein synthesis (Paik and Cohen, 1960; Metzenberg et al., 1961; Weber, 1967; Cohen, 1970; Dodd and Dodd, 1976). As discussed in Chapter 6, transcription from the genes encoding these enzymes is upregulated in the liver during natural and TH-induced metamorphosis, which in turn leads to higher levels of translation to produce the corresponding enzymes.

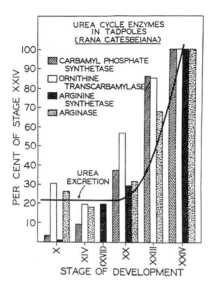

Figure 4.11 The urea cycle. Two enzymes, carbamyl phosphate synthetase I and ornithine transcarbamylase, are localized in the mitochondria; the other three, argininosuccinate synthetase, argininosuccinate lyase, and arginase, are in cytoplasm. Based on Cohen (1970) and Atkinson et al.(1994).

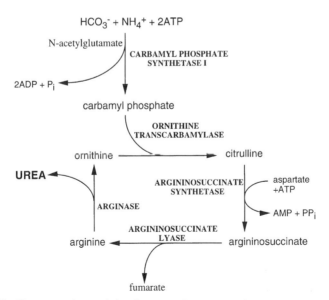

Figure 4.12 Urea excretion and development of enzymes of ornithine urea cycle in the metamorphosing tadpole. Arginine synthetase represents the overall synthesis of arginine from citrulline and aspartate and involves argininosuccinate synthetase and argininosuccinate lyase (see 4.11). Thus the activity for the arginine synthetase reflects the activity of the rate-limiting enzyme. Reprinted from Brown and Cohen (1958) with permission from The Johns Hopkins University Press.

5

Mechanism of Thyroid Hormone Action

5.1 INTRODUCTION

The action of thyroid hormone during development is regulated at multiple levels. This is in part because of the presence of numerous TH binding proteins in the plasma, cytosol, and nucleus. Owing to the existence of the many cytosolic proteins and many known effects of TH action at nongenomic levels (Davis and Davis, 1996), TH was considered to affect metamorphosis through its cytosolic actions. However, since first proposed in the 1960s, the model of nuclear TH action has gained strong support and it is now generally believed that the biological effects of TH are mainly mediated through gene regulation by nuclear thyroid hormone receptors (TRs) (Tata and Widnell, 1966; Tata, 1967; Oppenheimer, 1979). This chapter reviews some recent findings on the function of TRs.

5.2 THYROID HORMONE AND ITS BINDING PROTEINS

Both of the naturally occurring thyroid hormones, 3,3′,5,5′-tetraiodothyronine (T_4), commonly known as thyroxine, and 3,3′,5-triiodothyronine (T_3), are synthesized in the thyroid gland, although T_4 can be converted to T_3 in other organs as well (Fox, 1983; Dodd and Dodd, 1976). Their biosynthysis involves first the activation of the thyroglobulin gene in the thyroid gland. The gene product, thyroglobulin, then undergoes a series of post-translational processing, including iodination and tyrosine condensation, to produce T_4 (Fig. 5.1). T_4 can be either secreted into the circulating plasma or converted into T_3 through deiodination by 5′-deiodinase in the thyroid gland. The T_3 and T_4 secreted from the thyroid gland are subsequently carried by the plasma to different organs/tissues where they exert their biological effects.

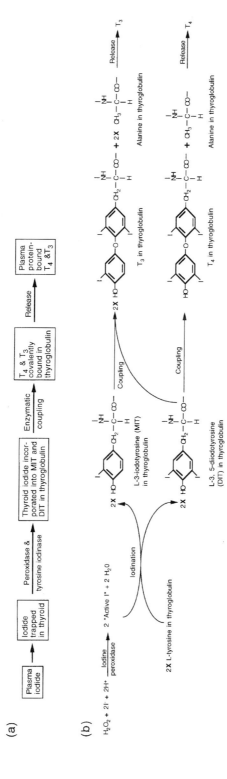

Figure 5.1 A biosynthetic pathway of T_3 and T_4. B. Enzymatic reactions in the biosynthesis of T_3 and T_4. Based on Fox (1983).

T_4 can also be converted into T_3 in various target tissues by 5′-deiodinase (St. Germain, 1994; St. Germain and Galton, 1997). In addition, both T_3 and T_4 can be inactivated through the action of 5-deiodinases, producing T_2 and reverse T_3, respectively. Thus the ratios of T_3 to T_4 and the cellular levels of T_3 and T_4 can vary from tissue to tissue, depending upon the levels of various deiodinases. Two different 5′-deiodinases have been identified (St. Germain, 1994; St. Germain and Galton, 1997). They have different enzymatic properties and tissue distributions in mammals. A 5′-deiodinase and a 5-deiodinase from *Rana catesbeiana* (Davey et al., 1995; Becker et al., 1995) and a 5-deiodinase from *Xenopus laevis* (St. Germain et al., 1994) have been cloned and found to have distinct developmental regulation patterns in different organs, supporting the idea that thyroid hormone levels can vary in different organs/tissues.

Upon secretion from the thyroid gland into the serum, thyroid hormone encounters serum thyroid hormone binding proteins, resulting in little free thyroid hormone in the serum (Table 5.1; Jorgensen, 1978; Barsano and DeGroot, 1983; Galton, 1983; Benvenga and Robbins, 1993). The serum TH binding proteins, including serum albumin, serve as carriers to bring TH to target tissues, where TH enters the cytoplasm (Fig. 5.2).

The mechanisms by which TH is taken up by the cells are unknown at present. Both T_3 and T_4 are hydrophobic at neutral pH. Thus it is possible that some TH may enter cells through passive diffusion through the cell membrane. Another route is to enter cells with the serum TH binding proteins as a complex. Lipoproteins and other TH transport proteins can affect the efflux rate and cellular TH levels in tissue culture cells (Benvenga and Robbins, 1998). In addition, there is evidence that active pathways exist to import and export TH to and from the cytoplasm (Blondeau et al., 1988; Oppenheimer et

TABLE 5.1 Cytosolic and Plasma Thyroid Hormone Binding Proteins

Protein	Location	Reference
Myosin light chain kinase	Cytosolic	Hagiwara et al., 1989
Pyruvate kinase, subtype M1	Cytosolic	Parkison et al., 1991
Pyruvate kinase, subtype M2	Cytosolic	Kato et al., 1989; Shi et al., 1994
Prolyl 4-hydroxylase, β-subunit	Cytosolic	Cheng et al., 1987; Pihlajaniemi et al., 1987; Geetha-Habib et al., 1988; Noiva and Lennarz, 1992
Aldehyde dehydrogenase	Cytosolic	Yamauchi and Tata, 1994
Transthyretin	Plasma	Yamauchi et al., 1993
T_4-binding globulin	Plasma	Jorgensen, 1978; Benvenga and Robbins, 1993
Serum albumin	Plasma	as above
Lipoproteins	Plasma	as above

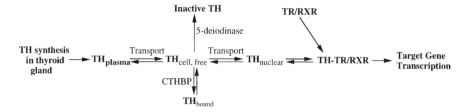

Figure 5.2 *A proposed molecular pathway for TH to regulate cellular gene transcription. TH synthesized in the thyroid gland is circulated through the plasma to target tissues, where it is transported through yet unknown mechanisms into cells. The cellular concentrations of TH are further modulated through the binding by CTHBPs, inactivation by 5-deiodinase, and possibly other mechanisms. Within the target tissue T_4 can be converted to T_3 through 5'-deiodination. Although both T_4 and T_3 can bind to TRs, T_3 has a stronger affinity and is likely to be the predominant TH regulating gene transcription. The cytoplasmic TH is then transported into the nucleus. The nuclear TH can bind to TRs, which are most likely bound to TREs in target genes as heterodimers with RXRs (9-cis-retinoic acid receptors), and regulate transcription.*

al., 1987; Robbins, 1992; Ribeiro et al., 1996; Benvenga and Robbins, 1993). This is first suggested by the cloning and characterization of a yeast gene involved in exporting glucocorticoid (Kralli et al., 1995). Furthermore, FK506, a known inhibitor of mdr 1(multi-drug resistance) transporter can increase cellular levels and potency of steroid hormones in mammalian cells, implicating the existence of hormone export pathways (Kralli and Yamamoto, 1996). Direct evidence for the existence of a TH-exporting pathway comes from a study on a mutant mammalian cell line that requires higher levels of TH added to the culture medium to achieve similar responses as other cell lines (Ribeiro et al., 1996). This defect in TH responses can be remedied by blocking export pathways with the drug verapamil, a known inhibitor of mdr/P-glycoprotein transporters, arguing that the defect in the cell line is due to increased levels of TH export (Ribeiro et al., 1996). In addition, studies on amino acid transport in mammals have provided evidence that some amino acid transporters, e.g., those involved in tryptophan, may transport TH as well (Kamp and Taylor, 1997). Given the conservation in TH action (see Sections 5.3 and 5.4), both the export and import pathways are expected to be present in amphibians. Their exact roles in amphibian development and physiology will require the cloning and characterization of the genes involved.

Within the cytoplasm, TH encounters another group of proteins (Fig. 5.2). These are the cytoplasmic TH binding proteins (THBP, Table 5.1; Barsano and DeGroot, 1983; Cheng, 1991). In general, these proteins are multifunctional proteins that also serve other roles in the cytoplasm. They bind TH with dissociation constants of about $1-100$ nM. Two genes encoding cytosolic TH binding proteins have been cloned from *Xenopus laevis*. The first one is the frog homolog of a mammalian and avian cytosolic aldehyde dehydrogenase (Yamauchi and Tata, 1994) and the second one is the homolog of the mammalian M2 pyruvate kinase (Shi et al., 1994). Interestingly, the human M2 pyruvate kinase functions as the enzyme in the tetramer form but as a TH binding protein in its monomer form (Ashizawa and Cheng, 1992). The binding of TH shifts the equilibrium toward the monomer form and thus inhibits the kinase activity. It is therefore possible that TH may be able to affect the function of different cytosolic binding proteins. In fact, it had been assumed that TH regulated cellular events through its action in the cytoplasm until the discovery of nuclear TH binding proteins or TH receptors (TRs). The exact roles of these cytosolic TH binding proteins in the TH signal transduction pathway are still unknown. They may serve as chelators to limit the cellular free TH concentration. Alternatively, they may serve as carrier proteins in the cytoplasm to facilitate the transport of TH to the nucleus and/or as buffer proteins to maintain a sufficient level of cellular TH (Fig. 5.2).

The biological effects of TH are mainly mediated by TRs (Fig. 5.2). TRs are high-affinity TH-binding proteins with dissociation constants of less than 1 nM. They are localized in the nucleus in both the presence and absence of TH (Oppenheimer, 1979; Galton, 1983). Using photoaffinity labeling, Samuels and colleagues (Perlman et al., 1982) showed that TRs are chromatin-associated. These and other studies implied that TRs mediate the biological effects of TH by regulating gene expression in a TH-dependent manner. This idea was proved with cloning and functional studies of TRs in 1986 (Sap et al., 1986; Weinberger et al., 1986).

5.3 THYROID HORMONE RECEPTORS

The first TR genes were cloned in 1986 in chicken and human (Sap et al., 1986; Weinberger et al., 1986). Subsequently, TR genes have been cloned in many vertebrate species (Mangelsdorf et al., 1995), including four genes in *Xenopus*, two TRα genes and two TRβ genes (Fig. 5.3; Brooks et al., 1989; Yaoita et al., 1990). Alternative splicing of the TRβ transcriptions gives rise to two different isoforms for each TRβ gene (Fig. 5.3). In higher vertebrates, there is only one TRα and one TRβ gene. As in *Xenopus*, alternative splicing yields two TRβ isoforms (Lazar, 1993).

TRs belong to the superfamily of nuclear hormone receptors, including receptors for glucocorticoid, estrogen, retinoic acid, and vitamin D (Evans, 1988; Mangelsdorf et al., 1995; Tsai and O'Malley, 1994; Yen and Chin, 1994). Members of this family share many structural features. In general, each can be divided roughly into five domains, A/B, C, D, E, and F, respectively, from the amino to the carboxyl terminus (Fig. 5.3, Krust et al., 1986). The DNA binding domain (domain C) is located in the amino half of the protein and is the most highly conserved domain among different receptors (Fig. 5.4, Table 5.2). The large ligand binding domain (domain E) is in the carboxyl half of the protein and is conserved among TRs in different species. The rest of domains vary in sizes and sequences among different nuclear receptors (Fig. 5.4). The *N*-

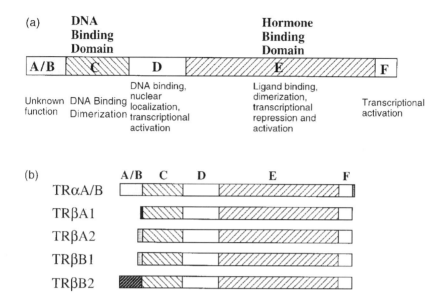

Figure 5.3 (A) *Domain structure and function of TRs (Krust et al., 1986; Tsai and O'Malley, 1994). The A/B domain has transcription activation properties for a number of nuclear receptor, especially GR (Tsai and O'Malley, 1994). However, no definite function can be assigned to it for TRs, at least, in amphibians. (B) Different isoforms of Xenopus laevis TRs. Only a single form of TRαA and TRαB has been cloned but two isoforms are known for each of the two TRβ genes. TRαA/B have three extra amino acids at the carboxyl terminal end compared to TRβ (see Fig. 5.4). TRβA1 has a very short A/B domain and TRβA2 and TRβB1 have identical A/B domains. TRβB2 has a very different A/B domain that is similar in length to A/B domains of TRαA/B. For simplicity, the "C" domain includes all the amino acids encoded by the two exons encoding the Zn^{2+} fingers of the DNA binding domain (i.e., up to the third amino acid of TRβA1 at the amino terminal side; see Fig. 5.4). Based on Yaoita et al. (1990).*

terminal A/B domain is highly variable in sequence and length, the shortest being the TRβ in *Xenopu laevis* (Fig. 5.3B, Yaoita et al., 1990). In general, this domain contains a transactivation function (AF), although its role in TR function is unclear. Another transactivation function domain is the AF-2 domain, which is located at the very C-terminus (F domain and part of the E domain). At least for TRs, this region also influences hormone binding (Tone et al., 1994).

Between the DNA and ligand binding domains is the D domain, or variable hinge region. This region often contains a nuclear localization signal and influences both DNA binding and transactivation (Giguere et al., 1986; Godowski et al., 1988; Hollenberg and Evans, 1988; Picard and Yamamoto, 1987; Guiochon-Mantel et al., 1989; Zechel et al., 1994; Lee and Mahdavi, 1993; Uppaluri and Towle, 1995; Puzianowska-Kuznica et al., 1996). How the D

```
GR    MDPKDLLKPS SGSPAVRGSP HYNDKPGNVI EFFGNYRGGV SVSVSASCPT   50

GR    STASQSNTRQ QQHFQKQLTA TGDSTNGLNN NVPQPDLSKA VSLSMGLYMG  100

GR    ESDTKVMSSD IAFPSQEQIG ISTGETDFSL LEESIANLQA KSLAPDKLIE  150

GR    ISEDPGGFKC DISAQPRPSM GQGGSNGSSS TNLFPKDQCT FDLLRDLGIS  200
                                                        MTMP    4
ER
GR    PDSPLDGKSN PWLDPLFDEQ EAFNLLSPLG TGDPFFMKSE VLSEGSKTLS  250

RXRα              MSSAAM DTKHFLPLGG RTCADTLRCT TSWTAGYDFS   36
ER    LPNKTTGVTF LHQIQ~~ELE TLTRPPLKIS LERPLGEMYV ENNRT~IFNY   54
GR    LEDGTQRLG  DHAKDMLLP SADRPISQVK TEK       EDYI ELCTP~VVNE 295

RXRα  SQ VNSSSLS SSGLRGSMTA PLLHPSLGNS GLNNSLGSPT QLPSPLSSPI   85
ER    PE GTTYDFA AAAAPVYSS~ S~SYAASSET FGSS~~TGLH T~NNVPP~PV  103
GR    EKFGPVYCVG NFSGS~LFGN KSSAI~VHGV STSGGQMYHY D~NTATI~QQ  345

TRαA              MDQ NLSGLDCLSE PDEKRWPDGK RKRKNSQCMG   33
RXRα  NGMGPPFSVI SPPLGPS~AI PSTPGLGYGT GSPQIHSPMN SVSSTEDIKP  135
ER    VFLAKLPQLS PFIHHHGQQV PYYLESEQGT FAVREAAPPT FY~SSSDNRR  153
GR    DVKPVFNLGS PGTSIAEGWN RCH~SGNDTA ASPGNVNFPN  ~SVFSNGY  393

TRαA  KSGMSGDSLV SLPSAGYIPS YLDKDEP**CVV CSDKATGYHY RCITCEGCKG**   83
TRβA1              ME~~~~~ ~~~~~~~L~~~ ~G~~~~~~~~ ~~~~~~~~~~   37
RXRα  PP~IN~ILK~ PMHPS~AMA~ F  TKHI~**AI** ~G~RSS~K~~ **GVYS**~~~~~~  183
VDR     ~EFMAAT TSIADTDMEF DKNVPRI~G~ ~G~~~~~F~F **NAM**~~~~~~~   47
ER    Q~~RERMSSA NDKGPPSME~ TKET RY~**A**~ ~~~Y~S~~~~ **GVWS**~~~~~A  202
GR    S~PGIRSDAS PSP~TSSTST GPPP KL~**L**~ ~~~E~S~C~~ **GVL**~~GS~~V  442

TRαA  FFRRTIQKNL HPSYSCKYDG CCIIDKITRN QCQLCRFKKC IAVGMAMDLV  133
TRβA1 ~~~~~~~~~~ ~~~~~~~~E~ K~V~~~V~~~ ~~~E~~~~~~ ~~~~~~T~~~   87
RXRα  ~~K~VRKDL  TYTCRDSK   D~M~~~RQ~~ R~~Y~~YQ~~ L~M~~       226
VDR   ~~~~SMKR   KAMFT~PFN~ D~R~T~DN~~ H~~S~~L~R~ VDI~~MKEFI   95
ER    ~~K~S~~G   ~ND~M~PATN Q~T~~~NR~K S~~A~~LR~~ YE~~~MKGGI  250
GR    ~~K~AVEG   QHN~L~AGRN D~~~~~~R~K N~PA~~YR~~ LQA~~        485

TRαA  LDDGKRVAKR KLIEENRQRR RKEE             MIKTLQQRPE  167
TRβA1 ~~~N~~L~~~ ~~~~~~~EK~ ~~D~             IQ~S~V~K~~  121
RXRα           ~~ EAVQ~E~~~G K  ~           RNENEVESSN  250
VDR   ~T~EEVQR~~ QM~NKRKSEE ALK~SMRPKI SDEQQKMIDI LLEAHRKTFD  145
ER    RK~RRGGRLL KH KRQKEEQ EQKNDVDPEL QELQE  LQA LLQQQ~   D  294
GR                                        N  LEA RKTKKK   I  496
                                            *  *

TRαA  PSSEEWELIR IVTEAHRSTN AQGSHWKQRR KFLPEDIGQS PMASMPD    214
TRβA1 ~TQ~~~~~~Q V~~~~~~A~~ ~~~~~~~~L~ ~~~~~~~~~A ~IVNA~E    168
RXRα  SAN~DMPVEK ~L ~~EHAVE PKTETYTEAN MG~APNSPSD ~VTNICQ    296
VDR   TTYSDFNKF~ PPVREN~DPF RRITRSSSVH TQGSPSEDSD VFTSS~~SSE 195
ER    PELQELQELQ ALLQQQQDPS EIRTASIWVN PSVKSMKLSP VLSLTAEQLI 344
GR    KGIQ QSTT ATARESPETS MTRTL   VP ASV AQLTP T      LI   532
       *    *    *       *    **    **    *  *        **
```

```
TRαA                       GD  KVDLEAFSEF  TKIITPAITR  VVDFAKKLPM     246
TRβA1                      GG  ~~~~~~~~Q~  ~~~~~~~~~~  ~~~~~~~~~~     200
RXRα                       A   A~KQL~T                 L~EW~~RI~H     314
   VDR  HGFFSASLFG QFEYSSMGGK  SGE~SMLPHI  ADLVSYS~QK  IIG~~~MI~G     245
   ER   SALMEAEAPI VYSEHDSTKP  LSEASMMTLL  ~NLADRELVH  MINW~~RV~G     394
   GR   SLLEVIEPEV LYSGYDSSIP  DTTRRLM~SL  NMLGGRQVVS  A~RW~~AI~G     582
         *    *    **    **  *            *   *   *   *     *   *


TRαA    FSELTCEDQI ILLKGCCMEI  MSLRAAVRYD  PDSET    LT  LSGEMAVKRE     293
TRβA1   ~C~~P~~~~~ ~~~~~~~~~~  ~~~~~~~~~~  ~E~~~    ~~  ~N~~~~~T~G    247
RXRα    ~~~~PLD~~V ~~~RAGWN~L  LIASFSH~SI  AVKDG    IL  ~ATGLH~H~N    361
   VDR  ~RD~IA~~~~ A~~~SSVI~V  IM~~SNQSFS  L~DMSW   TC  G~EDFKY~VD    293
   ER   ~VD~~LH~~V H~~ECAWL~~  LMVGLIW~ S  VEHPGK   ~S  FAPNLLLD~N    441
   GR   ~RN~HLD~~M T~~QYSW~FL  ~VFALGW~ S  YKQTNGSILY  FAPDLVITED    631
             *        *           * *  *               *** *


TRαA    QLKNGGLGVV SDAIF DLGR  SLAAFNLDDT  EVALLQAVLL             MS   334
TRβA1   ~~~~~~~~~~ ~~~~~ ~~~V  ~~SS~S~~~~  ~~~~~~~~~~             ~~   288
RXRα    SAHSA~V~AI F~RVLTE~VS  KMRDMQM~K~  ~LGC~R~IV~             FN   403
   VDR  DVTQA~HNME LLEPLVKFQV  G~KKLD~HEE  ~HV~~M~ICI             L~   335
   ER   ~GRCVEGL~E IFDMLVTTAT  RFRMMR~RGE  ~FIC~KSIIL  LNSGVYTFL~      491
   GR   RMH LPFMQE RCQEMLKIAG  EMSSLQISYD  ~YLCMKVL~~  M              CT  673
                *          *                * *


TRαA    SDRTGLICTD KIEKCQETYL  LAFEHYINHR  KHNIPH    F  WPKLLMKVTD    381
TRβA1   ~~~P~~ASVE R~~~~~~GF~  ~~~~~~~~Y~  ~~~~A~    ~  ~~~~~~~~~~    335
RXRα    P~SK~~SNPL EV~ALR~KVY  ASL~A~CKQK  YPEQPG    R  FA~~~LRLPA    450
   VDR  P~~P~~QDKA LV~SI~DRLS  STLQT~~LCK  H PPPGSRLL  YA~MIQ~LA~    384
   ER   ~TLES~EDT~ L~HIILDKII  DTLV~FMAKS  GLSLQQQQRR  LAQ~~LILSH    541
   GR   IPKE~~KSHA LF~EIRM~~I  KELGKA~VK~  EG~SSQNWQR  FYQ~TKLLDS    723
             *     *  *   *        *       *   *   *


TRαA    LRMIGACHAS RFL HMKVEC  PTEL FPPLF  LEVFEDQEV                   418
TRβA1   ~~~~~~~~~~ ~~~ ~~~~~~  ~~~~ ~~~~~  ~~~~~~~~~                   369
RXRα    ~~S~~LKCLE HLF FF~LIG  D~PI   DTFL M~ML~APHQM  T                488
   VDR  ~~SLNEEHSK QYR SISFLP  EHSMKLT~LM  ~~~~S~EIP                   422
   ER   I~HMSNKGME HLY SMKC    KNVVPLYD~L  ~~MLDAHRIH  TPKDKTTTQE    588
   GR   MHEVAENLLA FCFLSFLD    KSMSIEFPDM  LSEIISNQIP  KYSSGNLKKL    771
                *                *            *


   ER   EDSRSPPTTT VNGASPCLQP  YYTNTEEVSL  QSTV                       622
   GR   LFHQK                                                         776
```

Figure 5.4 Sequence comparison among Xenopus nuclear receptors: TRs (Yaoita et al., 1990), RXRα (Blumberg et al., 1992), VDR (vitamin D receptor, Li et al., 1997c), ER (estrogen receptor, Xing and Shapiro, 1993), and GR (glucocorticoid receptor, Gao et al., 1994). Gaps are introduced to allow maximal alignment. The "~"s indicate residues that are identical to the corresponding ones in TRαA. The DNA binding domain (bold letters) is highly conserved among different receptors. TRs, VDR, and RXR belong to the same nuclear receptor subfamily but ER and GR belong to a different subfamily. Consistently, there are many residues (indicated by "*") in the carboxyl half of the proteins that are conserved between ER and GR but different in the other receptors.

TABLE 5.2 Homology Comparison of *Xenopus* Nuclear Receptors[a]

Identity\similarity	TRα	TRβA1	RXRα	VDR	ER	GR
TRα	100	96	64	64	59	59
TRβA1	91	100	65	65	58	58
RXRα	53	55	100	56	58	53
VDR	52	53	44	100	59	50
ER	56	56	53	45	100	65
GR	48	44	52	39	59	100

[a] The percentage of amino acid identity (in italics) and similarity (roman) are calculated for only the DNA binding domain (see Fig. 5.4). All nuclear receptors have similar levels of sequence identities with TRs in the DNA binding domain, but TRs share higher levels of amino acid sequence similarity with RXR 19-cis retinoic acid receptor or retinoid X receptor and VDR (Vitamine D receptor) than ER (estrogen receptor) and GR (glucocorticoid receptor). This may reflect the fact that TR, RXR, and VDR belong to the same nuclear receptor subfamily whereas ER and GR belong to another subfamily (Tsai and O'Malley, 1994; Mangelsdorf et al., 1995).

domain influences receptor function is unknown. Its variable length and sequence may facilitate conformational change of the receptor during DNA binding and/or transcriptional activation, thus also facilitating the determination of the specificity in transcriptional regulation by the receptors.

5.3.1 Thyroid Hormone Binding by Thyroid Hormone Receptors

Specific binding of thyroid hormone by TRs requires the minimum thyroid hormone binding domain, which spans the *C*-terminal domain of about 250 amino acids (Forman and Samuels, 1990; Lazar, 1993). This domain shares only a low level of homology among different nuclear receptors. However, it is highly conserved throughout evolution. Thus TRs from species as diverse as frogs and humans are more than 90% identical in this region (Yaoita et al., 1990). Such conservation accounts for the essentially identical high affinities of TRs for T_3 (with K_d in the subnano molarity range) (Sap et al., 1986; Weinberger et al., 1986; Davey et al., 1994; Puzianowska-Kuznicka et al., 1996).

The TH-binding domain contains nine heptad repeats (Fig. 5.5; Forman and Samuels, 1990). Deletion and mutational analyses have shown that these repeats are not only important for TH-binding but also for dimerization of the receptor with itself or other members of the nuclear receptor superfamily (the thyroid and retinoid receptor subfamily) and transcriptional regulation (Forman and Samuels, 1990; Tsai and O'Malley, 1994). The crystal structure of the domain bound with a ligand reveals that the holo-ligand binding domain consists of 12 α-helices with the ligand buried inside (Fig. 5.5; Wagner et al., 1995). The last helix (helix 12) encompasses the carboxyl terminus where the AF-2 domain (F domain) is located.

Although the sequence of the ligand binding domain for different nuclear receptors diverges extensively, the domain seems to maintain a similar overall

THYROID HORMONE RECEPTORS

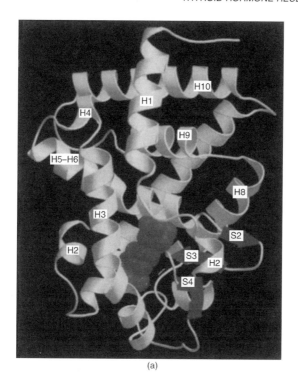

(a)

```
                 H1                        H2  S1
QRPEPSSEEW ELIRIVTEAH RSTNAQGSHW KQRRKFLPED IGQSPMASMP  213

               H3              H4           H5-H6
DGDKVDLEAF SEFTKIITPA ITRVVDFAKK LPMFSELTCE DQIILLKGCC  263

  H5-H6      S2     S3     S4      H7                H8
MEIMSLRAAV RYDPDSETLT LSGEMAVKRE QLKNGGLGVV SDAIFDLGRS  313
                              R1                R2,3

            H9                                H10
LAAFNLDDTE VALLQAVLLM SSDRTGLICT DKIEKCQETY LLAFEHYINH  363
          R4-6                             R7,8

              H11                           H12
RKHNIPHFWP KLLMKVTDLR MIGACHASRF LHMKVECPTE LFPPLFLEVF  413
           R9

EDQEV                                                  418
```
(b)

Figure 5.5 (a) Stereo drawing of the TR ligand binding domain, with the secondary structure elements labeled. The hormone (magenta) is depicted as a space-filling model; α-helices and coil conformations are yellow, β-strands are blue. Reprinted with permission from Nature (378:690–697) Copyright (1995) Macmillan Magazines Limited. (b) Structural organization of the ligand binding domain of Xenopus laevis TRαA. Based on the data of Wagner et al. (1995), there are 12 α-helices (bold letters, H1–H12) and 4 β-strands (italic letters, S1–S4) in the TH-binding domain. This region also contains nine heptad repeats (Forman and Samuels, 1990; underlined, R1–R9) that are important for receptor dimerization with itself or other members of the retinoid-thyroid receptor subfamily.

structure. First, all nuclear receptors contain the nine heptad repeats, although the exact sequences differ (Forman and Samuels, 1990). Second, the crystal structure of the holo-TH binding domain is remarkably similar to the holo-RA (retinoic acid) binding domain of RARγ (RA receptor γ, Renaud et al., 1995). It also resembles that of steroid hormone receptors (Tanenbaum et al., 1998; Moras and Gronemeyer, 1998). Both TR and RARγ ligand binding domains contain 12 α-helices and have the ligand binding pocket buried inside. Interestingly, the apo-RA binding domain of RXRα (Bourguet et al., 1995) has a slightly different structure. Although it has the 12 α-helices in similar locations as in the RARγ and TRα, its last helix (helix 12) stretches away from the rest of the protein, in contrast to the ligand-bound RARγ and TRα, where the helix 12 wraps tightly against the rest of the ligand binding domain with the ligand buried inside. This structural difference argues that upon ligand binding, a conformational change occurs in the ligand binding domain that leads to the inward folding of the helix 12. Indeed, biochemical studies using spectroscopy, enzymatic analysis, and conformation-specific antibodies have provided direct evidence for conformational changes in TRs induced by TH binding (Ichikawa et al., 1990; Bhat et al., 1993; Toney et al., 1993; Yen et al., 1993; Tsai and O'Malley, 1994; Yen and Chin, 1994). Such conformational changes may be responsible for the ligand-dependent transcriptional regulation by nuclear receptors.

Of the two natural thyroid hormones, T_3 binds to TRs with about 5- to 10-fold higher affinities than T_4 (Jorgensen, 1978). Both T_3 and T_4 can activate TH-dependent target genes and induce precocious metamorphosis (Dodd and Dodd, 1976; Weber, 1967; Shi and Brown, 1990). However, it is not clear whether T_4 functions directly through TRs or is first converted through the action of 5'-deiodinases to T_3, which then activate TRs. Several studies suggest that the latter is likely the in vivo pathway for T_4 to exert its biological effect. First, 5'-deiodinases are widely present in different tissues and organs (St. Germain, 1994; Davey et al., 1995; Becker et al., 1997; St. Germain and Galton, 1997). Active mechanisms appear to exist that transport the hormones into and out of the cell, thus allowing cells to respond to extracellular T_3 even if the cells themselves lack any 5'-deiodinase to convert T_4 to T_3. Finally, studies with inhibitors of 5'-deiodinases can inhibit the biological effects of T_4 (Becker et al., 1997), implying that at least in certain cases T_3 is the active thyroid hormone whereas T_4 is inactive or much less active. Thus further biochemical studies are needed to understand the structural differences between the T_4- and T_3-bound TRs. Such differences may result in distinct interactions between TRs and their co-factors (see below) to lead to possible different biological effects.

5.3.2 DNA Binding by Thyroid Hormone Receptors

The DNA binding domain of TRs mediates the specific recognition of the thyroid hormone response elements (TREs) present in TH response genes (Tsai and O'Malley, 1994; Yen and Chin, 1994). The DNA binding domain consists of two adjacent Zn^{2+} fingers, each of which contains two histidine and two cysteine residues that coordinate a Zn^{2+} ion in a tetrahedral configuration (Fig. 5.6; Rastinejad et al., 1995). This coordination of the Zn^{2+} ions by the two Zn^{2+} fingers determines the overall structure of the DNA binding domain. The primary sequences of this domain share considerable homology among the members of the nuclear receptor superfamily (Fig. 5.4; Table 5.2), which accounts for the similarity in the DNA sequences that they recognize.

TRs can bind to DNA as monomers, homodimers, and heterodimers formed with other members of the thyroid–retinoid receptor subfamily (Forman and Samuels, 1990; Lazar, 1993; Tsai and O'Malley, 1994; Mangelsdorf et al., 1995).

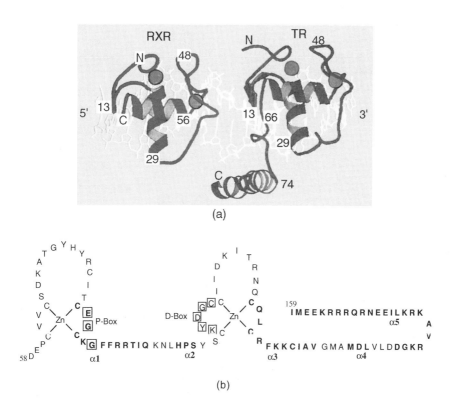

Figure 5.6 (a) *The overall architecture of RXR (red) and TR (blue) DNA binding domain on a direct repeat sequence of AGGTCA with 4-bp separation (yellow). The 5' and 3' ends of the DNA are indicated. The amino acids 1–74 of TR correspond to amino acids 61–134 of Xenopus TRαA (see B). Reprinted with permission from* Nature *(375: 203–211) Copyright (1995) Macmillan Magazines Limited. (b) Structural organization of the DNA binding domain. The sequence from amino acids 58 to 159 of Xenopus TRαA corresponds to the region whose structure was determined by X-ray crystallography as shown in (a), and contains 5 α-helices (α1–α5, bold letters). The P-box (residues in shaded boxes) and D-box (residues in boxes) are important for DNA recognition and dimerization, respectively (Tsai and O'Malley, 1994).*

Biochemical studies have shown that the optimal binding site is the hexanucleotide AGGTCA. This hexanucleotide itself can be recognized specifically by a TR monomer. TR homodimers and heterodimers can recognize a palindromic sequence made of two inverted repeats of this hexanucleotide or of two direct repeats of the hexanucleotide separated by 4 bp. Two direct repeats separated by 1, 2, 3, or 5 bp can be bound by TR dimers with drastically reduced affinities.

TR monomer–DNA complexes are much less stable than the TR dimer–DNA complexes. The best DNA-binding dimers are the ones formed between TRs and RXRs (9-*cis*-retinoic acid receptors) (Fig. 5.7; Heyman et al., 1992; Kurokawa et al., 1993; Leid et al., 1992; Marks et al., 1992; Perlmann et al., 1993; Yu et al., 1991; Zhang et al., 1992). These heterodimers not only bind specifically to TREs with high affinities but also regulate the transcription of TH response genes with high specificity both in vitro and in tissue culture cells (also see Fondell et al., 1993; Wong et al., 1995, 1997a; Wong and Shi, 1995; Puzianowska-Kuznicka et al., 1997). These observations suggest that TR-RXR heterodimers are the true mediators of the biological effects of TH (Mangelsdorf and Evans, 1995).

Figure 5.7 Strong binding of a TRE requires the presence of both TRs and RXRs. The mRNAs encoding indicated receptors were injected into Xenopus laevis oocytes. After overnight incubation to allow the synthesis of the receptors, protein extracts were prepared and used to test their ability to bind the ^{32}P-labeled TRE present in the Xenopus TRβA promoter (Ranjan et al., 1994) by the gel mobility shift assay. The binding was carried out with individual extracts or mixtures of different combinations of TR- and RXR-containing extracts. Adapted from Wong and Shi (1995) with permission from The American Society for BioChemistry and Molecular Biology, Inc.

Studies of TH response gene promoters have revealed that most of them have TREs consisting of two direct repeats of sequences highly similar to AGGTCA (Umesono et al., 1991; Naar et al., 1991; Ranjan et al., 1994). Biochemical studies as well as transcriptional analysis of mutant TREs and TRs show that the binding of such a TRE by a TR/RXR heterodimer involves the recognition of the 5' repeat by the RXR and the 3' repeat by the TR of the heterodimer (Perlmann et al., 1993; Kurokawa et al., 1993). Such an occupation of the TRE by the heterodimer has been directly visualized in an X-ray structure of a cocrystal of a TRE made of two direct repeats separated by 4 bp and a heterodimer consisting of the DNA binding domain of a TR and an RXR (Fig. 5.6, Rastinejad et al., 1995). The crystal structure provides evidence for specific DNA–amino acid contacts that appear to determine the binding specificity. The residues that are involved in such contacts are from the same regions that have been shown by biochemical experiments to be critical for DNA binding specificity (Tsai and O'Malley, 1994).

Furthermore, specific interactions between the TR and RXR moieties are also present in the crystal, supporting the biochemical data that implicate the involvement of the DNA binding domain (Perlmann et al., 1993), in addition to the ligand binding domain (Forman and Samuels, 1990), in heterodimerization between TRs and RXRs.

The importance of TR/RXR heterodimers in mediating TH signaling are reflected not only in their ability to recognise specifically a TRE but also in the effects of DNA on TR/RXR heterodimers to interact with transcription cofactors. As discussed in detail below, unliganded TR/RXR heterodimers are capable of repressing the transcription of TH-activated genes. Such repression involves very likely the participation of corepressors like SMRT (silencing mediators of receptors of thyroid hormone, Chen and Evans, 1995) or N-CoR (nuclear receptor co-repressors; Horlein et al., 1995). TRs can interact with SMRT and N-CoR in the absence of TH in solution and TH inhibits this binding. Interestingly, for N-CoR to bind to TR in a DNA–TR complex, the complex must be formed between a TRE and TR/RXR heterodimer (Zamir et al., 1997a). TR monomer or homodimer–TRE complexes show little or no binding to N-CoR as analyzed by the gel mobility shift assay. Similar studies also indicate that TRE sequences can influence the interaction between DNA-bound TR and the coactivator SRC-1 (Takeshita et al., 1998). Thus specific receptor dimers are important for DNA sequence recognition and the binding site in turn can influence the interactions between the receptors and downstream mediators.

5.4 MECHANISM OF TRANSCRIPTIONAL REGULATION

TH can not only upregulate but also downregulate gene expression in target tissues or cells. It is generally believed that both effects of TH are mediated by TRs at the transcriptional level. Thus, depending upon the target promoters and/or cell types, TH-bound TRs can either activate or repress transcription. The vast majority of the known TH response genes are those that are upregulated by the hormone and most studies of receptor function have been on these upregulated genes. Relatively little is known about how liganded TRs repress transcription. The discussion focuses only on the mechanisms involved in regulating the transcription of TH-upregulated genes.

Similar to DNA binding by TRs, RXRs are critical for optimal transcriptional response to TH. Earlier transcription studies using transient transfection assays showed that overexpression of TRs alone could activate a reporter promoter bearing a TRE when TH is present. However, RXRs are present in many types of cells. In fact, co-overexpression of TRs and RXRs leads to more greatly enhanced activation of the reporter gene than when TRs alone are overexpressed (see references in the preceding section). Furthermore, studies on amphibian metamorphosis have demonstrated a coordinated expression of TR and RXR genes during organ transformations (Wong and Shi, 1995). In addition, both TRs and RXRs are required for efficient transcriptional regulation of a TRE-containing reporter in frog oocytes or endogenous TH-response genes in frog embryos (Wong et al., 1995; Wong and Shi, 1995; Puzianowska-Kuznicka et al., 1997). These studies provide strong evidence for a physiological role of RXRs in TH signal transduction.

5.4.1 Activation versus Repression

Transcriptional activation by TH requires the binding of TRs, most likely as heterodimers with RXRs, to TREs present in the regulatory regions of the TH-response genes. The binding of TREs by TR–RXR heterodimers is, however, independent of TH both in solution and in chromatin (Figs. 5.7, 5.8; Perlman et al., 1982; Wong et al., 1995). Thus TH appears to bind to TRs and trigger conformational changes in TRs that activate the receptor function. Indeed, biochemical studies such as protease digestion and spectroscopic analysis have demonstrated TH-induced conformational changes in TRs (Baht et al., 1993; Toney et al., 1993; Yen et al., 1993). How such changes lead to transcriptional activation remains to be elucidated.

Interestingly, cultured cell transfection experiments have revealed that in the absence of TH, TRs repress the transcription of TRE-containing promoters although in the presence of TH, the TRs enhance the transcription from these promoters (Baniahmad et al., 1992; Brent et al., 1993; Damm et al., 1989; Glass et al., 1989; Graupner et al., 1989; Sap et al., 1989; Wong et al., 1995). Thus unliganded TRs seem to function as repressors. Such a suggestion has been substantiated by in vitro studies. Roeder and his colleagues (Fondell et al., 1993) established an in vitro transcription system using partially purified basal transcription factors and RNA polymerase II, which could accurately transcribe a TRE-containing promoter. Addition of TRs to such a transcription reaction led to the repression of the promoter activity and this repression could be relieved by T_3. However, no transcriptional activation above the basal level

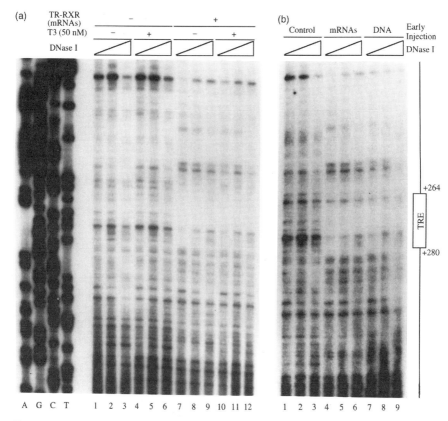

Figure 5.8 TRβ/RXRα heterodimer can bind to a TRE in a T_3-independent manner and even when the TRE is preassembled into chromatin in vivo. (A) TR/RXR can bind to a TRE independent of T_3. Groups of 30 oocytes were injected with (+, lanes 7–12) or without (−, lanes 1–6) TRβ/RXRα mRNAs (2.7 ng/oocyte) 6 hr before single-stranded pTRβA was injected and treated with (+, lanes 4–6, 10–12) or without (−, lanes 1–3, 7–9) T_3 as indicated. The single-stranded plasmid was replicated into the double-stranded form and chromatinized in the presence of TRβ/RXRα (see Fig. 5.9). After overnight incubation, the oocytes were collected and processed for PCR-mediated DNase I footprinting analysis with increasing amounts of DNase I (400 U/ml, 600 U/ml, and 800 U/ml, respectively). (Lanes A, G, C, T) Sequencing ladders of pTRβA using the same end-labeled primer used for the PCR-mediated DNase I footprinting. The diagram at right of Part B illustrates the position of the TRE in TRβA promoter. Note that the protection of the region further upstream (top of the gel) also depends on the presence of the TRβ/RXRα. The nature of this protection is unknown. (B) TR/RXR heterodimer binds to the TRE preassembled into chromatin in vivo. Oocytes were preinjected with TRβ/RXRα mRNAs (lanes 7–9) or single-stranded pTRβA promoter construct first (lanes 7–9) followed by the injection of single-stranded pTRβA promoter construct or mRNAs, respectively, 6 hr later. Control indicates no mRNA injection (lanes 1–3). DNase I footprinting was done as in (A). Adapted from Wong et al.(1995) with permission from Cold Spring Harbor Laboratory Press.

activity obtained in the absence of TRs was observed when both T_3 and TRs were present in this system, suggesting that some components necessary for activation are missing in the in vitro system.

5.4.2 Coactivators and Corepressors

Increasing evidence suggests that both transcriptional repression by unliganded TRs and activation by T_3-bound TRs involve other TR-interacting cofactors. Many such factors have been isolated based on their ability to interact with TRs in the presence or absence of T_3 or under both conditions (Halachmi et al., 1994; Baniahmad et al., 1995; Chen and Evans, 1995; Horlein et al, 1995; Le Douarin et al., 1995; Lee et al., 1995a,b; Onate et al., 1995; Burris et al., 1995; Zamir et al., 1996, 1997b; Kamei et al., 1996; Chakravarti et al, 1996; Chen et al., 1997; Li et al., 1997a; Takeshita et al., 1997; Fondell et al., 1996; Yuan et al., 1998; Koenig, 1998; Chen and Li, 1998; Rachez et al.,1998).

Two corepressors have been studied extensively. These are two large proteins (SMRT, 1495 amino acids, and N-CoR, 2453 amino acids, respectively) that can interact with unliganded TRs but not with T_3-bound TRs, suggesting that they may be involved in transcriptional repression by unliganded TRs (Chen and Evans, 1995; Horlein et al., 1995). Consistent with this idea, GAL4–SMRT fusion protein strongly represses the transcriptional activity of a reporter gene downstream of GAL4 DNA binding sites (Chen and Evans, 1995). Both SMRT and N-CoR appear to be members of a related family, sharing considerable structural and sequence similarity. Both interact with the D domain of TRs. Interestingly, a mutation in the D domain of TR that abolishes the ability of TR to suppress transcription also fails to interact with SMRT, again pointing out the potential involvement of SMRT in TR-mediated repression (Chen and Evans, 1995). In addition to TR, these corepressors also interact with other members of the nuclear hormone superfamily.

The transcription coactivator proteins interact with nuclear receptors in the presence of the ligand. Often they can interact with many different nuclear hormone receptors and enhance the activities of these receptors when cotransfected with the receptors into mammalian tissue culture cells (Onate et al., 1995; Torchia et al., 1998). Further evidence for a role of coactivators in nuclear hormone receptor function has come from studies with gene knockout in mice. The inactivation of SRC-1 in mice leads to partial resistance to steroid hormones (Xu et al., 1998).

Coactivator–receptor interaction involves the ligand binding domain of the hormone receptor and LXXLL motifs present in many coactivators (Le Douarin et al., 1995; Feng et al., 1998; Nolte et al., 1998; Darimont et al., 1998; McInerney et al., 1998). The core domain of the carboxyl terminal activation domain (AF-2) plays a critical role in this interaction. This domain consists of a part of the E and F domains and is highly conserved among nuclear hormone receptors. The crystal structure of the holo-ligand binding domain of retinoic acid receptor (RAR)-γ shows that this core domain is a part of an α-helix that folds back towards the core of the ligand binding domain (Renaud et al., 1995). In contrast, in the apo-ligand binding domain of the human RXRα (Bourguet et al., 1995), this helix is more extended whereas the rest of the structure is highly similar to holo RAR-γ. These results suggest that the ligand-induced change in the α-helix folding may be responsible for the recruiting of coactivators such as SRC-1. A similar mechanism may be employed by TRs as the structure of the ligand binding domain is conserved (see Section 5.3.1 and Wagner et al., 1995). In support of this, mutational and structural studies of TR have shown that ligand binding induces the formation of a hydrophobic groove within the ligand binding domain through the folding of the carboxyl terminal α helix (helix 12) and thus facilitates binding to LxxLL motifs in coactivators (Feng et al., 1998; Darimont et al., 1998).

5.4.3 Role of Chromatin

Most of the functional studies of hormone receptors were carried out in vitro or by transient transfection experiments in tissue culture cells. However, genomic DNA in eukaryotic cells is associated with histones and other nuclear proteins and assembled into chromatin. Growing evidence indicates that chromatin structure plays important roles in regulating gene transcription (Kornberg and Lorch, 1995; Lewin, 1994; Svaren and Horz, 1993; Workman and Kingston, 1998; Wolffe, 1998). It is known that transcriptionally active chromosome domains have distinct structure and protein compositions compared to repressed chromatin. In addition, transcriptional activation is often accompanied by chromatin reorganization (Lee and Archer, 1994; Mymryk and Archer, 1995; Svaren et al., 1994; Truss et al., 1995; Zaret and Yamamoto, 1984; Almer et al., 1986; Archer et al., 1992; Struhl, 1996; Wolffe, 1994, 1997). One of the best studied examples is the nucleosome remodeling following glucocorticoid induction of MMTV promoter (Archer et al., 1991, 1992; Pina et al., 1990; Truss et al., 1995). In this case, the binding by glucocorticoid receptor (GR) appears to modify the nucleosome, allowing the binding of the transcription factor NFI, which in turn activates the promoter. Thus to understand the mechanism of TR action, it is important to use properly chromatinized templates. One way is to establish permanent cell lines with the TRE-containing promoters of interest inserted into the genome. Another approach is to utilize the yeast model system. However, in this case, the unliganded TRs fail to repress basal activity; instead they activate the promoter slightly even when T_3 is absent (Lee et al., 1994b; Sande and Privalsky, 1994; Uppaluri and Towle, 1995). The third, experimentally more feasible way is to make use of the *Xenopus* oocyte transcription system.

The *Xenopus* oocyte offers a unique system to study transcription owing to its large storage of basal transcription factors as well as many other proteins important for early embryogenesis including histones for chromatin assembly. Furthermore, it is easy to introduce exogenous proteins into oocytes by micro-injecting either purified proteins or their mRNAs into the cytoplasm and to introduce promoter containing DNA into the nucleus for promoter function studies. The plasmid DNA injected into the nucleus is assembled into chromatin (Almouzni et al., 1990). Interestingly, the types of chromatin formed differ depending upon the forms of the injected DNA. When double-stranded promoter-containing plasmid DNAs are used, they are chromatized in 5–6 hr with less well defined nucleosome arrays such that the transcription from the promoters is at high levels (Fig. 5.9A). In contrast, when single-stranded plasmid DNAs are used, they are quickly replicated (1–2 hr) and assembled into chromatin in a replication-coupled chromatin assembly pathway, mimicking the genomic chromatin assembly process in somatic cells. The resulting templates produce much lower levels of transcriptional activity (Fig. 5.9A). Thus by using different forms of promoter-containing DNA, it is possible to study the transcriptional regulation under different chromatin conditions.

Xenopus oocytes have only a very low level of endogenous TR that is insufficient to affect the transcription of a TRE-containing promoter (Eliceiri and Brown, 1994; Wong and Shi, 1995; Wong et al., 1995). However, when exogenous *Xenopus* TRs and RXRs are co-introduced into the oocytes, they can repress the transcription from the double-stranded template containing a TRE (Fig. 5.9B) (TRs can function less efficiently in the absence of injected RXRs) (Wong and Shi, 1995; Wong et al., 1995). Transcriptional repression of the promoter occurs even when the single-stranded DNA is injected to allow the promoter DNA to be assembled into chromatin in the replication-coupled chromatin assembly pathway. However, in this case, efficient transcriptional repression by the unliganded TR–RXR requires the presence of unliganded TRs during replication coupled chromatin assembly. Currently, it is unclear how the unliganded TRs and the chromatin assembly process cooperate to effect maximal repression of the promoter. It is possible that the binding by TR–RXR heterodimer may help to establish a more ordered and more repressive state of nucleosomes near the promoter region. However, current biochemical methods have failed to detect any differences in chromatin structure when the receptors are introduced before, as opposed to after, chromatin assembly.

Independent of whether the double- or single-stranded promoter DNA is used, the addition of T_3 leads to transcriptional activation (Wong et al., 1995). The final transcriptional activity is essentially identical when either the single- or double-stranded promoter DNA is used (Fig. 5.9B), indicating that thus T_3-bound TRs can overcome any repression incurred by chromatin. Therefore, the central question for transcriptional activation is how the TR–RXR heterodimer overcomes the repression by chromatin. To this end, it has now been shown that T_3 binding to TRs bound to chromatized templates causes the disruption of the ordered chromatin formed during replication-coupled chromatin assembly (Fig. 5.10). Furthermore, this chromatin disruption occurs even when transcription is blocked by α-amanitin (Fig. 5.10). Thus T_3-bound

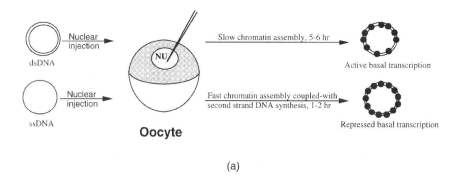

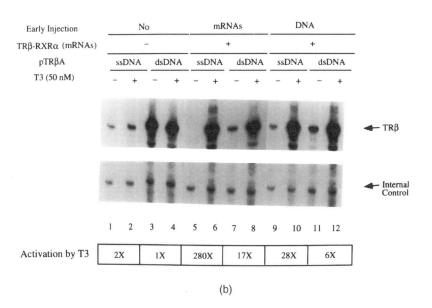

Figure 5.9 (A) *Schematic diagram showing the difference in chromatin assembled from plasmid DNA templates microinjected into frog oocyte nucleus (NU) and the resulting transcriptional activities of the template. Double-stranded DNA, dsDNA; single-stranded DNA, ssDNA. (B) Replication-coupled chromatin assembly in the presence of unliganded TR/RXR heterodimer is required for efficient repression of transcription from the TRβA promoter. Groups of 20 oocytes were injected without (−) or with (+) TRβ/RXRα mRNAs (0.9 ng/oocyte), either 6 hr before (early injection mRNAs) or 6 hr after (early injection DNA) the injection of either ssDNA or dsDNA of the pTRβA. The oocytes were then treated for 16 hr with (+) or without (−) T_3 as indicated. The RNA transcribed from the TRβA promoter was analyzed by using the primer extension assay. The mRNA level in each case was quantified by densitometery and normalized against the internal control. Fold of transcriptional activation by T_3 was quantified by comparing the mRNA level in the presence of T_3 with that in the absence of T_3. Adapted from Wong et al. (1995) with permission from Cold Spring Harbor Laboratory Press.*

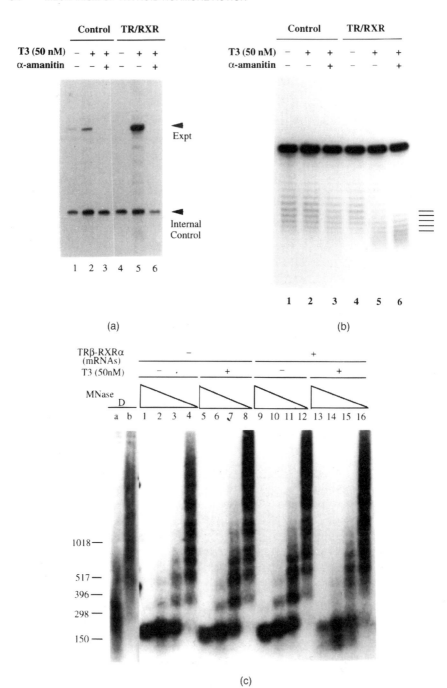

Figure 5.10 The transcriptional activation by liganded TR/RXR from the repressive chromatin assembled by the replication-coupled pathway is accompanied by extensive chromatin disruption. (A) The transcriptional activation by liganded TR/RXR can be inhibited by α-amanitin.

TR–RXR heterodimers can disrupt chromatin structure through an active process. Interestingly, this disruption requires at least the presence of the intact C-terminal activation domain (parts of E and F domains) as mutations in TR that abolish transcriptional activation by T_3 and also eliminate chromatin disruption (Wong et al., 1997a). On the other hand, mutant promoters that fail to direct the specific transcription can still direct chromatin disruption by TR–RXR heterodimers in the presence of T_3 as long as they contain a TRE (Wong et al., 1997a). These results suggest that chromatin disruption and transcription process are separable but both require the binding of transcriptionally active TR–RXR heterodimers.

The ability of TR–RXR heterodimers to bind to a TRE present in a nucleosome as suggested by the transcription studies in oocytes and in vivo footprinting experiments is also supported by in vitro biochemical studies (Wong et al., 1995, 1997b). The TRE-containing region of the *Xenopus* TRβA gene can be reconstituted into a nucleosome in vitro. Interestingly, the nucleosome is rotationally positioned with the major groove of the TRE exposed. Upon the addition of TRs and RXRs to the nucleosome, a triple complex is formed containing the DNA, histones, and a TR–RXR heterodimer. However, the addition of T_3 to this triple complex does not appear to affect the rotational positioning of the DNA around the histone octamer. Therefore, TR–RXR heterodimer binding even in the presence of T_3 is insufficient to disrupt the nucleosome. Alternatively, the disruption can not be detected by the DNase I footprinting assay used for the in vitro experiment (for in vivo disruption assay on plasmid DNA, the micrococcal nuclease digestion is used).

Groups of 20 oocytes were injected with ssDNA (100 ng/µl, 23 nl/oocyte) and TR/RXR mRNAs (100 ng/µl, 27 nl/oocyte) and treated with or without hormone as indicated. α-Amanitin was co-injected with ssDNA at a concentration of 10 µg/µl. The transcription was analyzed by primer extension using CAT primer, and the internal control is the primer extension product from an unknown endogenous mRNA as described (Wong et al., 1995). (B) The DNA topology assay indicates that liganded TR/RXR also induces extensive chromatin disruption and that this chromatin disruption is not the by-product of processive transcription. The injections were the same as in (A). The DNA was purified from each group and the topological status of the DNA was analyzed using a chloroquine agarose gel. The top band in each lane represents the nicked form of plasmid. (C) Chromatin disruption in vivo by the binding of liganded TR/RXR heterodimer can also be detected by micrococcal nuclease (MNase) digestion assay. Oocytes were injected with (+) or without (−) TRβ/RXRα mRNAs 6 hr before injection of single-stranded pTRβA and treated with (+) or without (−) T_3 as indicated. After an overnight incubation, they were processed for MNase digestion with decreasing amounts of MNase (7.5, 2.5, 0.83, 0.24 units, respectively). After MNase digestion, DNA was purified and analyzed on a 1.5% agarose gel in 1 X TBE before transfer to a filter. The filter was hybridized with a randomly primed radiolabeled DNA fragment of the TRβA promoter region that contained the TRE (from +218 to +314). Note that strong nucleosome disruption was present in the TRE region (the mono- to trinucleosome region, lanes 13–15). The positions of mono-, di-, and trinucleosomes are indicated at left, and the size markers (in base pairs) are shown at right. (Lane D, a and b): The digestion of naked DNA with micrococcal nuclease. Adapted from Wong et al. (1995, 1997a) with permission from Oxford University Press and Cold Spring Harbor Laboratory Press.

The effect of TR–RXR heterodimers on chromatin structure bears similarities to that of glucocorticoid receptors (GRs). GR plays a critical role in the transcription of MMTV late promoter (Archer et al., 1992; Hager et al., 1993; Pina et al., 1990). This promoter requires the participation of a number of different transcription factors, such as NFI, and is organized in an ordered nucleosome array (Richard-Foy and Hager, 1987). Four nucleosomes are assembled at and around the promoter with many important transcription factor binding sites (binding sites for NFI and GR, etc.) located within the second nucleosome. Such an organization prevents the binding by NFI. In the presence of its ligand, GR can bind to its recognition site within the nucleosome and alter it such that NFI can now bind to its recognition site, thus activating the promoter transcription (Archer et al., 1991, 1992; Pina et al., 1990; Truss et al., 1995).

Both TR and GR belong to the superfamily of nuclear hormone receptors and both are capable of disrupting chromatin structure. However, a distinct difference exists between the two receptors. Although TR can bind to a TRE in the absence of T_3, GR is normally localized in the cytoplasm in the absence of glucocorticoid. The addition of the ligand causes GR to translocate into the nucleus, bind to its recognition site, and activate transcription. Thus its chromatin disruption activity may be associated directly with DNA binding and the role of the hormone is to facilitate the DNA binding. In contrast, DNA binding alone by a TR–RXR heterodimer does not disrupt chromatin structure. It is the T_3 binding to the TR that triggers a change in the TR, allowing it to alter the chromatin and changing the TR from a repressor to an activator.

The mechanisms underlying transcriptional activation-associated chromatin disruption are under intense investigation. Studies from yeast to mammals have suggested the involvement of the SNF/SWI family of proteins in chromatin remodeling (Yoshinaga et al., 1992; Coté et al., 1994; Imbalzano et al., 1994; Kwon et al., 1994; Tsukiyama et al., 1994, 1995; Tsukiyama and Wu, 1995; Varga-Weisz et al., 1995; Owen-Hughes et al., 1996; Steger and Workman, 1996; Svaren and Horz, 1996). In particular, SWI/SNF complexes have been shown to be required for transcriptional activation by GR in yeast and mammalian cells (Yoshinaga et al., 1992; Muchardt and Yaniv, 1993; Ostlund Farrangs et al., 1997; Fryer and Archer, 1998). Similar protein complexes may be involved in chromatin disruption by liganded TR/RXR. In addition, as described below, recent studies have also suggested a role of histone deacetylases and acetyltransferases in chromatin remodeling and transcriptional activation by nuclear receptors.

5.4.4 Participation of Histone Acetyltransferases and Deacetylases

Histone acetylation has long been implicated to influence gene expression (Allfrey et al., 1964; Pogo et al., 1966; Roth and Allis, 1996; Wolffe, 1996; Wolffe and Pruss, 1996; Grunstein, 1997; Struhl, 1998). Histone acetylation occurs at the lysine residues on the amino-terminal tails of the histones, leading to the neutralization of the highly positively charged histone tails and reduced affinity toward DNA (Hong et al., 1993). These changes alter the nucleosomal confirmation and chromatin accessibility, allowing easier excess of transcription factors to chromatin templates (Norton et al., 1989; Lee et al., 1993; Vettese-Dadey et al., 1996). Consequently, histone acetylation can increase gene transcription. Consistent with such an idea, acetylated core histones have been shown to associate preferentially with transcriptionally active chromatin (Sealy and Chalkley, 1978; Vidali et al., 1978; Hebbes et al., 1988). Further support of this has come from recent demonstrations that many diverse transcription factors and cofactors have histone acetylase (or acetyltransferase) activity and that transcriptional corepressors are complexed with histone deacetylases (Wade and Wolffe, 1997; Pazin and Kadonaga, 1997; Struhl, 1998; Torchia et al., 1998).

Of particular interest to TR function is the identification of several coactivators as histone acetyltransferases. Among them are CBP/p300 (Ogryzko et al., 1996), SRC-1 (Spencer et al., 1997), ACTR, which belongs to the same family of coactivators as SRC-1 and TIF2/GRIP-1 (Chen et al., 1997), and PCAF (Blanco et al., 1998). These coactivators appear to exist as multimeric-complexes containing different coactivators with histone acetyltransferase activities (Chen et al., 1997; McKenna et al., 1998). Thus, consistent with the role of histone acetylation in gene expression, liganded TR/RXR heterodimers may activate gene transcription through the recruitment of this coactivator histone acetylases. The opposite appears to be true for transcriptional repression by unliganded TR/RXR. This is because the corepressors N-CoR and SMRT, which bind to unliganded but not T_3-bound TR/RXR, have been shown to form a complex containing histone deacetylases (Heinzel et al., 1997; Nagy et al., 1997). This deacetylase complex formation involves the binding of NCoR or SMRT to the transcriptional repressor Sin3A, which in turn interacts with histone deacetylase.

Studies on *Xenopus* TR/RXR regulation of the *Xenopus* TRβA gene have provided direct evidence for a role of histone acetylation in promoter activation (Wong et al., 1998b). As described earlier, a double-stranded plasmid containing the TRβA promoter is highly transcribed when injected into oocyte nucleus and this transcription can be repressed by unliganded TR/RXR (Fig. 5.11, lane 3). The addition of a specific inhibitor of histone deacetylase, TSA (Trichostatin A), can reverse this repression, mimicking the addition of T_3 while having no effect on the transcription in the absence of TR/RXR (Fig. 5.11, lane 2), indicating the involvement of histone deacetylase in the repression by TR/RXR. In addition, replication-coupled chromatin assembly of the single-stranded promoter injected into oocyte nucleus also represses gene transcription to a very low basal level (Fig. 5.11, lane 6), which can be further repressed by unliganded TR/RXR (Fig. 5.11, lane 8). TSA treatment relieves both types of repression (lanes 7 and 9), again just like the addition of T_3.

In contrast to the deacetylase-blocking experiments, overexpression of the catalytic subunit RPD3 of a frog histone deacetylase leads to transcriptional

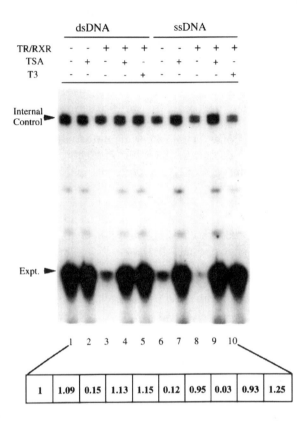

Figure 5.11 The histone deacetylase inhibitor TSA releases the transcriptional repression instigated by both chromatin and unliganded TR/RXR. Groups of oocytes were first injected with (+) or without (−) TR/RXR mRNAs and then injected with ds DNA or ss DNA of pTRβA as indicated. The oocytes were treated with (+) or without (−) TSA (5 ng/ml) or T_3 (50 nM) overnight. RNA was then prepared from the injected oocytes and the transcription from TRβA promoter was analyzed by primer extension (Expt.). The internal control represents the primer extension product derived from the endogenous storage pool of histone H4 mRNAs that serves as an RNA isolation and primer extension control. Levels of transcription from pTRβA were quantitated by phosphorimaging and normalized against the internal control. The level of transcription from control ds DNA of pTRβA was designated as 1 (lane 1) and the other lanes were compared with it. Reprinted from Wong et al. (1998b) with permission from Oxford University Press.

repression of the promoter and this repression can be reversed by the expression of TR/RXR in the presence of T_3 or the addition of TSA (Fig. 5.12; Wong et al., 1998b). Thus these two sets of complementary experiments strongly support a role of histone acetyltransferases/deacetylases in transcriptional regulation by TR/RXR.

MECHANISM OF TRANSCRIPTIONAL REGULATION 99

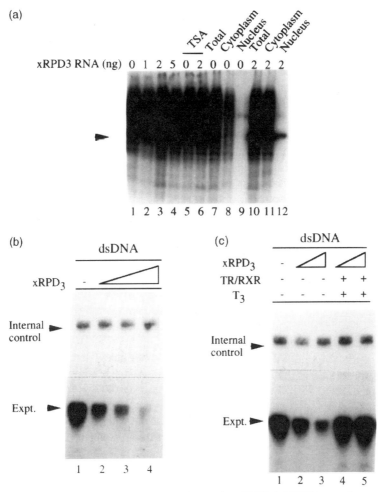

Figure 5.12 Expression of Xenopus histone deacetylase RPD3 (xRPD3) in oocytes represses transcription from TRβA promoter, and liganded TR/RXR overcomes the repression by xRPD3. (A) Expression of xRPD3 and its localization in oocytes. Groups of oocytes were co-injected with different amounts of in vivo transcribed xRPD3 mRNA as indicated and [^{35}S] methionine in the absence or presence of Trichostatin A (TSA) (5 ng/ml). After a 14-h incubation, protein extracts were made from total, cytoplasmic, and nuclear fractions as indicated and an amount of extract equivalent to one oocyte was separated by 10% SDS-PAGE. The [^{35}S] methionine-labeled xRPD3 was visualized by autoradiography (arrowhead). (B) Expression of the histone deacetylase xRPD3 represses transcription from the TRβA promoter. Groups of oocytes were injected with dsDNA of pTRβA and an increasing amount of xRPD3 mRNA (lanes 1-4: 0, 0.5, 1, and 2 ng, respectively). The oocytes were incubated overnight and the levels of transcription were analyzed by primer extension. (C) Liganded TR/RXR can relieve the repression established in the presence of xRPD3. Groups of oocytes were injected with ds DNA of pTRβA, and with or without increasing amounts of xRPD3 mRNA as indicated (0.5 ng, lanes 2 and 4, and 1 ng lanes 3 and 5). 14 hr later the oocytes were injected with or without TR/RXR mRNAs and treated with (+) or without (−) T_3 for 14 hr, before the levels of transcription were analyzed by primer extension. Reprinted from Wong et al. (1998b) with permission from Oxford University Press.

5.4.5 A Putative Model of Thyroid Hormone Receptor Action

The cumulative information of transcriptional regulation by nuclear receptors has clearly indicated a complex, multistep, multicomponent nature of the underlying mechanism. A potential model for TR/RXR function is outlined in Figure 5.13. In the absence of TH, TR/RXR recruits a corepressor and its associated deacetylase complex to the promoter, leading to histone deacetylation and transcriptional repression. Upon TH binding, the corepressor complex is dissociated and a coactivator complex is recruited to the promoter. This recruitment may lead to histone acetylation (Utley et al., 1998), chromatin disruption, and transcriptional activation.

Although the studies so far are supportive of an important role of histone acetylation in transcriptional activation, other pathways are likely involved. First of all, histone acetyltransferases can also acetylate other proteins such as general transcription factors (Imhof et al., 1997) and other transcription factors like P53 (Gu and Roeder, 1998). Thus they may also affect transcription independent of histone acetylation. Furthermore, there is evidence that corepressors such as N-CoR can interact with basal transcription factors and inhibit transcription independent of their ability to recruit deacetylases. In addition, chromatin disruption as detected by micrococcal nuclease digestion or plasmid DNA supercoiling assay appears to be necessary but not sufficient for transcriptional activation by liganded TR/RXR (Wong et al., 1997a). On the other hand, overexpression of RPD3 or TSA treatment has little effect on chromatin structure based on these assays despite their strong influence on transcription (Wong et al., 1998b). Finally, there are many other TR-interacting proteins of yet unknown function, and they are likely to affect transcription through distinct mechanisms. Thus further studies on the different cofactors and characterization of the nature of chromatin disruption and histone modification are needed to clarify the exact mechanism governing transcriptional regulation by TR/RXR.

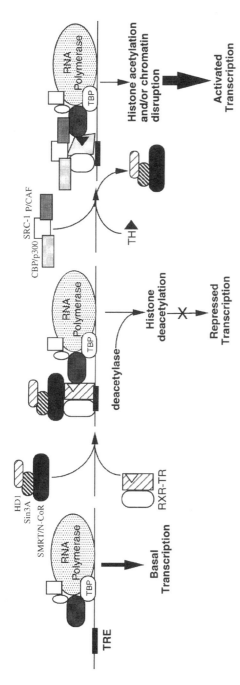

Figure 5.13 A model for transcriptional regulation by TR. Thyroid hormone receptor is presumed to form a heterodimer with RXR. The heterodimer binds to the TRE in a target gene. In the absence of TH, the heterodimer represses gene transcription likely through the recruitment of a corepressor complex containing the corepressor N-CoR (Horlein et al., 1995) or SMRT (Chen and Evans, 1995). The corepressor interacts with Sin3A, which in turn recruits a histone deacetylase (HD1) (Heinzel et al., 1997; Nagy et al., 1997), which can deacetylate histones as indicated, thus affecting transcription. Upon binding by TH, a conformational change takes place in the heterodimer, which may be responsible for the release of the corepressor complex and the recruitment of a coactivator complex containing coactivators such as SRC-1 (Onate et al., 1995), or CBP/p300 (Kamei et al., 1996; Chakravarti et al., 1996), and P/CAF (Blanco et al., 1998). There is evidence that SRC-1, CBP/P300, and P/CAF exist in a multimeric complex (Chen et al., 1997). Transcriptional activation is also associated with chromatin disruption (Wong et al., 1995, 1997a), which may be due to the recruitment of chromatin remodeling factors by TR/RXR or due to the action of the histone acetyl transferase activity of the coactivators (Ogryzko et al., 1996; Spencer et al., 1997; Blanco et al., 1998; Chen et al., 1997). This chromatin disruption may be necessary for transcriptional activation by TR/RXR. In addition to TBP (TATA box binding protein) and RNA polymerase, some other basal transcription factors are also depicted in the figure.

6

Gene Regulation by Thyroid Hormone

6.1 INTRODUCTION

The ability of TH to regulate gene transcription through the heterodimers of TRs and RXRs implys that TH controls amphibian metamorphosis by activating and repressing gene expression within the metamorphosing organs. The consequences of such gene regulation are to affect the expression of downstream genes and cell fate, and eventually to cause organ specific metamorphosis. That is, thyroid hormone induces a cascade of gene regulation to effect tissue transformation during metamorphosis (Shi, 1994).

Three models of the gene regulation cascade are possible with regard to the genes that are directly regulated by TR/RXR heterodimers (Fig. 6.1). In the first scenario (*A*), TR/RXR heterodimers activate or repress, in the presence of TH, a set of so-called direct response or immediate early response genes. These direct response genes are ubiquitous and thus themselves do not impose tissue specificity. However, their activation or repression in turn affect, through a cascade of events, the expression of downstream genes, that is, the so-called indirect or late TH-response genes. This second wave of gene regulation is tissue specific and responsible for subsequent tissue specific changes during metamorphosis.

The second extreme scenario (Fig. 6.1*B*) is that the liganded TR/RXR heterodimers regulate in a tissue-specific manner, specific direct response genes in different tissues, thus imposing tissue specificity at the first step of the gene regulation cascade. In this case, one would assume that TR/RXR heterodimers function together with tissue-specific factors to regulate the expression of different genes in different tissues.

The third and more likely possibility (Fig. 6.1*C*) is that some of the direct response genes are ubiquitous whereas others are tissue specific. Thus some

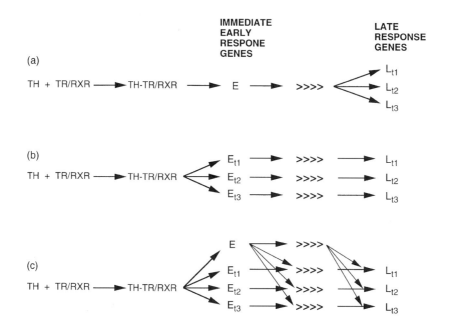

Figure 6.1 Possible gene regulation cascades leading to T_3-induced metamorphosis. (A) TR/RXR heterodimers bind to TH and thus activate/repress a set of direct response genes (E), which are ubiquitous. These early genes then affect other downstream ubiquitous genes in a cascade of events, leading to the activation of late respone genes that are tissue specific, for example, Lt1–Lt3 for late genes in tissue types 1–3. (B) Same as (A) except the early response genes are tissue specific, for example, Et1–Et3 for early genes in tissue types 1–3. (C) In this scenario, some early genes are tissue specific but others are not. At various step of the cascade, the ubiquitous genes participate in the regulation of tissue-specific ones to facilitate tissue-specific transformations.

tissue specificity is imposed at each step of the gene regulation cascade to achieve the eventual cell type specific transformation during metamorphosis.

Regardless what the exact model is, the cascade model suggests that to understand how TH induces metamorphic transformations in different organs, it is important to identify and characterize the genes at the various steps of the cascade. Earlier biochemical studies have suggested that many genes encoding various proteins associated with terminally differentiated cells are likely to be among the indirect response genes. But little information was available on the direct TH response genes. The following section focuses largely on the identification and characterization of such genes. For the purpose of the discussion, the genes that are regulated within 24 hr by the treatment of premetamorphic tadpoles with exogenous TH are referred to as early response genes, whereas those requiring more than 24 hr TH treatment to alter their expression are referred to as late response genes. Some of the early response genes are direct response genes but others are not. However, it should be pointed out that some late response genes may also require TR/RXR for their gene regulation, and thus in a sense are direct response genes. However, their regulation requires a proper regulation of genes in a prior step(s) of the cascade and therefore is delayed. For clarity, these genes are considered as indirect TH response genes or simply as late TH response genes.

6.2 METHODS FOR ISOLATION OF THYROID HORMONE RESPONSE GENES

As organs undergo metamorphic transformations both naturally or upon TH treatment, their gene expression profiles change, which is reflected by changes in the levels and types of various mRNAs present within the cells. By definition, any genes whose mRNA levels change during any stage of metamorphosis are TH response genes. Various methods are available to isolate these genes with varying degrees of sensitivities and efficiencies. Those based on differences in mRNA levels fall into three categories: differential hybridization, subtraction, and differential display. The following sections briefly summarizes these approaches.

6.2.1 Differential Hybridization

Differential hybridization (Sarget and Dawid, 1983; Sargent, 1987) was the first of the three approaches developed to isolate genes with different levels of their mRNAs in two different types of cells or states of a tissue. For this purpose, mRNAs are isolated from the two tissues or cell samples (+ and −, respectively). A cDNA library is constructed from + mRNA if genes to be isolated are expressed preferentially or only in the + sample. This library is then plated onto bacterial plates and duplicated filters are prepared for Southern blot hybridization. Two hybridization probes are made by reverse-transcribing the + and − mRNA in the presence of ^{32}P-labeled dNTP(s) and are used to hybridize the duplicated filters separately. The resulting signals from the two duplicated filters are then compared. The colonies that give different signals on the two duplicate filters represent genes that are expressed potentially at different levels in the two samples. Those colonies are then isolated and their differential expression in the two samples is verified by Northern blot hybridization.

This differential screening method is experimentally simple but not very sensitive. Genes that are expressed at low levels are poorly represented in the cDNA library and in the probes; thus they are unlikely to be isolated through this approach. However, more abundant, differentially expressed genes can be identified easily. Indeed, by using this method, Shi and Brown (1990) compared the gene expression pattern in *Xenopus laevis* tadpoles at stage 44 (feeding stage) to that in tadpoles at stage 49. This resulted in the isolation of a late

TH- response gene, the pancreatic trypsin gene, but failed to obtain any early TH response genes, which are likely expressed at lower levels (Shi and Brown, 1990; Shi, 1994).

6.2.2 Subtractive Differential Screen

A major disadvantage of the differential hybridization method is its low sensitivity toward genes expressed at low levels. To overcome this drawback, various PCR base methods have been developed to remove selectively genes that are expressed at similar levels in the two samples being compared while enriching and amplifying the genes whose mRNA levels differ in the two samples. One such method that has been applied in amphibian metamorphosis is schematically outlined in Figure 6.2 (Wang and Brown, 1991), the method involves first synthesizing cDNAs from the two (+ and − for with and without TH treatment, respectively) samples by reverse transcription using a dTn primer. The cDNAs are then converted into the double-stranded form and restricted into size suitable for PCR amplification by digesting them with one or two restriction enzymes that have 4 bp recognition sites to produce blunt-ended fragments. The fragments are then ligated to PCR linkers containing restriction enzyme recognition sites and amplified by PCR. The amplified cDNAs (+ and − cDNAs for the two samples, respectively) are used as starting materials for the subtractive differential screen.

To obtain genes that are preferentially expressed in the + sample, a small amount of the + cDNA (tracer) is mixed with a large excess of the − cDNA (driver), which has been tagged with photobiotin. The mixture is denatured by

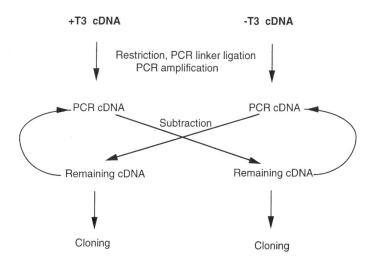

Figure 6.2 *Schematic of the PCR-based subtractive differential screen as reported by Wang and Brown(1991).*

heating and allowed to rehybridized by cooling. The + cDNA fragments derived from genes that are expressed equally or more abundantly in the − sample are hybridized preferentially with their biotinylated complements in the − cDNA sample. On the other hand, the + cDNA fragments derived from genes preferentially expressed in the + sample are hybridized more likely to their nonbiotinylated complements from the + cDNA. After hybridization, the biotinylated DNA, together with its hybridized complement, is then removed through the binding of the biotin to streptavidin followed by phenol/chloroform extraction. The remaining cDNA is then reamplified by PCR. This results in the preferential enrichment of fragments derived from genes that are more abundantly expressed in the + sample.

By using the − cDNA as the tracer and + cDNA as the driver, this subtraction procedure results in a − cDNA sample enriched with fragments derived from genes that are preferentially expressed in the − samples.

The above subtraction procedure can be repeated until there is little cross-hybridization (as judged by dot-blot hybridization) remaining between the + and − cDNAs. The cDNAs are then cloned into a plasmid vector through the restriction sites in the PCR linker and a differential hybridization is carried out using the same two cDNA samples as probes. The colonies in the + or − cDNA library that hybridize strongly only with the probe derived from the + or − cDNA, respectively, are isolated and subjected to Northern blot hybridization to verify the regulation of the corresponding genes in the original + and − RNA samples.

In theory, any genes that are differentially expressed in the two samples can be isolated with this method. However, genes that are expressed at high levels and/or have large differences in their mRNA levels between the two samples are preferentially enriched first. Their cDNA fragments are among the first to be isolated. These fragments can be removed from the enriched cDNAs by including them (after PCR amplification) in the driver for another round of subtraction. Their removal allows the less abundant fragments to be enriched and isolated through subsequent cloning and differential hybridization. The procedure can be repeated until few new differentially expressed cDNA fragments can be obtained from the next round of subtraction and cloning.

This method is clearly more sensitive than the simple differential hybridization described above. Furthermore, as most genes give rise to multiple short PCR fragments, an approximate estimate of the complexity, that is, how many, of the differentially expressed genes exist between the two samples, can be obtained based on the frequencies of multiple nonoverlapping fragments that are isolated for individual genes, just like that for a genetic screen (Wang and Brown, 1991, 1993).

On the other hand, some disadvantages still exist. First, because only short cDNA fragments are isolated from this screen, a full-length cDNA library must be constructed and screened with the short cDNA fragments to obtain the longer or full-length cDNA clones in order to determine the identities of the isolated genes. In addition, the genes isolated are still biased toward more abundant and more dramatically regulated (between the two samples) genes. Finally, the procedure is fairly tedious and lengthy, and extra care has to be taken to reduce PCR artifacts.

Regardless of the potential problems, this method has been successfully used to isolate genes regulated by TH in *Xenopus laevis*. Specifically, TH-regulated genes have been isolated from tadpole tail (Wang and Brown, 1991), hind limb (Buckbinder and Brown, 1992), intestine (Shi and Brown, 1993), brain (Denver et al., 1997), and a *Xenopus* tissue culture cell line (Kanamori and Brown, 1993). All these screens involved comparing gene expression differences in samples with or without 1 day of T_3 treatment of premetamorphic (stage 52–54) tadpoles or tissue culture cells.

6.2.3 Differential Display

The most recent and relatively straightforward method is the differential display (Liang and Pardee, 1992). This involves first synthesizing cDNAs from a subset of the mRNA population from the samples being compared by using a partially degenerate primer. A subset of this cDNAs is then PCR-amplified by using the primer used for cDNA synthesis and a 5'-primer that is again partially degenerate. The amplified products from the two samples are then displayed on a sequencing gel (a denaturing polycrylamide gel). Bands showing different intensities between the two samples (lanes on the gel) are potentially derived from mRNAs of different abundances in the two starting samples. They can be cut out from the gel, eluted into a buffer, and PCR amplified. The amplified products can be used as probes for Northern blot hybridization or other assays to verify the regulation of the corresponding genes between the two samples. If verified, they can be used to screen a full-length cDNA library to obtain the full-length cDNAs for further characterizations.

This approach is probably the simplest of all. Furthermore, it is possible to compare the expression profiles in more than two samples on a single gel. However, it requires a secondary screen to obtain a full-length cDNA clone, just like the subtractive differential screen. In addition, it compares only a small subset of the RNAs each time. Thus many primer sets (for cDNA synthesis and PCR) have to be employed to ensure that most, if not all, regulated genes are isolated. Finally, compared to the subtractive differential screen, the method is less sensitive because it uses only one round of amplification and thus is more biased toward more abundant genes. Currently, no report is published on the use of this method in the isolation of genes involved in amphibian metamorphosis.

6.2.4 Other Methods

Genes regulated during metamorphosis can also be obtained through other approaches. For example, the activities of many enzymes, such as the urea cycle enzymes in the liver, are known to be regulated during metamorphosis.

They can be isolated by screening cDNA libraries with heterologous cDNA probes if they are available. In fact, most of the urea cycle enzyme genes in *Rana catesbeiana* have been cloned this way and shown to be regulated at the mRNA level during metamorphosis (see below).

Another approach is to isolate proteins that are differentially expressed. The proteins can be sequenced and/or used to generate antibodies. The protein sequences can be used to design degenerate oligonucleotides to screen cDNA libraries to obtain the corresponding gene. Alternatively, the antibodies can be used to screen expression libraries to isolate the cDNA clones. A successful example is the cloning of the *Rana catesbeiana* tail collagenase gene (Oofusa and Yoshizato, 1994).

6.3 EARLY AND LATE THYROID HORMONE RESPONSE GENES

6.3.1 Early Thyroid Hormone Response Genes

Systematic isolation of early TH response genes, that is, those changing their mRNA levels within 24 hr of TH treatment of premetamorphic tadpoles has been carried out for four different organs of *Xenopus laevis*, the tail, hind limb, intestine, and brain. The tail represents one extreme case, where the entire organ resorbs, and the hind limb is another, where the new organ is developed de novo. On the other hand, the intestine and brain are typical of most organs, which are present both in pre- and postmetamorphic amphibians but drastically remodeled during metamorphosis in order to serve their roles in frogs. The screen for TH response genes was carried out by using the subtractive differential screen on RNAs isolated from the respective organs of premetamorphic tadpoles (stages 52–54) of *Xenopus laevis* with or without 1 day T_3 treatment (Wang and Brown, 1991; Buckbinder and Brown, 1992; Shi and Brown, 1993; Denver et al., 1997). The following provides a brief summary of the screening results.

6.3.1.1 Intestine Intestinal remodeling represents an important organ transformation that occurs during metamorphosis. It involves specific degeneration of larval (tadpole) tissues and concurrent development of adult ones. In *Xenopus laevis*, the most dramatic period is around stages 60–64, when extensive cell death occurs in the primary epithelium while secondary epithelial cells proliferate and differentiate (Chapters 3 and 4). Anatomically, the length of the small intestine reduces by as much as 90% from stage 58 to stage 64 and the epithelium changes from a simple tubular structure to a multiply folded form (Chapter 3; Marshall and Dixon, 1978; Ishizuya-Oka and Shimozawa, 1987a). These processes are controlled by TH and can be induced by TH even in organ cultures. To identify genes that may play a role in the remodeling process, a subtractive differential screen was conducted using intestinal RNAs isolated from stage 54 premetamorphic tadpoles that had been treated for 18 hr with or without 5nM T$_3$, close to the endogenous plasma T$_3$ concentration at the metamorphic climax (stage 62, Leloup and Buscaglia, 1977). A total of 22 T$_3$ upregulated and one downregulated genes was isolated (Table 6.1). Most of the genes respond to T$_3$ treatment very quickly (within a few hours) and their regulation by T$_3$ appears to be independent of new protein synthesis (Tables 6.1 and 6.2, Fig. 6.3), suggesting that they are direct TH response genes and represent the first period of gene regulation induced by TH.

The identities of many of the upregulated genes have been revealed through sequence analysis (Table 6.2). In agreement with the gene regulation cascade model above, several of the direct response genes encode transcription factors and thus are likely involved in directly activating or repressing transcription of intermediate and/or late TH response genes. In addition, several genes encoding proteins varying from a transmembrane protein to extracellular enzymes are also found to be among the early response genes in the intestine. These results suggest that TH simultaneously induces many intra- and extracellular events, which in turn cooperate to effect intestinal remodeling.

The first indication that these genes are likely to play important roles in tissue remodeling comes from their dramatic developmental regulation in the intestine during metamorphosis (e.g., Fig. 6.4*A*). Furthermore, the same expression patterns can be reproduced for most of these genes when premetamorphic tadpoles are treated with 5nM T$_3$ (Fig. 6.4*B*), a treatment that has been shown to induce intestinal remodeling, for example, intestinal length reduction and epithelial folding (Fig 2.4; Shi and Hayes, 1994). Developmentally, these early response genes fall into three general classes. The genes in the first class, for example, TRβ, a basic leucine-zipper (TH/bZip) motif-containing transcription factor, the extracellular matrix-degrading metalloproteinase stromelysin-3, and a nonhepatic arginase gene (Table 6.2), are expressed strongly only at the climax of intestinal remodeling (around stages 60–62). Much lower levels of expression of these genes are present in pre- or postmetamorphic intestine. The second class of genes, for example, NFI family of transcription factors and a transmembrane protein (Table 6.2), are activated during metamorphosis and their expression remains high in postmetamorphic frog intestine. Finally, two genes including the Na$^+$/PO$_4^-$ cotransporter (Table 6.2 and 6.4) are in the third class. They are expressed at high levels immediately before or after the climax of metamorphosis but minimally at the actual climax.

6.3.1.2 Tail The resorption of the tadpole tail is one of the last changes to be completed during metamorphosis. Although some events such as the resorption of the tail fin occur earlier, the reduction in tail length is minimal until about stage 62 in *Xenopus laevis* (Chapter 3; Nieuwkoop and Faber, 1956; Dodd and Dodd, 1976), when endogenous TH is at its peak concentration (Leloup and Buscaglia, 1977). However, tail resorption can be induced in premetamorphic tadpoles by TH treatment. Of the genes that are regulated by TH within the first day of treatment, 15 upregulated and 4 downregulated ones have been isolated (Wang and Brown, 1991, Wang and Brown, 1993). The

TABLE 6.1 Early Thyroid Hormone Response Genes in Limb, Tail, Intestine, and Brain of *Xenopus laevis*

Parameter	Hind Limb[a]	Intestine[b]	Tail[c]	Brain[d]
Morphological changes	De novo development	Remodeling	Resorption	Remodeling
Major events	Stages 51–56: cell proliferation and morphogenesis (digit formation)	Stages 58–62: cell death in primary epithelial, length reduction, development of the connective tissue and muscles	Stages 62–66: cell death, removal of degenerated tissues by phagocytosis etc.	Stages 56–66: degeneration of certain nervous structures used by tadpoles and development of ones used by frogs, cell death of larval neurons, and differentiation of adult neurons
	Stages 56–66: limb growth and differentiation	Stages 60–66: proliferation and differentiation of cells of the secondary epithelial, connective tissues, and muscles		
Number of upregulated genes isolated	14	22	15	20
Number of upregulated direct response genes[e]	3	15	11	12
Magnitude of T_3 regulation[f] (average)	3- to 12-fold (5)	3- to >20-fold (≥9)	6- to >20-fold (≥11)	2- to >10-fold (≥4)
Number of downregulated genes isolated	5	1	4	14

[a] TH treatment: 24 hr at 5 nM T_3 on stage 54 tadpoles (Buckbinder and Brown, 1992).
[b] TH treatment: 18 hr at 5 nM T_3 on stage 54 tadpoles (Shi and Brown, 1993).
[c] TH treatment: 24 hr at 100 nM T_3 on stage 54 tadpoles (Wang and Brown, 1991, 1993).
[d] TH treatment: 24 hr at 5 nM T_3 on stage 52–54 tadpoles (Denver et al., 1997). Only the diencephalon was dissected from the brain for the subtractive differential screen.
[e] Based on the resistance of the regulation of these genes to protein synthesis inhibition.
[f] The numbers refer to the factors of activation after respective treatment (a–d). The slightly higher magnitude of activation for the tail genes could be due to the higher concentration of T_3 used.

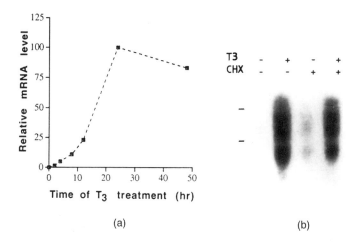

Figure 6.3 Fast and direct response of an early gene to T_3. (A) Kinetics of T_3 induction of the Na^+/PO_4^{-3} cotransporter gene (IU24) to T_3 (Table 6.2; Shi and Brown, 1993; Ishizuya-Oka et al., 1997b). (B) T_3 induction of IU24 is independent of new protein synthesis. Tadpoles at stages 52–54 were treated with 5nM T_3 for indicated periods (A) or 12 hr in the presence or absence protein synthesis inhibitors cycloheximide and anisomycin (CHX). Total RNA was then isolated from the intestine and subjected to Northern blot analyses with a IU24 cDNA probe. The mRNA levels were quantified by densitometric scanning and plotted in (A). The smear in (B) was due to the presence of multiple mRNA species and partial RNA degradation (also see 6.4). Modified after Shi and Brown (1993) with permission from the American Society for Biochemistry and Molecular Biology, Inc.

majority of these early upregulated genes appear to be direct response genes (Tables 6.1 and 6.2). Among the genes identified to date by their sequences are TRβ genes, three other putative transcription factors (two containing a basic leucine-zipper motif and the other a zinc finger protein), and genes encoding metalloproteinases and a type III deiodinase. Some of these genes have been independently isolated from the intestinal screen discussed above (Table 6.2). Developmentally, all but one of the upregulated genes are expressed strongly only during tail resorption. The exception is the 5-deiodinase, whose expression peaks around stages 60/61, immediately before massive tail resorption takes place (for more discussion on this gene, see Chapter 8).

All of the downregulated genes show a dramatic reduction in their expression as the tail resorbs (Furlow et al., 1997). These developmental profiles suggest that the genes participate in the resorption process. For example, as tail resorption takes place, the activities of many macromolecule-degrading enzymes increase dramatically (Chapter 4, Weber, 1967; Dodd and Dodd, 1976). The activation of the collagenase gene (Table 6.2) is likely to be responsible for the well-documented increase in collagen-degrading activity in the resorbing tail fin (Chapter 4, Gross, 1966).

TABLE 6.2 Early TH Upregulated Genes of Known Identity

Gene Name	Homologous Genes	Tissue Source	Response to TH[a]	Possible Function
Tail 1/3	Zinc finger (BTEB)	Tail[b], brain[c]	Direct	Transcription factors[b,c,d,f,g,m]
Tail 7	bZip (Fra-2)	Tail[b]	ND	
IU 16/33	NFI	Intestine[d]	Direct	
Tail 8/9	TH/bZip (E4BP4)	Tail[b], intestine[d]	Direct	
TRβ		Tail[b], intestine	Direct	
Tail 11	Collagenase-3	Tail[b]	ND	Matrix metalloproteins[b,d,f,h]
Tail 14	Stromelysin-3	Tail[b], intestine[d]	Direct	
Tail 13	FAPα	Tail[b]	ND	Proteases[f]
Tail D	N-Aspartyl dipeptidase	Tail[b]	ND	
Tail C	Fibronectin	Tail[b]	ND	ECM protein[f]
Tail 14	Integrin α-1	Tail[b]	Direct	ECM receptor[f]
Tail 15	Type III deiodinase	Tail[b], brain[c]	Direct	TH inactivation[b,c,i]
Xh20	Protein disulfide isomerase	Brain[c]	Direct	Protein isomerization[c]
Xh1	Cytochrome c oxidase subunit	Brain[c]	ND	Oxidative phosphorylation[c]
IU22	Nonhepatic arginase	Intestine[d]	ND	Proline biosynthesis, etc.[c]
IU24	Na+/PO$_4^-$ cotransporter	Intestine[d]	Direct	PO$_4^-$ transport[d,k]
IU12	Transmembrane protein	Intestine[d]	ND	Amino acid transport[l]
IU27	Sonic hedgehog	Intestine[d]	Direct	Morphogen[d,n]
P	Rat clathrin B chain	Limb[e]	ND	Vesicular intracellular transport[e]
B	Heat-shock protein	Limb[e]	Direct	Rapid cell growth[e]
I	Heat-shock protein	Limb[e]	ND	
M	Heat-shock protein	Limb[e]	ND	
N	Heat-shock protein	Limb[e]	ND	
H	Yeast MCM3, mouse P1	Limb[e]	ND	
J	Mouse eIF-4A	Limb[e]	ND	

[a] A direct response indicates that the regulation is resistant to protein synthesis inhibition; ND, not determinable.
[b] Wang and Brown (1993).
[c] Denver et al. (1997).
[d] Shi and Brown (1993).
[e] Buckbinder and Brown (1992).
[f] Brown et al. (1996).
[g] Ishizuya-Oka et al. (1997a).
[h] Patterton et al. (1995).
[i] St. Germain et al. (1994).
[j] Patterton and Shi (1994).
[k] Ishizuya-Oka et al. (1997b).
[l] Liang et al. (1998); Torrents et al. (1999).
[m] Puzianowska-Kuznicka and Shi (1996).
[n] Stolow and Shi (1995).

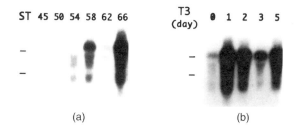

Figure 6.4 Expression of the early T_3 response gene Na^+/PO_4^{-3} cotransporter (IU24) gene (Table 6.2) in the intestine during development and T_3-induced metamorphosis. Total RNA was isolated from the intestine of tadpoles at different stages during natural metamorphosis (A) or at stages 52–54 and treated with 5nM T_3 for indicated number of days (B). The RNA was analyzed by Northern blot. Note that the gene is first upregulated by the beginning of metamorphic climax (stage 58) or after 1–2 days of T_3 treatment, then downregulated at the climax (stage 62) or after 3 days of T_3 treatment, and finally reactivated as the adult intestine develops. Modified after Shi and Brown (1993) with permission from the American Society for Biochemistry and Molecular Biology, Inc.

6.3.1.3 Hind Limb Hind limb development is one of the earliest events to occur during metamorphosis (Chapter 3, Nieuwkoop and Faber, 1956). In *Xenopus laevis*, the hind limb buds are detected by stage 48, when there is little endogenous TH. However, further development requires TH. This is initiated when endogenous plasma TH concentration starts to rise around stage 54 (Leloup and Buscaglia, 1977) and is essentially complete by the time the TH concentration reaches a peak at climax (stage 62). A subtractive differential screen isolated 19 genes, 14 of which were upregulated and five were downregulated in the hind limb within 1 day by TH (Table 6.1; Buckbinder and Brown, 1992). With the exception of a few genes, the activation of the upregulated genes occurs only after at least 12 hr of TH treatment, suggesting that the response to TH is delayed compared to the direct response genes. In general, the magnitude of their upregulation by TH is considerably less than that of the genes isolated from the tail and intestine (Table 6.1). Similarly, their expression levels during hind limb development (stages 54–66) vary only moderately compared to those for genes in the tail and intestine. Sequence analysis revealed that many of these genes encode proteins that are expected to be involved in cell proliferation and tissue growth (Table 6.2). This is in agreement with their potential functions during limb development, for hind limb morphogenesis takes place mostly before stage 56 and after that the hind limb undergoes rapid growth and maturation.

6.3.1.4 Brain Like the intestine, metamorphosis of the amphibian brain also involves selective elimination of certain larval cells through cell death and development of adult tissues through cell proliferation and differentiation (Chapter 3, Kollros, 1981; Gona et al., 1988; Tata, 1993; Hoskins, 1990). These

TABLE 6.3 Some Known Late TH Response Genes in *Xenopus laevis* and *Rana catesbeiana*

Gene	Tissue	Frog Species	Regulation by TH	References
Carbamyl-phosphate synthase I	Liver	R. catesbeiana	Up	Morris (1987); Helbing et al. (1992); Galton et al. (1991)
Argininosuccinate synthase	Liver	R. catesbeiana	Up	Morris (1987)
Argininosuccinate lyase	Liver	R. catesbeiana	Up	Iwase et al. (1995)
Arginase	Liver	X. laevis R. catesbeiana	Up	Xu et al. (1993); Helbing and Atkinson (1994); Atkinson et al. (1994); Iwase et al. (1995)
Ornithine carbamyltransferase	Liver	R. catesbeiana	Up	Helbing et al. (1992)
N-CAM	Liver	X. laevis	Up	Levi et al. (1990)
Albumin	Liver	R. catesteiana	Up	Schultz et al. (1988); Moskaitis et al. (1989)
α1-microglobulin	Liver	X. laevis	Down	Kawahara et al. (1997)
Myosin heavy chains	Limb	X. laevis R. catesbeiana	Up or down	Buckbinder and Brown (1992); Dhanarajan et al. (1988); Sachs et al. (1997b)
Tropomyosin	Limb	R. catesbeiana	Up	Dhanarajan et al. (1988)
M-Line protein	Limb	R. catesbeiana	Up	Dhanarajan and Atkinson (1981)
α-Actinin	Limb	R. catesbeiana	Up	Dhanarajan and Atkinson (1981)
Keratins	Epidermis	X. laevis R. catesbeiana	Up	Mathisen and Miller (1989) Ketola-Pirie and Atkinson (1988, 1990)
Magainin	Skin	X. laevis	Up	Reilly et al. (1994)
Trypsin	Pancreas	X. laevis	Down	Shi and Brown (1990)
Intestinal fatty acid-binding protein	Intestine	X. laevis	Down	Shi and Hayes (1994)
Mdr (multidrug resistence)	Intestine	X. laevis	Down	Zucker et al. (1997)
Gelatinase A	Intestine	X. laevis	Up	Patterton et al. (1995)
Collagenase-4	Tail	X. laevis	Up	Stolow et al. (1996)

include the elimination of the Mauther neurons and motor neurons that innervate tail muscles (Kollros, 1981) and development of neurons of the retina in the diencephalon (Hoskins, 1990). A similar subtractive differential screen was carried out using RNA isolated from the region of the brain encompassing the diencephalon, of stage 52–54 *Xenopus laevis* tadpoles with or without 20 hr treatment with 5 nM T_3 (Denver et al., 1997). This has resulted in the isolation of 34 cDNAs from TH-regulated genes, including 20 upregulated and 14 downregulated ones (Table 6.1). As in the tail and intestine, the regulation of most of these genes by TH are independent of new protein synthesis, although their magnitudes of TH-regulation are somewhat lower than those of the intestinal and tail genes. Developmentally, the upregulated brain genes can be grouped into three classes. The first includes genes that begin to be upregulated during prometamorphosis (stages 57/58) and continue to be expressed at high levels in frogs. The second class of genes are those that are upregulated around the same time as the first class but are then downregulated by the end of metamorphosis. The final class of genes are expressed at high levels only at the climax of metamorphosis (around stages 62). This distribution pattern of the upregulated genes is similar to that observed in the intestine except that no brain gene has been found to be first upregulated, then downregulated at the climax, and finally reactivated at the end of metamorphosis. This probably reflects the lack of synchronized degeneration of a major cell type like the larval intestinal epithelial cells during metamorphosis. On the other hand, sequence analyses indicate that like the genes isolated from other organs, the brain genes encode diverse groups of proteins that can affect both intra- and extracellular events to coordinate brain transformation (Table 6.2; Denver et al., 1997).

Most of the early response genes are not organ specific. Many of the genes isolated from tail and intestine could be activated by TH treatment in other tissues, including the hind limb and brain, even though their mRNA levels do not change appreciably during natural limb development (Wang and Brown, 1993; Shi and Brown, 1993; Table 6.2). Similarly, some of the limb genes were regulated in the intestine (Buckbinder and Brown, 1992), although their activation by TH in the intestine is to a lesser extent than that of the genes isolated from the intestinal screen. In addition, many of the early response genes also respond to T_3 with similar kinetics in tissue culture cells and conversely many TH response genes isolated from the tissue culture cells are also TH response genes in tadpole tissues (Kanamori and Brown, 1993). An exception is that many of the brain genes appear to be organ specific (Denver et al., 1997).

The brain screen is also unique in that it isolates a relatively large number of downregulated genes. For an unknown reason, all three other organs described above were found to have many fewer early response genes that were downregulated by TH within 24 h (Table 6.1). It is possible that the mRNAs of the downregulated genes are very stable and their levels change slowly even though the corresponding genes have been repressed by TH. Consequently, the differential screen would not identify them. However, no new downregulated genes were found in the tail when a differential screen was performed using stage 54 tadpoles treated with TH for 2 days, a treatment known to be sufficient to induce tail resorption (Wang and Brown, 1993, also see Section 6.4). Thus the treatment produced proper regulation of all genes that are crucial for tail resorption and at least in the tail, there are likely only a small number of early downregulated genes. At present, much less is known about the downregulated genes. It is difficult to assess their importance during metamorphosis.

6.3.1.5 Other Early Response Genes Although systematic search for early TH response genes has been carried out only for the above four organs in *Xenopus laevis*, various other approaches have revealed the existence of other early TH response genes in different organs. For example, many of the genes isolated from the tail and intestine have been shown to be early response genes in the hind limb and brain as well (Wang and Brown, 1993; Shi and Brown, 1993; Denver et al., 1997). In addition, studies of the C/EBPα gene in *Rana catesbeiana* as a possible regulator of liver cell differentiation have shown that it is upregulated by TH within a few hours (Chen et al., 1994), indicating that it is an early TH response gene. Furthermore, its upregulation precedes that of genes of the urea cycle (Chen and Atkinson, 1997), which are late TH response genes (see below), suggesting a possible role for C/EBPα in urea cycle gene expression (see Chapter 7). Finally, another early response gene is the collagenase-1 gene in *Rana catesbeiana*. It was discovered first through purification of the collagenase, which was known to be upregulated during tail resorption, and subsequent cloning and expression analysis (Oofusa and Yoshizato, 1991; Oofusa et al., 1994). This collagenase gene has since been shown to be a direct response gene containing a TRE (Oofusa and Yoshizato, 1996).

6.3.2 Late Thyroid Hormone Response Genes

Although TH apparently controls the metamorphic transition in all tissues/organs, different tissues undergo drastically different changes. These changes take place over a period of about 1 month in *Xenopus laevis* and undoubtedly involve alterations in the expression of many tissue specific genes. Similar changes in the expression of these genes are expected if premetamorphic tadpoles are treated with TH. Some of these genes respond to TH quickly, that is, the early response genes discussed above. Others change their expression only after at least 1 day of TH treatment. These latter genes are thus responding to TH indirectly and require the synthesis of new proteins for their regulation. These genes are considered to be the late TH response gene according to the gene regulation cascade model above. Many such genes have been identified over the years (Table 6.3). For example, as the herbivorous tadpole metamorphoses into a carnivorous frog, the digestive system is remodeled. This is accompanied by drastic changes in the expression of the intestinal fatty acid binding protein (IFABP) gene (Shi and Hayes, 1994) and the exocrine-specific genes encoding pancreatic enzymes such as trypsin (Table 6.3; Shi and Brown, 1990). Both the trypsin and IFABP genes are expressed only in their respective larval and adult organs (see Fig. 6.5 for IFABP). However, their mRNA levels are repressed to a minimum at the climax of metamorphosis, when the exocrine pancreatic or primary intestinal epithelial cells undergo degeneration. This suppression can be induced in premetamorphic tadpoles by prolonged (>1 day) treatment with TH (Fig. 6.6). At least for the IFABP gene, this downregulation appears to be an instance of gene repression or selective degradation of IFABP mRNA, because it occurs before cell death in the primary epithelium in which the gene is expressed (Shi and Hayes, 1994). Both the trypsin and IFABP genes are reactivated as the adult exocrine pancreatic (Shi and Brown, 1990) or intestinal epithelial cells differentiate (Fig. 6.6, also see Ishizuya-Oka et al., 1994, 1997c).

One of the best-studied systems is the activation of genes encoding the urea cycle enzymes in the liver upon the transition from ammonotelism to ureotelism in many anuran amphibians. The activities of these enzymes has long been known to be coordinately upregulated during metamorphosis or by TH

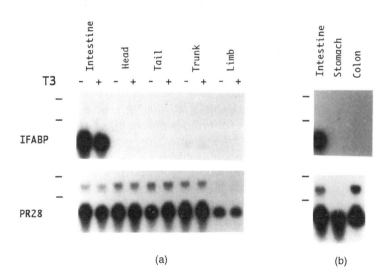

Figure 6.5 Tissue-specific expression of the late T_3 response gene: the intestinal fatty acid binding protein gene (IFABP). (A) IFABP is expressed only in tadpole intestine. Total RNA from indicated organs of stage 52—54 tadpoles treated with ($+T_3$) or without ($-T_3$) 5nM T_3 for 1 day was subjected to Northern blot analysis. Note that IFABP is expressed only in the intestine and that its mRNA level is not significantly altered by 1 day of T_3 treatment, unlike the early genes (e.g., see 6.3). (B) IFABP is expressed only in the small intestine but not in stomach or colon of juvenile frogs. Adapted from Shi and Hayes (1994) with permission from Academic Press.

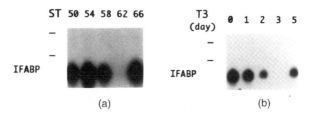

Figure 6.6 Similar expression profile of IFABP during natural and T_3-induced metamorphosis. Northern blot analysis of total RNA from the intestine of tadpoles at different stages (A) or stages 52–54 and treated with 5nM T_3 for indicated number of days (B). Note that IFABP is expressed in premetamorphic or postmetamorphic (stage 66 or after 5 days of treatment) animals but not at the climax (stage 62 or after 3 days of treatment) when there are few differentiated epithelial cells, which express IFABP. Adapted from Shi and Hayes (1994) with permission from Academic Press.

treatment (Chapter 4; Cohen, 1970; Dodd and Dodd, 1976). It has been shown more recently that the genes encoding the five enzymes in the urea cycle, that is, carbamyl phosphate synthetase I, argininosuccinate synthetase, argininosuccinate lyase, ornithine transcarbamylase, and arginase genes, are activated by prolonged TH treatment of premetamorphic *Rana catesbeiana* tadpoles (Table 6.3). In addition to the urea cycle enzymes, a few other genes have also been shown to be late TH response genes in the liver (Table 6.3; Atkinson et al., 1996).

In the developing hind limb, the genes of the connective tissue and muscles, such as the collagen and myosin heavy chain genes, are highly expressed as the organ differentiates (A. Kanamori, Y.-B.Shi, and D.D. Brown, unpublished; Buckbinder and Brown, 1992; Sachs et al., 1997b). These genes can also be activated in the hind limb when premetamorphic tadpoles are treated with TH. However, such activation occurs only after prolonged treatment, suggesting that these genes are late TH response genes.

Like the hind limb, the maturation of frog skin is also associated with the activation of adult-specific genes. Of these genes, the epidermal keratin gene can be activated by TH in premetamorphic tadpoles (Table 6.3, Mathisen and Miller, 1989). Furthermore, a short treatment of larval skin with TH can result in the activation of the adult keratin gene several days later even if the skin explant is subsequently maintained in TH-free medium, demonstrating that TH induces an irreversible gene regulation cascade that leads to the change in keratin expression (Miller, 1996). Another family of genes that is highly expressed in the adult skin are the magainin antimicrobial peptide genes (Bevins and Zasloff, 1990). The granular glands of adult frog skin are the predominant sites for synthesis and storage of these polypeptides. At least two of the genes in this peptide family have been shown to be activated at the climax of metamorphosis in *Xenopus laevis* (Reilley et al., 1994). As one would expect from the causative role of TH in metamorphosis, prolonged treatment of tadpoles (e.g., 7 days) with TH leads to premature activation of these genes

(Table 6.3). Although it is unclear what is the shortest treatment period required for the activation of these genes, they are probably late TH response genes as they are expressed in the differentiated cells in the granular glands.

More recently, Amano (1998) reported the isolation of a number of genes that were upregulated in the intestine of *Xenopus laevis* tadpoles after 4 days of TH treatment by using a PCR-based subtractive screening method that is slightly different from the one shown in Figure 6.2. Several distinct classes of genes were isolated. They encode transcription factors, collagens, components of the ubiquitin proteasome pathway, morphogenetic and growth factors, and so on. These genes can potentially participate in the regulation of larval epithelial apoptosis and/or adult cell proliferation and differentiation. Currently, detailed kinetic data are not available to determine whether they are early or late TH response genes, although they appear to be late TH response genes based on their isolation from tadpoles treated with TH for 4 days (Amano, 1998).

6.4 TISSUE-DEPENDENT VARIATION IN THE GENE EXPRESSION PROGRAM INDUCED BY THYROID HORMONE

The tissue-specific transformations during metamorphosis would argue that TH likely controls a different gene regulation cascade in each tissue. The systematic analysis of the early response genes in three contrasting tissues and available information on some late TH response genes have indeed revealed drastically different gene regulation profiles for the intestine, tail, brain, and hind limb of *Xenopus laevis*.

Among the changes in all organs, the total regression of the tail appears to be the simplest response to TH to explain (Brown et al., 1995; Kanamori and Brown, 1996). Although it consists of a variety of cell types, they are all destined to undergo cell death and resorption in the presence of TH. To get a sense of the complexity of the gene expression program induced by TH in tail resorption, an estimate of total number of genes regulated by TH within the first 48 hr was made based on how often non-overlapping small PCR fragments from a single gene were isolated during a differential screen (Wang and Brown, 1993). Within the first 24 hr after the addition of T_3, approximately 25 genes are predicted to be activated by TH, of which 15 have been isolated. Although the accuracy of the gene number is limited by the difficulty to take into account the size of the mRNA, abundance, and the magnitude of upregulation of an individual gene by TH, it clearly indicates that only a finite number of upregulated genes exist in the resorbing tail within the first 24 hr. Furthermore, a similar estimate predicts that only about 35 genes are induced within the first 48 hr by TH. A 48 hr treatment has been shown to be sufficiently long for tail resorption to occur even if the tail is subsequently maintained in a medium containing protein synthesis inhibitors (Fig. 6.7; Wang and Brown, 1993). On the other hand, blocking the RNA and/or protein

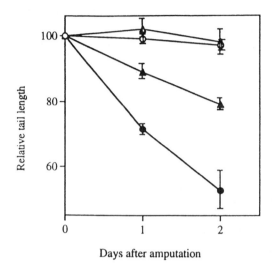

Figure 6.7 *Effect of TH pretreatment on cultured tails. Tails were amputated from stage 54 tadpoles that had been pretreated with 100 nM T_3 for various times up to 48 hr, then cultured in Steinberg's saline in the presence or absence of 10 µg/ml cycloheximide plus 100 nM anisomycin. Cultured tails from animals pretreated with TH for up to 24 hr were identical to controls. Control tails cultured in the presence (△ – △) or absence (○ – ○) of the inhibitors; tails pretreated with TH for 48 hr and then cultured in the presence (▲ – ▲) or absence (● – ●) of the inhibitors. The standard deviation of each measurement is indicated by the vertical bar. Reprinted from Wang and Brown (1993) with permission from the American Society for Biochemistry and Molecular Biology, Inc.*

synthesis at the onset of hormone treatment is known to inhibit tail resorption in organ cultures (Tata, 1966). These observations suggest that the 35 or so upregulated genes together with the downregulated genes are sufficient to effect tail resorption (Wang and Brown, 1993). It is important to realize, however, that there are possibly other genes that are crucial for the resorption process but are not included in the estimated 35 genes due to their low abundance (<10 copies of the mRNA per cell) and/or low magnitude of their upregulation by TH (<six-fold by 24 hr of 100nM T_3 treatment) (Wang and Brown, 1993).

In sharp contrast to the tail, the hind limb appears to have a much more complex gene regulation program (Buckbinder and Brown, 1992). It has been estimated that more than 120 genes are upregulated by TH within 24 hr. Although most of the 14 isolated genes are not likely to be direct TH response genes, many direct TH response genes isolated from the tail and intestine are also upregulated in the hind limb by T_3 (Table 6.2; Buckbinder and Brown, 1992; Shi and Brown, 1993; Wang and Brown, 1993). These results indicate the presence of at least two periods of gene upregulation within the first 24 hr, which is then followed by a third period that involves genes required for cell proliferation as reflected by a significant increase in cells in S phase of the cell cycle (Buckbinder and Brown, 1992). Finally, as cells differentiate after 3 days or longer, another period of gene regulation leads to the activation of genes of the differentiated adult cells such as myosin heavy chain gene (Table 6.3).

Intestinal remodeling, on the other hand, involves changes that are similar in part to both tail resorption and limb development. The tadpole intestine comprises mostly primary epithelium. One of the major events during intestinal remodeling is the degeneration of the primary epithelium. Thus it might be anticipated that at least some early response genes are common in the tail and intestine. In fact, many of the early response genes described above were found to be regulated in both organs (Tables 6.1, 6.2). Furthermore, several genes were independently isolated from both tissues. The total number of genes that are upregulated by TH within the first 24 hr was also estimated to be similar to that in the tail (Shi and Brown, 1993), contrasting with the large number of genes involved in the hind limb. Finally, the magnitude of the activation of the intestinal and tail genes by TH or during metamorphosis in these tissues is similar and considerably greater than that of the genes isolated from the hind limb (Table 6.1).

Although the early gene regulation program shares similarity between the intestine and tail, intestinal remodeling resembles limb development in a few aspects. The latter two organs eventually differentiate into adult-type tissues in response to TH. It is therefore not surprising that the gene regulation cascade is more complex in the intestine and limb than that in the tail. Like hind limb development, intestinal remodeling involves multiple periods of gene activation and repression. Many early response genes are eventually repressed toward the end of metamorphosis whereas others remain to be expressed at relatively high levels (Shi and Brown, 1993). On the other hand, the genes that are specific to differentiated epithelial cells, such as the IFABP gene, are repressed as the epithelium undergoes cell death. Eventually, for example, after several days of hormonal treatment, many of these epithelial genes are expected to be reactivated, such as the IFABP gene (Fig. 6.6), as secondary or adult epithelial cells differentiate to form the adult organ.

Unlike the limb, intestine, and tail, the brain appears to have a different, maybe more complex gene regulation program. As pointed out earlier, many of the genes isolated from the brain are organ specific in their regulation by T_3 and a relatively large number of downregulated genes have been isolated (Denver et al., 1997). In addition, many of the genes isolated from the tail, intestine, and limb screens are also regulated by T_3 in the brain. Thus the brain may have a relatively large number of genes regulated by T_3. This may reflect that many different cell types are present in the brain and each may undergo a different change at different developmental stages.

Another level of complexity in the gene expression program is present in all tissues. That is, the direct TH response genes identified so far belong to diverse groups, for example, genes encoding transcription factors and extracellular matrix-degrading enzymes. This diversity could be in part due to the fact that all these different organs themselves are very complex, consisting of many different cell types. Different cells could respond to TH differently by altering the expression of different genes, for example, the activation of stromelysin-3 genes occurs in only the fibroblastic cells of the intestine (Patterton et al., 1995; Ishizuya-Oka et al., 1996), while that of the TH/bZip gene takes place in all cell types within the intestine although temporally different in different cell types (Table 6.2; Ishizuya-Oka et al., 1997a). On the other hand, a given gene, such as stromelysin-3 (Table 6.2), could respond to TH in a similar manner in a specific cell type in different organs, thus allowing many of the same genes to be regulated by TH in organs that undergo drastically different transformations. These complex gene expression profiles contrast sharply with a much simpler but possible mechanism in which TH triggers a tissue-specific gene regulation cascade through sequential activation of a series of tissue-specific genes (Fig. 6.1*A*, *B*). Instead, the current information indicates that multiple intra- and extracellular processes involving many ubiquitous as well as cell type-specific genes are concurrently activated by TH in a given organ (Fig. 6.1*C*). It is currently unclear how TH induces such diverse developmental stage- and cell type-specific gene expression. As discussed in Chapter 8, temporal regulation of TR genes and genes that can affect cellular free TH levels appears to play an important role in regulating tissue response to TH. In addition, different TR-interacting cofactors (see Chapter 5) may be involved in different organs/tissues. Finally, certain preexisting cell- specific factors are presumed to contribute to the cell type specific gene expression program induced by TH.

7

Functional Implication of Thyroid Hormone Response Genes: Transcription Factors in the Thyroid Hormone Signal Transduction Cascade

7.1 INTRODUCTION

The gene regulation cascade model implies that the early T_3 response genes will directly or indirectly regulate the expression of downstream genes in different metamorphosing tissues. The existence of direct, sequential gene regulation cascades during metamorphosis is supported by the isolation of many immediate early or direct T_3 response genes that encode transcription factor (TFs) (see Chapter 6). The characterization of most of these genes is still very limited. However, the studies so far have provided important information on the function of these transcription factors.

7.2 EXPRESSION OF GENES ENCODING TRANSCRIPTION FACTORS DURING METAMORPHOSIS

Genes encoding transcription factors belonging to several different families have been isolated from the subtractive differential screens as described in Chapter 6. These include members of the nuclear factor I (NFI) family and TFs containing Zn^{2+} fingers and leucine zippers, etc. Interestingly, most of these TF genes are expressed ubiquitously in different organs during metamorphosis. Furthermore, these genes are activated by T_3 even in the absence of new protein synthesis (Shi, 1994, 1996; Brown et al., 1995). Thus they are likely to participate in the very early steps of metamorphosis in many different organs.

Although the TF genes are expressed in different organs, they are regulated temporally in an organ-dependent manner such that high levels of their mRNAs are present in the organs at the climax of the transformations of the respective organs (Shi, 1996; Brown et al., 1996). Several TF genes, for example, those encoding TRβ, two leucine zipper-containing TFs, and a Zn^{2+} finger-containing TF are regulated similarly in different organs during metamorphosis (Wang and Brown, 1993; Brown et al., 1996; Berry et al., 1998a,b). The limited information available allows the discussion on a few TF genes. One of the better studied TF gene encodes a leucine zipper-containing protein (TH/bZip or tail gene # 8, Fig. 7.1, Wang and Brown, 1993; Shi and Brown, 1993; Brown et al., 1996; Ishizuya-Oka et al., 1997a). The gene can be induced by T_3 in all organs of a premetamorphic tadpoles (Fig. 7.2), and this induction is independent of new protein synthesis (Shi and Brown, 1993; Wang and Brown, 1993; Ishizuya-Oka et al., 1997a), suggesting that TH/bZip is a direct TH response gene. Its predicted protein sequence shares strong homology with a mammalian transcriptional repressor in the DNA binding domain encompassing the basic region and the leucine zipper (Landschulz et al., 1988) (Fig. 7.1; Brown et al., 1996). However, little homology exists outside the putative DNA binding domain. Thus, although TH/bZip is likely to be a sequence specific DNA binding transcription factor, it remains to be determined whether it is a transcriptional activator or repressor.

Analysis of the expression of the *Xenopus* TH/bZip gene has demonstrated that it is developmentally regulated in a tissue-dependent manner. It has little expression in the hind limb throughout metamorphosis (Wang and Brown, 1993). In the tail, the TH/bZip mRNA level remains low till stage 62. Subsequently, it is upregulated when massive tail resorption, as reflected by the rapid reduction in tail length, takes places (Fig. 7.3*A*; Wang and Brown, 1993; Brown et al., 1996; Ishizuya-Oka et al., 1997a). Finally, in the intestine, high levels of TH/bZip are present around stages 60–62 (Fig. 7.3*A*). This is earlier than in the tail. However, it coincides with the period when larval epithelial apoptosis and adult cell proliferation take place (McAvoy and Dixon, 1977; Ishizuya-Oka and Shimozawa, 1991, 1992a; Ishizuya-Oka and Ueda, 1996). Furthermore, the expression profiles are reproduced by TH treatment of premetamorphic tadpoles (Fig. 7.3*B*; Wang and Brown, 1993; Ishizuya-Oka et al., 1997a).

```
E4BP4   QLRKMQTVKKEQASLDASSNVDKMMVLNSALTEVSEDSTTGEDVLL....  47
        .|..|....|.:.....::      ....|.....|..::||
TH/bZip GLVPCPTPVDPPAPPQVPPTLEVYAPPPHEEISPSTSLISGSGLLLARSL  52
```

Basic Domain

```
E4BP4   ......SEGSVGKNKSSACRRKREFIPDEKKDAMYWEKRRKNNEAAKRSR   91
              ..:..:......:.:|  :|||.|||||||. ||||||||||||..||||
TH/bZip FGCRRTPSSPTSSTSPGNSRGRREFTPDEKKDDQYWEKRRKNNEVEKRSR  102
```

Leucine Zipper

```
E4BP4   EKRRLNDLVLENKLIALGEENATLKAELLSLKLKFGLJSSTAYAQEIQKL  141
        ||||  .||.||.::|||  ||||  |:||||.|::::||:.
TH/bZip EKRRAGDLALEGRVIALLEENARLRAELLALRFRLGLVR...........  141
              *        *          *        *

E4BP4   SNSTAVYFQDYQTSKSNVSSFVDEHEPSMVSSSCISVIKHSPQSSLSDVS  191
                :.::..::.  |...:  |||.  .....  ..:.||.:|::|..|
TH/bZip ........DPCEETRGGYSAQCGLHEPPPANPPPPPLPHSEDSGFSTPS  183

E4BP4   EVSSV...EHTQESSVQGSCRSPENKFQIIKQEPMELESYTREPRDDRGS  238
        .|.|   ::..|  ..|..  .:|:. . :.  |..|.:|         :.||.
TH/bZip VGSPVFLEDRVPEHEPQPMPHSALSYYGPINGETVE.........NPRGR  224

E4BP4   YTASIYQNYMGNSFSGYSHSPPLLQVNRSSSNSPRTSETDDGVVGKSSDG  288
        ...           :|:::.:
TH/bZip LET......LGDCYKS..................................  234
```

Repression Domain

```
E4BP4   EDEQQVPKGHIHSPVELKHVHATVVKVPEVNSSALPHKLRIKAKAMQIKV  338
                                                  ||||||:|:  |
TH/bZip ........L.....................LPHKLRFKGGA.....  245

E4BP4   EAFDNEFEATQKLSSPIDMTSKRHFELEKHSAPSMVHSSLTPFSVQVTNI  388
                        . : .|.. :::.:...:
TH/bZip ...............SGEEGSHSIMQMTRETG.J................  262

E4BP4   QDWSLKSEHWHQKELSGKTQNSFKTGVVEMKDSGYKVSDPENLYLKQGIA  438
        :|  ||   ..::|.|..  :  :.|  :.  .:.|  .....||   |:  .:|
TH/bZip ...GLASE..TATGFSQKPPPCPRGGWMAHSQDGVPPAASENSELRSQLA  307

E4BP4   NLSAEVVSLKRLIATQPISASDSG  462
        .|||||  |||::. |.  .  ::  :
TH/bZip SLSAEVEHLKRIFSQQVAAHGGHE  331
```

Figure 7.1 Xenopus *early T_3 response gene TH/bZip (Brown et al., 1996; Ishizuya-Oka et al.,1997a) shares homology with the human transcriptional repressor E4BP4 (Cowell et al., 1992). The identical amino acid residues are indicated by vertical bars between them, and dots between the Xenopus and human residues indicate similar amino acids. The dots in the sequences are gaps introduced for better alignment. Several functional domains defined in E4BP4 are bracketed (Cowell et al., 1992; Cowell and Hurst, 1994). The leucine residues in the leucine zipper of the DNA binding domain (the Basic plus Leucine Zipper domains) are indicated by asterisks. Note that the DNA binding domain is highly conserved while the rest of the protein, including the repression domain, is divergent.*

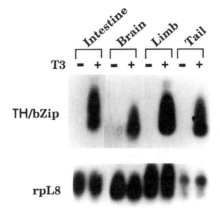

Figure 7.2 Ubiquitous induction of TH/bZip gene by T_3 in premetamorphic tadpoles. Xenopus-laevis tadpoles at stages 52–54 were treated in the presence (+) or absence (−) of T_3 for up to 24 hr. Poly $(A)^+$ RNA was isolated from the brain, hind limb (limb), tail, and intestine and copied into cDNA using a reverse transcriptase. The double-stranded cDNA was restricted to small fragments, ligated to a PCR linker, and amplified by PCR. Southern blot hybridization was performed on these cDNAs using TH/bZip or the control probe rpL8, which encodes the ribosomal protein L8 and is independent of T_3 (Shi and Liang, 1994). For details see Stolow and Shi (1995).

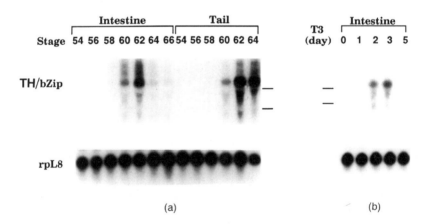

Figure 7.3 The Xenopus TH/bZip gene is highly expressed during both natural and T_3-induced metamorphosis. (A) Developmental expression of TH/bZip mRNA. Each lane had 10 μg RNA except the lane for the tail at stage 64, which had only 5 μg. (B) TH/bZip expression during T_3 treatment. Stage 54 tadpoles were treated with 5 nM T_3 for 0–5 days, and RNA was isolated from the intestine. 5 μg RNA was used per lane. The hybridization with rpL8 served as a loading control. Bars indicate the position of the 18S and 28S rRNA. From Ishizuya-Oka et al. (1997a).

More importantly, in situ hybridization reveals that the TH/bZip expression is first upregulated in the larval intestinal epithelial cells around stage 59/60, prior to any noticeable apoptosis in the larval intestinal epithelium (Ishizuya-Oka et al., 1997a). The expression is downregulated when cell death is actually taking place around stages 60–62. At these stages, the cells of the adult epithelium, connective tissue, and muscles are actively proliferating and express high levels of TH/bZip mRNA. As the adult cells differentiate, they downregulate TH/bZip expression. Such a regulation profile is similar to that of the TRβ genes (see Chapter 8), suggesting that TRβ may be involved in the upregulation of the TH/bZip gene. Regardless of the mechanism underlying the regulation, the tight associations of TH/bZip expression with cell type-specific changes imply that TH/bZip may participate in both apoptosis and cell proliferation. Such a role for TH/bZip is also consistent with its spatial expression profiles in the head and tail during metamorphosis (Berry et al., 1998a,b), where TH/bZip is found to be expressed in both dying and proliferating cells, although at higher levels in the proliferating ones. These results further imply that high levels of TH/bZip may not be compatible with high degrees of cell differentiation, at least in the tail and intestine. Thus in differentiated adult cells or in premetamorphic tadpoles, the TH/bZip mRNA levels are low.

Unlike the TH/bZip gene, the *Xenopus* NFI B and C genes represent another class of T_3-responsive TF genes (Puzianowska-Kuznicka and Shi, 1996). They are activated in the intestine during metamorphosis and remain highly expressed in postmetamorphic intestine (Fig. 7.4). They are also upregulated during tail resorption and limb development. However, their upregulation in the tail and hind limb occurs at different stages from that in the intestine. In the tail, it occurs slightly later, starting at stage 62, coinciding with the massive tail resorption. In the hind limb, high levels of NFI mRNAs are present at early stages, for example, stages 56–58, when limb morphogenesis, that is, digit formation, takes place. After stage 60, the mRNA levels are reduced somewhat to lower levels. In addition to these correlations of the mRNA levels with organ-specific metamorphosis, the protein levels of these NFI genes appear to be similarly regulated during metamorphosis. This has been shown by using gel mobility shift assay on an NFI consensus binding site followed by using specific antibodies to supershift the complexes containing a particular NFI (Fig. 7.5; Puzianowska-Kuznicka and Shi, 1996). Thus the NFIs may be involved in regulating the expression of both genes participating in cell death and those important for cell proliferation and differentiation, consistent with the diverse function of mammalian NFIs in transcription and replication (Cereghini et al., 1987; Apt et al., 1994; Nagata et al., 1982; Hay, 1985; Gronostajski et al., 1988).

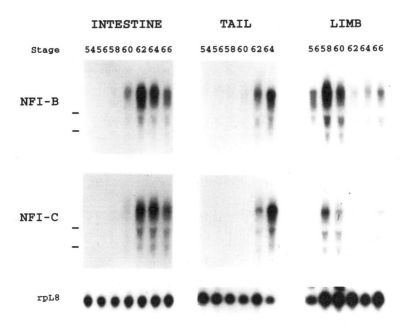

Figure 7.4 Northern blot analysis showing differential regulation of X. laevis NFI genes in the intestine, tail, and hind limb during metamorphosis. Ten micrograms of RNA was used per lane except for the tail at stage 64 and the hind limb at stage 56, which had only 5 μg RNA. Duplicate blots were probed with the coding regions of NFI-B1 and NFI-C1. After boiling off the probes, the filters were hybridized with rpL8 as a control for loading (Shi and Liang, 1994). The blots containing limb RNA were exposed for a longer period. The positions of 28S and 18S rRNA are indicated. Note that both genes had similar expression profiles. High levels of their mRNAs were present in the intestine during remodeling (stages 60–66), in the tail during resorption (stages 62–64), and in hind limb during and immediately after limb morphogenesis (stages 56–60; note that only half as much RNA was used for stage 56). Reprinted from Puzianowska-Kuznicka and Shi (1996) with permission from the American Society for Biochemistry and Molecular Biology, Inc.

7.3 BIOCHEMICAL AND MOLECULAR CHARACTERIZATION OF THYROID HORMONE-INDUCED TRANSCRIPTION FACTORS

Like thyroid hormone receptors, the TH-induced TFs are believed to regulate the transcription of target genes. Their specificity in gene regulation lies largely in their recognition of specific DNA sequences present in their target genes. Their interactions with the basal transcriptional machinery and/or other transcription factors participating in the regulation of the target gene transcription are also important for the specificity. Thus the understanding of the function of the T_3-induced transcription factor genes requires the determination of their DNA binding specificity, transcriptional activity, and target genes.

No systematic effort has been carried out to isolate target genes of the TH-induced TF genes or investigate their DNA binding specificity. However,

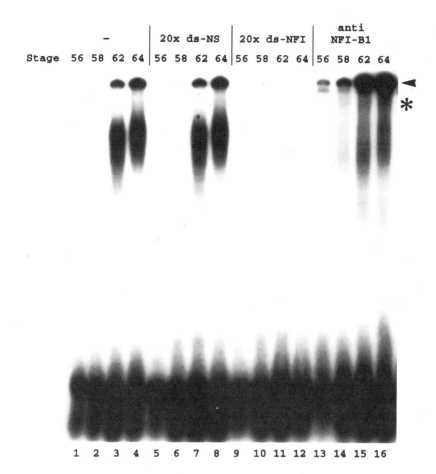

Figure 7.5 NFI binding activity is regulated similarly as the NFI mRNAs during development. Whole cell extracts were isolated from the intestine of tadpoles at different developmental stages and analyzed for binding to a ^{32}P-labeled double-stranded oligonucleotide containing a consensus NFI binding site (ds-NFI). Specific complexes were formed in the absence (lanes 1–4) or presence of a 20-fold excess of a nonspecific competitor (ds-NS) (lanes 5–8) but not in the presence of a 20-fold excess of the unlabeled ds-NFI (lanes 9–12). The addition of an anti-NFI-B1 antibody could supershift most of the complexes formed (lanes 13–16). The arrowhead and asterisk indicate the positions of the supershifted NFI-B1-DNA and NFI-C1-DNA complexes, respectively. Reprinted from Puzianowska-Kuznicka and Shi (1996) with permission from the American Society for Biochemistry and Molecular Biology, Inc.

their amino acid sequence homologies with known TFs in other animal species have provided valuable information about their biochemical and molecular properties. Thus the strong sequence conservation in the basic region and the leucine zipper region (the bZip domain) between the *Xenopus* TH/bZip and the mammalian bZip domain-containing transcriptional repressor implies that the *Xenopus* TH/bZip protein recognizes similar DNA sequences as the mammalian protein (Fig. 7.1; Brown et al., 1996; Ishizuya-Oka et al., 1997a). Similarly, the frog NFI B and C are the homologs of avian and mammalian NFI B and C, respectively, and are expected to bind to similar DNA sequences as the avian and mammalian proteins (Puzianowska-Kuznicka and Shi, 1996). Indeed, direct DNA binding experiments using the *Xenopus* proteins overexpressed in frog oocytes show that the frog proteins can recognize a consensus NFI binding site (Fig. 7.5).

The identification of some DNA binding sequences by a TF allows the investigation on its transcriptional properties. Various methodologies are available for such studies. These include (1) in vitro transcription studies with a reconstituted RNA polymerase II transcription system, the purified TF of interest, and a reporter promoter containing the TF binding site; (2) cotransfection of a reporter plasmid and a plasmid overexpressing the TF into model tissue culture cells; and (3) microinjecting a reporter plasmid into the nucleus of frog oocytes that contain overexpressed TF through microinjection of its mRNA into the oocyte cytoplasm (see Chapter 5), and so on. In all cases, the properties of the TF can be investigated after adding or expressing the wild type TF if none of this TF or insufficient quantities of it are present endogenously. Even if endogenous TF is present, useful information can be obtained by introducing exogenous mutated TF, such as dominant or chimeric TFs.

Using the oocyte system, the frog NFI B and C have been shown to be able to activate a promoter bearing a consensus NFI binding site, confirming their properties predicted from sequence homology analysis (Fig. 7.6; Puzianowska-Kuznicka and Shi, 1996). Similarly, *Xenopus* TRβA, which is encoded by one of the two TRβ genes and also a T_3-induced TF, has been shown to be able to regulate properly a target promoter, which happens to be its own, in the frog oocyte (Chapter 5, Wong and Shi, 1995; Wong et al., 1995, 1998a). Furthermore, the oocyte system has allowed the investigation of the role of chromatin in the transcriptional regulation by TRs and the effects of TRs on chromatin (Wong et al., 1995, 1997a, 1998b; also see Chapter 5), a study difficult to be carried out in vitro or in tissue culture cells through transient transfection. Thus the frog oocyte system will remain a very powerful system in which to study the molecular mechanisms by which the T_3-induced TFs regulate transcription.

7.4 TARGET GENES OF THE THYROID HORMONE-INDUCED TRANSCRIPTION FACTORS

As mentioned above, there is little information on the target genes of the TFs genes induced by TH during metamorphosis. The only possible exceptions are the TRβ genes in *Xenopus laevis*. This is because all of the early direct TH response genes isolated so far are presumably regulated by TRα and/or TRβ, the only TRs known in any vertebrate species. Thus to understand the gene

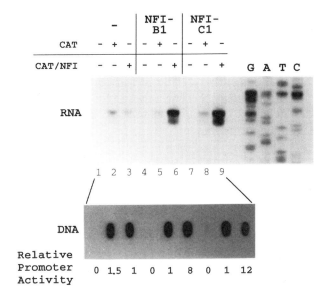

Figure 7.6 *Transcriptional activation by X. laevis NFIs in a reconstituted oocyte system. Control oocytes (−) or oocytes preinjected with the mRNA for NFI-B1 or NFI-C1 were injected with either one of two promoter vectors. The first vector (CAT) was a vector containing the SV40 promoter upstream of the CAT gene, and the second one (CAT/NFI) had two copies of the NFI binding site (ds-NFI, see Fig. 7.5) inserted into the CAT vector. After overnight incubation, the oocytes were homogenized and half of the oocyte homogenate was used for RNA analysis by primer extension (upper panel). The other half was used to quantify the injected DNA by slot blot analysis (lower panel). The relative promoter activity was determined by normalizing the primer extension signal with the DNA signal. The DNA sequencing ladder with samer primer was used as a size standard. Reprinted from Puzianowska-Kuznicka and Shi (1996) with permission from the American Society for Biochemistry and Molecular Biology, Inc.*

regulation cascade involving the TH-induced TF genes, it will be essential to isolate their downstream target genes.

The most direct way to isolate the target genes of a TF is to first generate antibodies against the TF (see Gould et al., 1990 for an example). Next, nuclei can be isolated from tissues of TH-treated premetamorphic tadpoles or tadpoles undergoing natural metamorphosis. The nuclei are then restricted with endonucleases or partially digested with micrococcal nuclease, and incubated with the anti-TF antibodies to immunoprecipitate the chromatin-bound TF together with the DNA fragments bound by the TF. The selected DNA fragments can be cloned and sequenced to identify the TF binding sites and the corresponding genes, which are the TF target genes. Although straightforward, this method will likely yield many DNA fragments that do not containing TF binding sites (false positives). Thus careful characterization of the selected binding sites will be required to confirm a target gene. In addition, some target genes may be missed.

Another way to obtain TF target genes is first to isolate TF binding sites in genomic DNA through binding site selection (Fig. 7.7; see Kinzler and Vogelstein, 1989, for an example). First the TF is overexpressed in *E. coli* or other systems and purified. Then genomic DNA is restricted to an average of 500 bp in length and ligated to PCR primers and amplified. The DNA is incubated with the purified TF. The DNA–TF complex is separated from unbound DNA by using gel mobility shift assay or immunoprecipitation to isolate DNA containing the binding sites. The DNA can be amplified again and the selection can be repeated until most or all DNA fragments contain TF binding sites. The DNA fragments can then be cloned and used to screen larger genomic DNA. The latter, more likely to contain transcribed regions, can be used as probes for Northern blot analysis to determine the expression profiles

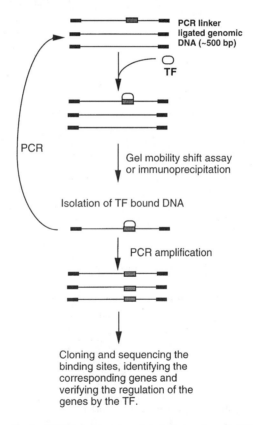

Figure 7.7 Schematic of a PCR-based approach to isolate genomic DNA fragments containing binding sites for a transcription factor (TF). The genomic DNA (lines) is digested to an average of 500 bp in length, ligated to a PCR linker (filled boxes), and amplified by PCR. The amplified DNA is then used in binding site selection as shown in the Figure. Open boxes, TF; shaded boxes, TF binding sites.

of the corresponding mRNAs. Those genes whose expression profiles are consistent with TF expression patterns (i.e., if the TF is a repressor, the genes should be highly expressed when the TF is low and repressed when TF is high; the opposite is true if TF is an activator) are very likely to be the desired target genes. Further studies of these candidate genes will include cloning of the promoters and investigating their regulation by the TF. In vivo genomic footprinting can also be carried out to see if the TF binding sites are occupied by the TF when the genes are expected to be regulated by the TF.

A final approach to identify the target genes of a TF is through a reverse analysis. That is, the promoter of a late TH response gene is analyzed to determine what transcription factors are involved in its regulation. If a transcription factor whose binding site is present in the promoter is expressed in a proper temporal and spatial manner that is consistent with a role for the transcription factor in the regulation of the late TH response gene, the late TH response gene is most likely to be a target gene of the transcription factor.

A gene regulation cascade involving a TH-regulated TF gene has been suggested from studies on the carbamyl phosphate synthetase (CPS) and ornithine transcarbamylase (OTC) genes of the frog *Rana catesbeiana* (Atkinson et al., 1998). As described in Chapter 6, the *Rana* CPS gene is a late TH-response gene participating in the urea cycle in the liver. As the tadpole is transformed into a frog, the animal changes from ammonotelism to urotelism and the urea cycle enzyme genes, including the CPS gene, are upregulated (Atkinson et al., 1994). The activation of the CPS and OTC genes by TH is delayed (Fig. 7.8) and appears to be indirect (Helbing and Atkinson, 1994; Helbing et al., 1992). Analysis of the CPS and OTC promoters reveals the existence of binding sites for the transcription factor C/EBPα (Chen et al., 1994;

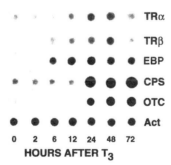

Figure 7.8 Sequential upregulation of Rana TRα, TRβ, C/EBPα (EBP), CPS, and OTC mRNAs in the liver of thyroid hormone (T_3)-treated tadpoles. RNA was isolated from the liver of these tadpoles at the times indicated and aliquots of it were used for dot-blot hybridization. Because the relative levels of actin (Act) mRNA do not change in the liver of T_3-treated tadpoles (Helbing et al., 1992), it is used as an internal control. Note that TRα is present even before the treatment. TRβ and C/EBPα are activated first, followed by the activation of CPS and OTC genes about 20 hr later. From Chen and Atkinson (1997).

Chen and Atkinson, 1997), a leucine zipper containing TF that is involved in hepatocyte cell differentiation (Friedman et al., 1989; Umek et al., 1991; Cao et al., 1991; Sladek and Darnell, 1992). Direct DNA binding experiments in vitro and transcriptional activation experiments in tissue culture cell transfection studies have shown that C/EBPα can regulate these genes through direct recognition of its binding sites present in these genes (Chen and Atkinson, 1997). Interestingly, C/EBPα gene is also upregulated during liver remodeling and its regulation by TH is fast, within a few hours (Fig. 7.8; Chen et al., 1994; Chen and Atkinson, 1997); just like many of the direct TH response genes isolated in *Xenopus laevis*. Thus TH may directly activate the C/EBPα gene. The C/EBPα in turn activate the CPS and OTC genes to establish a TF-mediated transcriptional regulation cascade to effect a part of the reprogramming of gene expression in the liver during metamorphosis. Similarly, roles for *Xenopus laevis* C/EBPα have also been suggested by its conserved developmental expression profiles during metamorphosis (Xu and Tata, 1992). The future challenge will be to test the validity of this and other possible gene regulation cascades and determine the interrelations among them and with other intra- and/or extracellular events triggered by TH during tissue remodeling.

8

Competence and Tissue-specific Temporal Regulation of Amphibian Metamorphosis

8.1 INTRODUCTION

Natural metamorphosis takes place at precise developmental stages. This process, however, can be precociously induced in premetamorphic tadpoles as early as stage 41 for *Xenopus laevis*, only 3 days after fertilization, by treatment with exogenous TH (Dodd and Dodd, 1976; Gilbert and Frieden, 1981; Tata, 1968). Younger tadpoles or embryos were found to be refractory to this TH induction. This observation suggests that the tadpole is competent to respond to TH only after specific developmental stages.

This stage-dependent phenomenon is also observed to be true if one examines the metamorphic transitions of different tadpole organs during natural development. Individual tissues undergo their unique metamorphic changes at distinct developmental stages. This is exemplified by the transformation of the hind limb, tail, and the intestine in *Xenopus laevis* (Chapter 3, Fig. 3.1). The earliest change to take place is hind limb development (Chapter 3; Nieuwkoop and Faber, 1956). The hind limb bud begins to grow around stage 48 and the most dramatic morphological differentiation takes place around stages 53–56. Subsequently, the hind limb undergoes mostly growth with little morphological change. In stark contrast, tail resorption is one of the last changes to occur during metamorphosis (Chapter 3; Nieuwkoop and Faber, 1956). Although some changes in the tail, such as tail fin resorption, take place a little earlier (Dodd and Dodd, 1976), drastic tail resorption as reflected by the reduction in tail length begins around stage 62 and its completion marks the end of metamorphosis. The intestine, on the other hand, begins its remodeling process around stage 58 when its connective tissue and muscle layers increase in thickness (Chapter 3; McAvoy and Dixon, 1977; Ishizuya-Oka and Shimozawa, 1987a). This is followed by the degeneration of the larval epithelium through programmed cell death (apoptosis) around stages 60–62 (McAvoy and Dixon, 1977; Ishizuya-Oka and Shimozawa, 1992a; Ishizuya-Oka and Ueda, 1996). Concurrently, secondary epithelial cells rapidly proliferate and differentiate toward the end of metamorphosis to form a much more complex frog organ (McAvoy and Dixon, 1977; Ishizuya-Oka and Shimozawa, 1987a).

The molecular mechanisms underlying such developmental stage-dependent phenomena are still unclear. On the other hand, each tissue is absolutely dependent on the presence of TH and its receptors to undergo metamorphic changes, suggesting that the regulation of TH and TR levels may be important for the tissue-specific temporal transformation of various tissues/organs. Indeed, biochemical and molecular studies as described below provide strong evidence to support such a mechanism.

8.2 EXPRESSION AND FUNCTION OF THYROID HORMONE AND 9-CIS RETINOIC ACID RECEPTORS IN PRE- AND METAMORPHOSING TADPOLES

Thyroid hormone receptors are the presumed mediators of the regulatory effects of TH during metamorphosis. Their function in vivo depends on the presence of RXRs, which form heterodimers with TRs to facilitate the transcriptional repression by unliganded TRs and activation by TH-bound TRs. Thus extensive studies have been carried out to determine the levels of TR and RXR in different organs at various stages of animal development and to investigate the developmental function of TR/RXR heterodimers. The following summarizes studies in *Xenopus laevis*. Similar observations have also been made in *Rana catesbeiana* (Schneider and Galton, 1991; Helbing et al., 1992; Davey et al., 1994).

8.2.1 Thyroid Hormone and 9-cis Retinoid Acid-Receptor Expression

8.2.1.1 Correlation with Organ Transformations As the mediators of the biological effects of TH, TRs and RXRs are expected to be present during amphibian metamorphosis. Indeed, the mRNAs of both TRα and TRβ are expressed during amphibian metamorphosis (Schneider and Galton, 1991; Shi et al., 1994; Eliceiri and Brown, 1994; Fairclough and Tata, 1997; Helbing et al., 1992; Yaoita and Brown, 1990; Kawahara et al., 1991; Tata, 1996). In

Xenopus, the TRα genes are activated shortly after tadpole hatching and their expression is maintained at high levels in tadpoles throughout metamorphosis (Fig. 8.1). Similarly, both RXRα and RXRγ are also expressed in premetamorphic as well as metamorphosing tadpoles (Fig. 8.1; Wong and Shi, 1995). The third member of the RXR family, the RXRβ gene, is also present in *Xenopus laevis* (Marklew et al., 1994), although its expression during metamorphosis has not been analyzed.

The TRβ genes, however, have little expression in premetamorphic tadpoles (Fig. 8.1). Their mRNA levels are drastically upregulated during metamorphosis, coinciding with the rising concentration of endogenous TH (Fig. 8.1; Yaoita and Brown, 1990). In fact, the TRβ genes are among the early TH response genes isolated from subtractive screens for TH-regulated genes (Chapter 6). Furthermore, their regulation by TH occurs within a few hours of TH treatment of premetamorphic tadpoles and is independent of new protein synthesis, suggesting that they are directly regulated by TH through the TRs themselves (Kanamori and Brown, 1992).

Analysis of the genes encoding *Xenopus* TRβ genes shows that both TRβA and TRβB genes span at least 70 kb of genomic DNA and consist of at least 12 exons each (Fig. 8.2; Shi et al., 1992). About half of these exons encode the 5′-untranslated region and alternative splicing of those exons gives rise to the

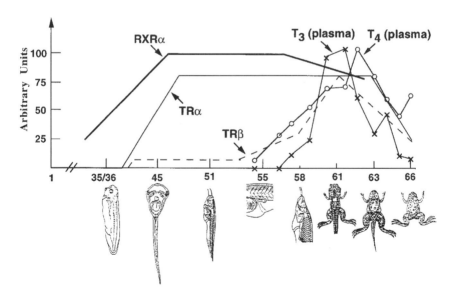

Figure 8.1 Correlation of the levels of endogenous TH and the mRNAs of TRα, TRβ, and RXRα genes with stages during Xenopus laevis development. The plasma concentrations (arbitrary units) of thyroid hormone T_3 (crosses) and T_4 (open circles) are from Leloup and Buscaglia (1977). The mRNA levels for TRα (solid line), TRβ (broken line), and RXRα (bold line) are based on published data (Yaoita and Brown, 1990; Kawahara et al., 1991; Wong and Shi, 1995) and for clarity, plotted on different scales.

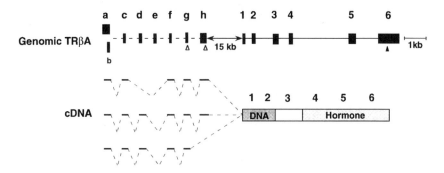

Figure 8.2 Organization of TRβ genes in Xenopus laevis. The two TRβ genes have similar structures and only TRβA is shown here. (Top) Genomic structure. The exons a–h encode mostly the 5'-untranslated region and exons 1–6 encode the protein-coding region from zinc finger 1 of the DNA-binding domain (exon 1) to the carboxyl terminus. The translation initiation and termination sites are indicated by open and solid triangles, respectively. The relative location of exons a and b are unknown and thus are shown at the same position. The exons are shown as solid bars and introns as lines. The dashed lines indicate introns of unknown sizes. (Bottom) Three examples of TRβA mRNAs derived from alternative splicing of the 5' exons. The 5' exons present in the mRNA are indicated by short bars underneath the corresponding exons in the genomic DNA. No evidence has been reported for the existence of alternative splicing involving exons 1–6 (Shi et al., 1992). Reprinted from Shi (1996) with permission from Academic Press.

multiple isoforms of TRβ mRNAs, leading to at least two TR proteins for each gene that differ only in a small region of the amino terminus (Fig. 8.2; Yaoita et al., 1990; Shi et al.,1992). Furthermore, each gene has two alternative promoters. One of the promoters is expressed at low but constitutive levels whereas the other is highly expressed only in the presence of TH (Kanamori and Brown, 1992; Shi et al., 1992). Analysis of the inducible promoter reveals that both genes contain at least a strong TRE consisting of two near perfect direct repeats of AGGTCA separated by 4 bp (Ranjan et al., 1994; Machuca et al., 1995; Wong et al., 1995, 1998a), which is likely responsible for the upregulation of the TRβ genes by TH during metamorphosis in different organs.

When the expression of the TR and RXR genes was analyzed in individual organs during *Xenopus* metamorphosis, a strong correlation was found between their mRNA levels and organ-specific transformation (Wang and Brown, 1993; Yaoita and Brown, 1990; Shi et al., 1994; Wong and Shi, 1995). In general, TR and RXR genes are coordinately regulated during metamorphosis and their expression is high in a given organ when metamorphosis takes place and low before or after metamorphosis. Thus the mRNA levels for these receptor genes (except for TRβ genes) are high in the hind limb around stages 54–56 when hind limb morphogenesis, for example, digit formation, takes place (Fig. 8.3; Wang and Brown, 1993; Wong and Shi, 1995). Similarly, in the tail, the genes are expressed at low levels until stage 62 when they are upregulated, coinciding with the rapid tail resorption. In the animal intestine, the mRNAs of both TRα and RXRα genes are present at intermediate levels compared to those in the tail and hind limb, and those of TRβ and RXRγ genes are upregulated during tissue remodeling (Shi et al., 1994; Wong and Shi, 1995). It should be pointed out, however, that some differences do exist in the expression profiles of these receptor genes (Wong and Shi, 1995). For example, as illustrated in Figure 8.3, TRβ gene expression does not change as drastically as that of the TRα genes during hind limb morphogenesis (stages 54–58), suggesting that TRα may play a more important role in this transition. In any case, these observations together suggest that TR/RXR heterodimers mediate the effects of TH during metamorphosis and that one mechanism for regulating stage-dependent tissue transformation is to modulate the levels of endogenous hormone receptors. The availability of these receptors may thus be required for the correct temporal regulation of changes in individual tissues.

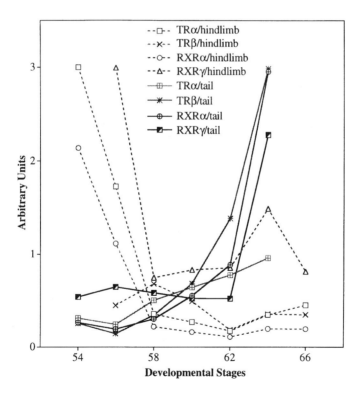

Figure 8.3 Coordinated regulation of TR and RXR genes during metamorphosis. With the exception of TRβ mRNA in the hind limb, the mRNA levels of the receptor genes are high during hind limb morphogenesis (stages 54–56) and tail resorption (stages 60–64) and low at other stages.

8.2.1.2 Cell Type-specific Expression of TRβ Genes Correlates with Apoptosis and Cell Proliferation in the Xenopus Intestine

The tadpole intestine offers a unique opportunity to investigate the role of TR genes, especially the TRβ genes, during metamorphosis. The *Xenopus* TRβ genes are direct TH response genes and have little expression before or after metamorphosis but are highly expressed during intestinal remodeling (Wong and Shi, 1995). Furthermore, the intestine consists of essentially three major types of tissues that are well separated spatially and easily identifiable (Fig. 3.1). These tissues within the intestine undergo distinct metamorphic changes at different stages. Thus a simple analysis of the TRβ gene expression in different cell types during metamorphosis may provide important clues on the role of TRβ in cell death or proliferation and differentiation.

In situ hybridization using a TRβ antisense RNA probe reveals a strong correlation of TRβ gene expression with cell type-specific changes in the *Xenopus* intestine (Figs. 8.4, 8.5; Shi and Ishizuya-Oka, 1997a). The TRβ mRNAs are absent or at very low levels prior to stage 55. They are first upregulated in the larval epithelium, to the maximal levels by stages 59/60, at the onset or immediately prior to larval epithelial cell death. Interestingly, the mRNA levels are downregulated as the cells undergo apoptosis (stages 60–62). The upregulation of the TRβ mRNAs occurs around stages 60–62 in the adult epithelium; connective tissue, and muscles. The genes are downregulated again in a sequential order in the adult epithelium, connective tissue, and muscles as their cells differentiate. In particular, the downregulation occurs last in the longitudinal muscle, which is also the last tissue to attain its adult form. Thus TRβ appears to be involved in the early stages of apoptosis and adult cell proliferation but is not required or is required at only very low levels for differentiated adult cells.

8.2.2 Functions of Thyroid Hormone Receptors in Frog Development

Two different approaches have been used to study the roles of TRs in developing *Xenopus laevis*. One is to test the function of endogenous TRs and the other is to determine the effects of exogenous TRs on animal development. By microinjecting exogenous genes directly into the caudal skeletal muscle of *Xenopus* tadpoles, De Luze et al. (1993) have demonstrated that the exogenous plasmids can survive for weeks or longer in the muscle cells. Furthermore, if a promoter is present in the microinjected plasmid, the promoter is transcribed and responsive to endogenous transcription factors. Thus when a reporter plasmid bearing a TRE (thyroid hormone response element) is injected into the muscle cells of premetamorphic tadpoles of *Xenopus laevis*, the endogenous TRs can activate the exogenous promoter when TH is added to the rearing water. This result indicates that functional TRs are present in the tadpoles to mediate the effect of TH on target gene transcription, consistent with the presence of TRs and RXRs in the tail of premetamorphic tadpoles (Wong and Shi, 1995; Eliceiri and Brown, 1994; Fairclough and Tata, 1997).

EXPRESSION AND FUNCTION OF THYROID HORMONE RECEPTORS   141

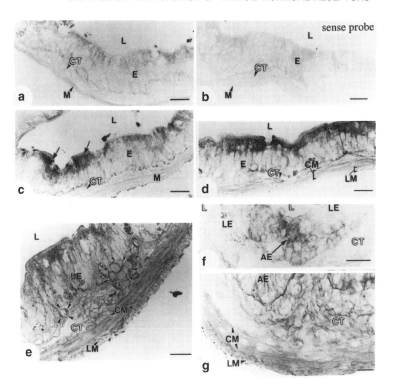

Figure 8.4 In situ hybridization reveals that TRβ genes are first activated in larval epithelial cells and then in the other cell types of the small intestine. (a) Stage 55. No signal could be detected with an antisense TRβ RNA probe. (b) Stage 55. Control hybridization with a sense probe. (c) Stage 57. TRβ gene are activated in the epithelium (arrows, E) facing the lumen (L) but not yet in the connective tissue (CT) or muscles (M). (d) Stage 59. TRβ genes are highly expressed in the larval epithelium. The TRβ signals can also be detected, although weakly, in the connective tissue and the circular muscle (CM) but not in the longitudinal muscle (LM). (e) Stage 60. The adult epithelial cells can now be detected as small islets (arrowheads) between the larval epithelium (LE) and the connective tissue and are expressing the TRβ genes. The signals in the larval epithelium are weaker than those at stage 59 in (d), whereas the signals in the connective tissue and circular muscles are stronger than those at stage 59. The signals in longitudinal muscle remain very week or negative. (f) Stage 61. The proliferating adult epithelium (AE) remains positive in TRβ expression but the degenerating larval epithelium is negative. (g) Stage 62. The TRβ mRNA signals in the longitudinal muscle are now at least as strong as those in the circular muscle. TRβ genes are also expressed in the adult epithelium and connective tissue. Larval epithelium degeneration is almost complete. Modified after Shi and Ishizuya-Oka (1997a) with permission from Karger and Basel.

Using this same method, Ulisse et al. (1996) have subsequently shown that dominant-negative mutant TRs made based on in vitro and tissue culture transfection studies can prevent the activation of the TH-dependent *Xenopus* TRβA promoter linked to a reporter injected into *Xenopus* tadpoles. Thus the exogenous mutant TRs can interfere with the function of endogenous TRs, possibly by forming heterodimers with endogenous RXRs to bind to the TRE

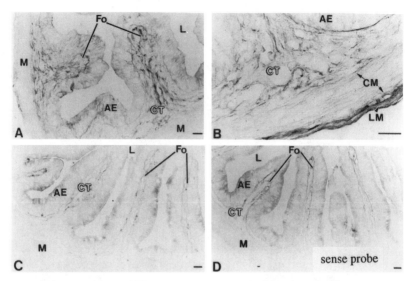

Figure 8.5 The repression of TRβ genes occurs later in the connective tissue and muscles than in the epithelium. (A) Stage 63. Multiple folds (Fo) are formed as adult epithelial cells begin to differentiate. The TRβ signals in the connective tissue (CT) and the muscles (M) are weaker than those at stage 62. The adult epithelium (AE) is essentially negative. (B) Lower region of the intestinal fold at stage 63. The longitudinal muscle (LM), which is the last tissue to attain its adult form, still expresses the TRβ genes at high levels. However, the signals are weak in the connective tissue and undetectible in the circular muscle (CM). (C) Stage 66. By the end of metamorphosis, the TRβ genes are repressed in all tissues of the intestine. (D) Stage 66. Control hybridization with a sense probe. Reprinted from Shi and Ishizuya-Oka (1997a) with permission from Karger and Basel.

in the TRβ promoter. This experiment again supports a functional role of endogenous TRs.

Puzianowska-Kuznicka et al. (1997) have made use of the lack of endogenous TR/RXR in early embryos to investigate the function of TR/RXR in development. By microinjecting mRNAs encoding *Xenopus* TRα and RXRα into fertilized eggs, they have overexpressed TRα and RXRα either individually or together into *Xenopus* embryos (Puzianowska-Kuznicka et al., 1997). The overexpression of individual receptors has little or no effects on embryo development in both the presence or absence of T_3. On the other hand, TRs/RXRs together have severe teratogenic effects on embryonic development if overexpressed at high levels in the absence of T_3. In the presence of T_3, even low levels of TRs/RXRs cause maldevelopment (Fig. 8.6). The phenotypes of the embryos in presence and absence of T_3 are distinct even though some similarities exist, consistent with the fact that TR/RXR heterodimers are transcription repressors in the absence of T_3 and activators when T_3 is present.

More importantly, the expression of several genes known to be regulated by T_3 during metamorphosis is specifically altered by the overexpressed TR/RXR

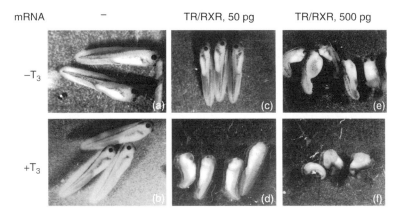

Figure 8.6 Overexpression of TRs and RXRs together results in dose- and T_3-dependent embryonic abnormalities. Embryos were injected with the indicated amounts of TR/RXR mRNAs and cultured for 2 days in the presence (+) or absence (−) of 100 nM T_3. Reprinted from Puzianowska-Kuznicka et al.(1997) with permission from American Society for Microbiology.

(Puzianowska-Kuznicka et al., 1997). The expression of two such genes is shown in Figure 8.7. The stromelysin-3 gene encodes a metalloproteinase that may participate in extracellular matrix remodeling and is a direct T_3 response gene in all tadpole organs examined (Wang and Brown, 1993; Patterton et al., 1995). The second gene, the *Xenopus* hedgehog gene, encodes a putative morphogen and is a direct T_3 response gene in the intestine but is not regulated by T_3 in most other organs examined (Stolow and Shi, 1995). Both genes are also expressed in early embryos when both TR and T_3 are not yet synthesized (Fig. 8.7; Patterton et al., 1995; Stolow and Shi, 1995). They are subsequently repressed upon the completion of tadpole organogenesis as tadpole feeding begins, and are reactivated in all (stromelysin-3) or certain (hedgehog) organs by T_3 during metamorphosis. The overexpression of TR or RXR alone has little effect on the expression of either gene, independently of T_3 (Fig. 8.7). However, coexpression of TR and RXR leads to a small but significant repression of the expression of the two target genes, especially the hedgehog gene, in the absence of T_3 and the addition of T_3 leads to the activation of the stromelysin-3 gene and only the reversal of the repression of the hedgehog gene (Fig. 8.7). As total embryo RNA is used for Northern blot analysis of the gene expression, it may not be surprising to see that the hedgehog gene is not upregulated by the overexpressed TR/RXR in the presence of T_3 because its upregulation by T_3 during metamorphosis is limited to a few organs. On the other hand, transcriptional repression likely involves different TR/RXR cofactors (Chapter 5) that may be present in all cell types to mediate the observed repression of the hedgehog gene by the overexpressed TR/RXR in the absence

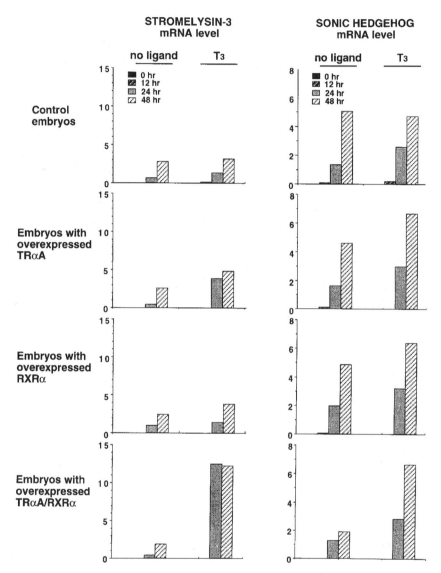

Figure 8.7 Overexpression of TR and RXR together but not alone in early Xenopus embryos leads to specific regulation of two T_3 response genes, the Xenopus sonic hedgehog and stromelysin-3 genes. Embryos were injected with indicated mRNAs (500 pg per embryo for each mRNA) and cultured in the presence or absence of 100 nM T_3. Total mRNA was isolated and analyzed by Northern blot hybridization. The quantification of the hybridization signals shows that the genes are repressed by the unliganded TR in the presence of RXR and the addition of T_3 leads to the reversal of the repression on both genes and strong activation of the stromelysin-3 gene, in agreement with the observation that stromelysin-3 is upregulated by T_3 ubiquitously in tadpoles (Wang and Brown, 1993; Patterton et al., 1995) whereas hedgehog is upregulated only in a few organs (Stolow et al.,1995). Based on Puzianowska-Kuznicka et al. (1997).

of the ligand. These results thus provide strong evidence to support the conclusions that TR/RXR heterodimers are the mediators of the regulatory effects of T_3 and that RXRs are required to mediate efficiently the effects of T_3 during metamorphosis, which was first suggested by the coordinated regulation of TR and RXR genes in different organs during metamorphosis (see previous section, Wong and Shi, 1995).

The upregulation of TRα and RXRα genes in post-hatching tadpoles (Fig. 8.1) correlates with the downregulation of several TH-response genes such as hedgehog (Stolow and Shi, 1995) and stromelysin-3 (Patterton et al., 1995). This and the ability of overexpressed unliganded TR/RXR to downregulate TH response genes precociously also suggest a role for TR/RXR in premetamorphic tadpoles. That is, the heterodimer represses the genes participating in larval organogenesis once larval development is complete (at the tadpole feeding stage, that is, stage 45, Fig. 8.1). Many of these genes are TH response genes and will be reused again late during metamorphosis to facilitate adult organ development and larval tissue resorption. Thus TR/RXR serves dual functions, that is, repressing these TH response genes in premetamorphic tadpoles when TH is absent and reactivating them during metamorphosis. Such a repression function of TR/RXR may be critical to ensure a proper period of tadpole development before transforming into a frog because continued expression of the TH response genes may trigger premature tissue transformation.

8.3 REGULATION OF CELLULAR THYROID HORMONE LEVELS

In addition to the regulation of the levels of hormone receptors, another possible mechanism for determining the spatial and temporal aspects of metamorphosis is the regulation of cellular TH levels in the animal. This could be achieved in several different ways, three of which are described here. First is the actual production of TH in the thyroid gland. Secondly, the levels of cellular TH could be affected by cellular proteins that bind TH, that is, cytosolic thyroid hormone binding proteins (CTHBPs). Finally, the metabolic conversion of T_4 to T_3 and/or the inactivation of both hormones may be an important step in regulating the levels of functional TH.

8.3.1 Circulating Thyroid Hormone

Thyroid hormone is synthesized in the thyroid gland in two forms, 3,5,3′,5′-tetraiodothyronine (T_4 or thyroxine) and its monodeiodinated form 3,5,3′-triiodothyronine (T_3). In *Xenopus laevis*, the gland first appears in the embryo as a median thickening of the pharyngeal epithelium at the time of tadpole hatching (stages 35/36) (Chapter 2, Dodd and Dodd, 1976). This rudiment then develops into a functional larval thyroid gland around stage 53.

In 1977, Leloup and Buscaglia reported a systematic quantification of plasma TH levels during *Xenopus laevis* development (Leloup and Buscaglia,

1977) (Chapter 2, also Fig. 8.1). They found detectable levels of T_4 in the larval plasma as early as stage 54, shortly after the formation of the functional thyroid gland. The levels of plasma T_3 appear to be lagging slightly behind T_4 during development. These low levels of endogenous TH can apparently trigger the initiation of metamorphosis with the morphogenesis of the hind limb as one of the first visible changes to take place around stages 53–56. Subsequently, the plasma concentrations of both thyroid hormones increase, reaching peak levels around stage 60, the climax of metamorphosis. Currently, the mechanisms underlying the developmental regulation of the plasma TH levels are unknown. However, many factors appears to be involved. These include the factors that influence the growth and maturation of the thyroid gland and hormonal clues that regulate thyroid hormone synthesis and release from the gland (Chapter 2; Dodd and Dodd, 1976; Gilbert and Frieden, 1981). Regardless of the regulation mechanisms, the high levels of TH during development correspond exactly to the period of metamorphosis, consistent with the causative role of TH during this tadpole-to-frog transition.

8.3.2 Cytosolic Thyroid Hormone Binding Proteins

In order to regulate cellular gene expression, the circulating TH in the plasma has to be transported through yet undefined mechanisms into target cells in different tissues (see Chapter 5). Upon entering the cytoplasm, it is likely to encounter cytosolic TH binding proteins (CTHBPs) (Chapter 5; Barsano and DeGroot, 1983). Many such cellular proteins have been characterized (Chapter 5; Shi et al., 1994; Ashizawa and Cheng, 1992; Yamauchi and Tata, 1994). In general, CTHBPs are multifunctional proteins that serve one or more other roles (e.g., pyruvate kinase, myosin light chain kinase, and disulfide isomerase) and bind TH with 10- to 100-fold weaker affinity than the nuclear TRs.

The exact roles of these CTHBPs in TH signal transduction is still unclear. These proteins could participate in TH import from the extracellular medium, intracellular TH metabolism, and transport to the nucleus, or serve as a buffer to modulate intracellular free TH concentrations. This last function is strongly supported by studies on the human M2 pyruvate kinase (Ashizawa and Cheng, 1992). As a monomer, it binds TH with high affinity and specificity whereas the homotetramer form functions as the M2 pyruvate kinase. Interestingly, overexpression of the monomer form of the protein leads to an inhibition of TH-dependent transcriptional activation by TR in a tissue culture cell line. Based on this observation, S.-Y. Cheng and co-workers suggest that the monomer form of M2 pyruvate kinase functions as a chelator of cellular TH, thus reducing cellular free TH concentration and inhibiting the effect of TH (Ashizawa and Cheng, 1992).

To investigate the roles of cellular thyroid hormone binding proteins during metamorphosis, several CTHBPs have been identified in *Xenopus laevis* (Yamauchi and Tata, 1994, 1997; Shi et al., 1994). Two of them have been cloned. The first one was isolated from frog liver owing to its ability to bind

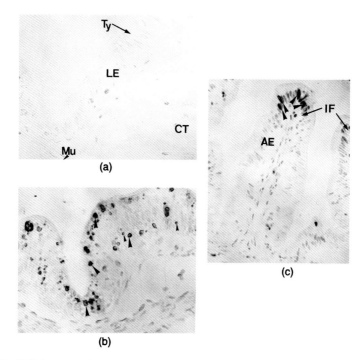

Figure 4.5 DNA fragmentation assay (TUNEL) detects apoptosis during intestinal remodeling. Intestinal cross-sections from Xenopus laevis tadpoles at different stages were analyzed by the TUNEL assay (Gavrieli et al., 1992), by which cells undergoing apoptosis are labeled. (A) No cell death was observed in premetamorphic tadpole intestine at stage 59 (X630). (B) Most of the larval epithelial cells were labeled at the metamorphic climax (stage 60). Note the apoptotic morphology of the dying cells. Small and large arrowheads refer to labeled nuclear fragments of apoptotic bodies and intact nuclei of the larval cells, respectively (X630). (C) Cell death (arrowheads) in postmetamorphic frog intestine (stage 66) was limited to the crests of intestinal folds (IF), where the fully differentiated epithelial cells degenerate after functioning for a finite period of time (X420). Ty; typholosole; LE; larval epithelium; AE; adult epithelium; Mu; muscle; CT, connective tissue. Reprinted from Shi and Ishizuya-Oka (1997) with permission from Plenum Press.

COLOR PLATES

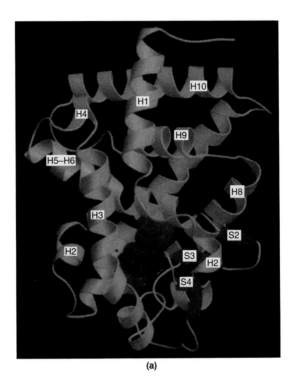

(a)

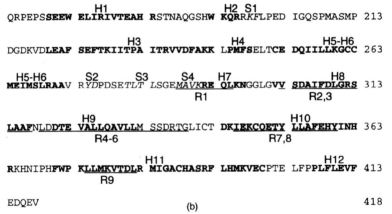

(b)

Figure 5.5 (a) Stereo drawing of the TR ligand binding domain, with the secondary structure elements labeled. The hormone (magenta) is depicted as a space-filling model; α-helices and coil conformations are yellow, β-strands are blue. Reprinted with permission from Nature (378:690–697) Copyright (1995) Macmillan Magazines Limited. (b) Structural organization of the ligand binding domain of Xenopus laevis TRαA. Based on the data of Wagner et al. (1995), there are 12 α-helices (bold letters, H1–H12) and 4 β-strands (italic letters, S1–S4) in the TH-binding domain. This region also contains nine heptad repeats (Forman and Samuels, 1990; underlined, R1–R9) that are important for receptor dimerization with itself or other members of the retinoid-thyroid receptor subfamily.

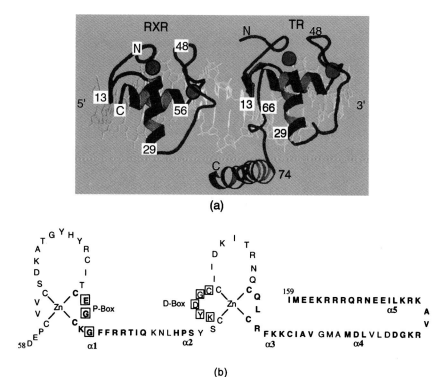

Figure 5.6 (a) The overall architecture of RXR (red) and TR (blue) DNA binding domain on a direct repeat sequence of AGGTCA with 4-bp separation (yellow). The 5' and 3' ends of the DNA are indicated. The amino acids 1–74 of TR correspond to amino acids 61–134 of Xenopus TRαA (see B). Reprinted with permission from Nature (375: 203–211) Copyright (1995) Macmillan Magazines Limited. (b) Structural organization of the DNA binding domain. The sequence from amino acids 58 to 159 of Xenopus TRαA corresponds to the region whose structure was determined by X-ray crystallography as shown in (a), and contains 5 α-helices (α1–α5, bold letters). The P-box (residues in shaded boxes) and D-box (residues in boxes) are important for DNA recognition and dimerization, respectively (Tsai and O'Malley, 1994).

COLOR PLATES

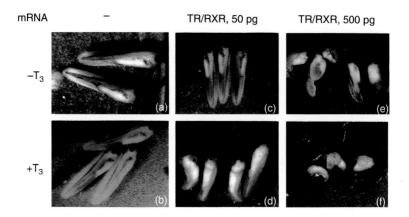

Figure 8.6 Overexpression of TRs and RXRs together results in dose- and T_3-dependent embryonic abnormalities. Embryos were injected with the indicated amounts of TR/RXR mRNAs and cultured for 2 days in the presence (+) or absence (−) of 100 nM T_3. Reprinted from Puzianowska-Kuznicka et al.(1997) with permission from American Society for Microbiology.

TH (Yamauchi and Tata, 1994). Partial peptide sequencing showed that it is likely to be the frog homolog of mammalian and avian cytosolic aldehyde dehydrogenase. Currently, it is unknown whether it is expressed during metamorphosis and how it is regulated. The second frog CTHBP gene was cloned based on its homology to the human M2 pyruvate kinase gene (Shi et al., 1994). The deduced amino acid sequence of the frog CTHBP is about 90% identical to the human protein, with all known important regions completely conserved. Thus, although the biochemical function of the protein has not been studied, it is likely to function both as a pyruvate kinase in the tetramer form and a CTHBP in the monomer form.

The potential role of the frog M2 pyruvate kinase gene (here referred to as xCTHBP) during metamorphosis is suggested by the tissue dependent developmental expression of its mRNA. Of the three *Xenopus laevis* organs analyzed, the intestine has very low but relatively constant levels of xCTHBP mRNA during development (Shi et al., 1994). In contrast, the xCTHBP expression in the tail and hind limb is drastically altered during metamorphosis. Low levels of xCTHBP mRNA are present in either the hind limb during morphogenesis (stages 54–56) or tail during resorption (stages 62–64) (Fig. 8.8). Interestingly, the xCTHBP mRNA levels increase dramatically in the growing hind limb. They are also high in premetamorphic tail but decrease precipitously in the resorbing tail (Fig. 8.8, compare the xCTHBP mRNA levels with the sizes of the organs). This suggests a potential role of the xCTHBP as chelator of intracellular TH to reduce free TH levels, thus participating in the regulation of the metamorphic transition of at least these two organs.

8.3.3 Deiodinases

Thyroid hormone is synthesized initially as T_4 or thyroxine in the thyroid gland (St. Germain, 1994). A fraction of the T_4 in turn is converted in the thyroid gland into T_3 by 5'-deiodinases, the more potent form of TH (Jorgensen, 1978; Weber, 1967). Both T_3 and T_4 are then secreted into the serum and carried to target tissues. The hormones are presumably transported into the cells either via carrier proteins or through energy-driven transport processes (Chapter 5, Robbins, 1992; Oppenheimer et al., 1987). Within the target cells, the levels of the hormones can be further modulated owing to local metabolism. This includes the conversion of T_4 to T_3 through 5'-deiodination to enhance the effect of TH and the inactivation of both forms through 5-deiodination, conjugation, deamination and oxidative decarboxylation (St. Germain, 1994). Among them, the best studied is the deiodination.

Two families of deiodinases have been discovered (St. Germain, 1994). These are the 5'- and 5-deiodinases. At least two different 5'-deiodinases have been identified that have different enzymatic characteristics. These enzymes appear to have different tissue distributions in mammals with the type I form high in the liver, kidney, and thyroid, and the type II form high in the brain, pituitary,

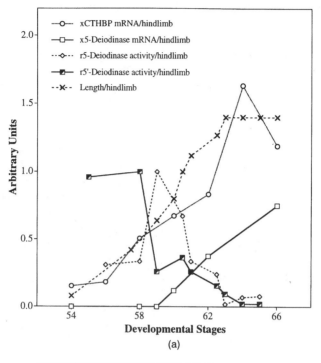

(a)

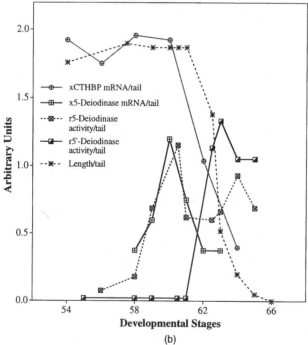

(b)

and brown adipose tissue of adult rat (St. Germain, 1994). Thus it seems likely that in developing animals, these activities are regulated in a tissue-specific manner.

A systematic survey of the 5'-deiodinase activity in various tissues during *Rana catesbeiana* development has indeed shown that it is regulated in a tissue- and developmental stage-dependent manner (Becker et al., 1997). The 5'-deiodinase activity could not be detected in the liver, kidney, or red blood cells, and only very low levels are present in the brain and heart. In the other tissues studied, high levels of 5'-deiodinase activity are present in a given organ when metamorphosis takes place but no or low levels are present at other stages. For example, the 5'-deiodinase activity is high in the hind limb during early stages of metamorphosis when limb morphogenesis takes place (Fig. 8.8*A*) but low subsequently when the limb merely increases in size. Similarly, in the tail, little 5'-deiodinase is present in pre- and prometamorphic tadpoles but high levels are present at the climax when the tail resorbs rapidly (Fig. 8.8*B*). A gene encoding the type II 5'-deiodinase has been cloned in *Rana catesbeiana* and found to share homologies with the type I 5'-deiodinase in rat (Davey et al., 1995). Its mRNA levels are regulated similarly as the 5'-deiodinase activity in various tissues during metamorphosis (Becker et al., 1997). Thus the type II deiodinase gene may play a role in regulating tissue T_3 levels to coordinate the temporal regulation of organ-specific tissue transformation.

In contrast to 5'-deiodinases, which convert T_4 to the more active form T_3, 5-deiodinases convert T_4 and T_3 to rT_3 (reverse T_3) and T_2, respectively (St. Germain, 1994). Because both rT_3 and T_2 have little affinity for TRs, the action of 5-deiodinases inactivates TH, thus inhibiting the effect of TH. A 5-deiodinase gene was recently cloned from *Xenopus laevis* owing to its upregulation in the tail by the TH treatment of premetamorphic tadpoles (St. Germain et al., 1994). Its sequence shares considerable homology in several regions with the rat type I 5'-deiodinase. However, it functions primarily as a 5-deiodinase, with its 5'-deiodinase activity about 600 fold weaker (St. Germain et al., 1994).

Figure 8.8 The temporal expression profiles of the Xenopus CTHBP (xCTHBP) and Xenopus 5-deiodinase genes and activities of the 5'- and 5-deiodinases in Rana catesbeiana *during tail and hindlimb metamorphosis. Note that both* Xenopus *genes are expressed at low levels during limb morphogenesis (stages 54–56) (A) and tail resorption (stages 62–66) (B). The expression of the two* Xenopus *genes increases in the hind limb during the limb growth period when there are few morphological changes (stage 58–66). On the other hand, the activities of the 5'-deiodinase are high during* Rana *metamorphosis in both hind limb and tail. Although the activities of 5-deiodinase are also high in the tail and hind limb around metamorphic stages, its peak levels are shifted away from the peak of metamorphosis in the two organs. The mRNA levels of the xCTHBP and 5-deiodinase in* Xenopus *are based on published data (Wang and Brown, 1993; Shi et al., 1994; St. Germain et al., 1994) and are calculated as mRNA levels per unit quantity of total RNA from the tail or hind limb. The deiodinase enzymatic activities in* Rana *are based on Becker et al. (1997). The lengths of the tail and hind limb are plotted as the ratios of the lengths of the respective organs to the lengths of the tadpole body (Shi et al., 1994; Atkinson et al., 1981).*

Interestingly, its expression during *Xenopus* metamorphosis is regulated in a tissue-specific manner that is distinct from that of the type II 5'-deiodinase in *Rana catesbeiana* (Fig. 8.8; Wang and Brown, 1994; St. Germain et al., 1994; Becker et al., 1997). It has little expression in the intestine. In the hind limb, it is not expressed until about stage 60 when its mRNA level is upregulated, paralleling the increase in limb size. In contrast, relatively high levels of its mRNA are present in the tail during premetamorphic stages. Immediately prior to rapid tail resorption (stages 58–61), the expression of the 5-deiodinase gene is suddenly upregulated several fold. Its mRNA levels then return to the premetamorphic levels as tail resorption takes place after stage 61.

The *Rana catesbeiana* homolog of the *Xenopus* 5-deiodinase has also been cloned (Becker et al., 1995). It shares 61% and 76% identities with the *Xenopus* gene at the cDNA and protein sequence levels, respectively. The gene is expressed in different tissues in *Rana catesbeiana* tadpoles, although no significant changes were found based on a PCR analysis (Becker et al., 1995, 1997). On the other hand, like the *Xenopus* 5-deiodinase gene, the *Rana* gene is also upregulated by TH treatment of premetamorphic tadpoles (Becker et al., 1995).

Unlike the 5-deiodinase mRNA, the 5-deiodinase activity is regulated in a tissue- and stage-dependent manner during *Rana* metamorphosis. Somewhat surprisingly, the profile of the 5-deiodinase is in general similar to that of 5'-deiodinase (Fig. 8.8; Becker et al., 1997), suggesting that either additional 5-deiodinase gene(s) exists in *Rana catesbeiana* and/or the gene is also under post-transcriptional regulation. Detailed comparisons of the 5- and 5'-deiodinase profiles in *Rana catesbeiana* reveal that the 5'-deiodinase activity correlates tightly with metamorphosis whereas the peak levels of the 5-deiodinase activity do not (Fig. 8.8; Becker et al., 1997). For example, in the hind limb, high levels of 5'-deiodinase activity are present at earlier stages when morphogenesis takes place. In the tail, high levels of 5-deiodinase activity are achieved at earlier stages before tail metamorphosis, whereas the 5'-deiodinase activity reaches peak levels when massive tail resorption takes place. Thus a proper balance of the deiodinase activities appears to play a role in coordinating tissue specific metamorphosis. Indeed, Huang et al., (1999) have shown recently that overexpression of 5-deiodinase in *Xenopus laevis* tadpoles by using transgenesis (see chapter 10 for more on this approach) leads to a delay in metamorphosis and resistance to TH treatment, consistent with its role to inactivate TH.

8.4 MOLECULAR BASIS FOR COMPETENCE AND TIMING OF TISSUE-SPECIFIC TRANSFORMATION

The development of amphibians proceeds through a complex genetic program that produces a larval form in the absence of a functional thyroid gland (Dodd and Dodd, 1976; Nieuwkoop and Faber, 1956). This phase of development is followed by metamorphosis, a process involving a complicated coordination of the transformations of individual tissues. The development of the functional thyroid gland seems to play a critical role in initiating this metamorphic process. This is because tadpoles prior to the formation of a functional thyroid gland (at stage 53 in *Xenopus*) can undergo precocious metamorphosis when exposed to exogenous TH (Chapter 2; Dodd and Dodd, 1976; Gilbert and Frieden, 1981; Tata, 1968), suggesting the only missing signal is the hormone itself. However, several levels of regulation are involved to ensure proper metamorphic transitions. Thus embryos and tadpoles younger than a critical stage (stage 41 for *Xenopus*) (Tata, 1968) do not respond to exogenous TH. Furthermore, different tissues undergo transformations at very different developmental stages. Therefore, like the metamorphic process itself, the mechanisms underlying these regulations are expected to be very complex.

At least two factors are critical in the signal transduction by TH during metamorphosis. These are the levels of the receptors and the concentrations of intracellular free TH. As summarized above, the expression of the receptor genes and those genes that can influence cellular free TH concentrations are regulated in a tissue-specific and developmental stage-dependent manner.

Assuming that in general the mRNA levels reflects the levels of the corresponding proteins, the observed regulation of these genes provides a molecular model for the competence of the tadpoles to respond to TH and tissue-specific developmental regulation of metamorphosis. Thus in embryos and tadpoles before stage 40 (in *Xenopus laevis*) neither the TRα nor TRβ genes, the only known TR genes in all animal species, are expressed. This results in a lack of functional TR/RXR heterodimers and is most likely responsible for the inability of embryos or young tadpoles to respond to exogenous TH. After stage 40, TRα and RXRα genes are expressed and this expression results in the formation of TR/RXR heterodimers, thus making the tadpoles competent now to respond to exogenous TH.

Interestingly, the endogenous TH synthesis is not detectable until around stage 54 (Fig. 8.1). It is unclear why the animal needs to have functional TR/RXR heterodimers before the formation of a functional thyroid gland. However, the onset of the expression of TR and RXR genes (Wong and Shi, 1995; Yaoita and Brown, 1990) coincides with the repression of the embryonic expression of at least two genes, the sonic hedgehog gene (Stolow et al., 1995) and stromelysin-3 gene (Patterton et al., 1995). The TH-response genes are also activated during metamorphosis and thus likely play important roles during both embryogenesis and metamorphosis. The expression profiles of TR and RXR genes during early development together with the ability of unliganded TR/RXR heterodimers to repress these early TH-responsive genes in embryos (see above; Puzianowska-Kuznicka et al., 1997), therefore, suggest that the activation of TR and RXR genes during late embryogenesis (around stages 40–45 in *Xenopus laevis*, just prior to the onset of tadpole feeding, Wong and Shi, 1995; Yaoita and Brown, 1990; Kawahara et al., 1991; Nieuwkoop and Faber, 1956), serves a role to repress the expression of the early TH-response genes such as sonic hedgehog and stromelysin-3 genes after they fulfill their embryonic roles. Such a repression of gene expression may be critical to ensure a proper period of tadpole development before changing into frogs, for continued expression of these early TH-responsive, metamorphosis-participating genes may trigger premature, deleterious tissue transformation.

The regulation of receptor gene expression in individual tissues during metamorphosis will also likely contribute to the timing of tissue-specific metamorphosis. In addition, the tissue-specific regulation of at least two types of genes, the CTHBP and deiodinase genes, could influence the metamorphic timing as well. That is, the expression of these genes could lead to tissue-specific regulation of the cellular free TH levels, even though all organs/tissues are exposed to the same levels of circulating TH synthesized in the thyroid gland. Thus the combination of the levels of TR/RXR heterodimers and available cellular TH could act as the causative factors determining the timing of the transformations of individual tissues.

For example, in the hind limb at stages 54–56, the xCTHBP and 5-deiodinase genes are repressed, but the 5′-deionase activity is high (based on studies in *Rana*, Fig. 8.8A). This should allow for the accumulation of relatively high levels of cellular free T_3, the more potent form of TH. The free T_3 could in turn interact with the very high levels of TR/RXR heterodimers to activate the limb morphogenic process. It should be pointed out that even though the plasma TH levels are low around stages 54–56, at about 1–2 nM T_4 and no detectable T_3, the dissociation constant of TR–TH complexes is even lower, about 0.1 nM or less (Weinberger et al., 1986; Sap et al., 1986; Puzianowska-Kuznicka et al., 1996). Thus it is possible to accumulate sufficient levels of cellular free T_3 to bind to TR/RXR heterodimers, thus activating the limb morphogenic program.

On the other hand, in the tail, the xCTHBP gene is expressed at very high levels up to approximately stage 62. Consequently, the cellular free TH levels could be limiting. Furthermore, when the plasma TH concentration rises between stages 56 and 62, the 5-deiodinase expression is also high (Fig. 8.8B). This could counteract any resulting increase in cellular TH by converting the cellular TH into inactive forms through 5-deiodination. Thus, despite the high levels of plasma TH during this period, the intracellular free T_3 is likely to be very low. This coupled with the relatively low levels of TR and RXR expression will effectively suppress tail resorption. After stage 60, the TR and RXR genes as well as 5′-deiodinase are upregulated and the xCTHBP and 5-deiodinase genes are downregulated, thus allowing the accumulation of high levels of TR/RXR and cellular T_3 to activate the tail resorption process.

Finally, in the intestine, the xCTHBP and deiodinase genes have very little expression throughout metamorphosis. The controlling factors appear to be the levels of the receptors, 5′-deiodinase, and plasma TH. Although the TRα and RXRα genes are expressed at relatively constant levels, the expression of TRβ and RXRγ genes and 5′-deiodinase (based on studies in *Rana*, Becker et al., 1997), and the plasma TH levels are upregulated between stages 58–66

(Leloup and Buscaglia, 1977; Wong and Shi, 1995). This period corresponds exactly to the period of intestinal remodeling. Furthermore, the upregulation of TRβ genes in the different intestine tissues correlates temporally with the tissue transformation processes (see above and Shi and Ishizuya-Oka, 1997a). Thus the tissue-specific temporal regulation of the receptor genes and intracellular free TH concentrations appear to be important molecular factors that help determine the competence of premetamorphic tadpoles to respond to exogenous TH and the developmental regulation of tissue-specific metamorphosis.

Although the mechanism of TH action proposed above for metamorphosis is based largely on the studies in two frog species, it is likely to be applicable for biological functions of other hormones of the nuclear receptor superfamily. However, compared to the regulation of amphibian metamorphosis by thyroid hormone, considerably less is known about the developmental and organ-specific regulation of these other hormones and their receptors. Interestingly, insects, like amphibians, also undergo a larva-to-adult transition in a hormone (ecdysone) controlled process (Gilbert and Frieden, 1981). The hormone ecdysone regulates insect metamorphosis through the ecdysome receptor, which is also a member of the nuclear hormone receptor superfamily (Gilbert and Frieden, 1981; Gilbert et al., 1996; Mangelsdorf et al., 1995; Chapter 11). Recent molecular and genetic investigations have provided strong evidence that a complex regulation of ecdysone titers and its receptor gene expression is required to effect the proper tissue transformations (Thummel, 1995; Gilbert et al., 1996; for more details see Chapter 11).

Thus, in conclusion, the regulation of TH synthesis, factors affecting free TH levels, and receptor expression provide the molecular basis for the observed temporal regulation of tissue-specific metamorphosis. Although the model is based largely on the studies in *Xenopus laevis* and *Rana catesbeiana*, it is likely to be true for other amphibians. However, the genes described here have not been cloned in other amphibian species. Thus it will be interesting to clone these genes in other amphibians and study their regulation and function.

Although the evidence available is consistent with such a model, it is clearly oversimplified. There are many other factors that could affect intracellular free TH levels. For example, the efficiency of cellular uptake of TH may vary in different cells. In addition, there may be active mechanisms to export cellular TH (Chapter 5). Furthermore, other enzymes in the TH metabolic pathway could also regulate cellular TH concentrations. Finally and most importantly, the proposed functions for genes discussed here remain to be tested in vivo.

9

Thyroid Hormone Regulation and Functional Implication of Cell–Cell and Cell–Extracellular Matrix Interactions During Tissue Remodeling

9.1 INTRODUCTION

The cells that constitute a tissue or organ do not exist or function in isolation. Instead they are in constant contact with each other or other cell types and/or with the extracellular matrix (ECM). These extracellular forms of communication are of extreme importance because not only do they influence the local cellular environment but they have an impact on the three-dimensional structure of the tissue and ultimately the entire organism. Without these types of extracellular interactions, the position and identity of each individual cell are affected. Communication with the extracellular environment helps maintain the status and function of each cell. A critical player in this communication is the extracellular matrix (Hay, 1991). The multitude of cell changes and movements that occur in amphibian metamorphosis undoubtedly depend on the influence of interactions among different cells and between cells and the ECM.

The extracellular matrix is composed of many proteins that form a complex networked structure that lies beneath epithelia and surrounds, to varying degrees, most other type of cells, especially the cells in the connective tissue (Hay, 1991; Timpl and Brown, 1996). The most common form of ECM is probably the connective tissue matrix, which consists of mostly collagen fibrils of 10 nm or wider. This matrix surrounds cells of the fibroblast family and other minor cell types in connective tissues like cartilage, skin, and tendon (Hay, 1991; Hukins, 1984). On the other hand, the ECM that is best studied with regard to its role on cell function is perhaps the basement membrane, which underlies and separates the epithelial and muscle cells from the collagen fibrils of the connective tissue. This special ECM, more properly called basal lamina, most commonly consists of laminin, entactin, type IV collagen, proteoglycans, and so on.

The ECM serves as a structural support media for the cells it surrounds but clearly this is not its only biological function. The ECM can interact directly with nearby cells through their cell surface receptors, especially the integrins (Damsky and Werb, 1992; Schmidt et al., 1993; Brown and Yamada, 1995). The ECM also serves as a reservoir for many signaling molecules such as growth factors, and ECM remodeling can alter the availability of these molecules (Vukicevic et al., 1992; Werb et al., 1996). In addition, because most cells are surrounded by ECM, cell–cell interactions clearly depend on the nature of the ECM. The interaction between the cell and ECM is thus a dynamic, multi-faceted one that influences cell shape, cell metabolism, and ultimately the fate of the cell. Several examples of cell behavior affected by the ECM are cell–cell and cell–matrix adhesion, cell movement, cell proliferation, and apoptosis (Hay, 1991; Ruoslahti and Reed, 1994). These behaviors have direct relevance not only to the maintenance of a cell's status but also to developmental processes and tissue remodeling. Notably, all of these cellular events occur during metamorphosis as different tissues undergo dramatic morphological changes. Therefore, it is reasonable to predict that the ECM will be involved in this particular developmental event.

9.2 EXTRACELLULAR MATRIX REMODELING AND CELL–CELL INTERACTION DURING INTESTINAL METAMORPHOSIS

The transformation of the larval intestine to the adult form in frogs represents an excellent model system to study how different cell types participate in adult organogenesis. As described in Chapter 3, the intestine consists of several well defined, spatially separated layers of tissues. The adult frog intestine is structurally similar to mammalian and avian adult intestine whereas the tadpole intestine is similar to that seen during embryonic development in higher vertebrates.

As in many other organs, the epithelium provides the basic physiological function of the intestine. However, several other major tissues exist in this organ. These include the connective tissue and muscles. The development of the epithelium and the other tissues is well coordinated. More importantly, interactions among different cell types appear to be an essential component of intestinal organogenesis (Dauca et al., 1990; Louvard et al., 1992; Mamajiwalla et al., 1992; Simon and Gordon, 1995; Simon-Assmann and Kedinger, 1993). Recombinant organ cultures of birds or mammals directly demonstrated that intestinal mesenchyme is capable of inducing intestine-specific morphogenesis and cytodifferentiation of heterologous embryonic endoderm (Haffen et al., 1983; Ishizuya-Oka and Mizuno, 1984, 1985). Furthermore, intestinal epithelial cells cultured in vitro proliferate and differentiate in the presence but not in the absence of co-cultured fibroblasts (Kedinger et al., 1987; Stallmach et al., 1989).

Similar to that in birds and mammals, the development of the adult intestinal epithelium during amphibian metamorphosis also requires the participation of the mesenchyme. There is little connective tissue in the tadpole intestine. However, prior to the epithelial transformation, the connective tissue increases in thickness (McAvoy and Dixon, 1977; Ishizuya-Oka and Shimozawa, 1987b). During adult epithelial development, extensive cell–cell contacts are also established between the epithelium and the connective tissue (Fig. 9.1; Ishizuya-Oka and Shimozawa, 1987b). More importantly, when intestinal metamorphosis is induced with T_3 in organ cultures, the development of the adult epithelium requires the presence of connective tissue (Fig. 9.1, Ishizuya-Oka and Shimozawa, 1992b). If the epithelium is cultured alone or if the posterior small intestine, which has little connective tissue, is cultured in vitro, T_3 treatment leads to only the apoptotic degeneration of the larval epithelial cells. On the other hand, co-culture of the larval epithelium with the connective tissue in the presence of T_3 results in first the apoptosis of the larval epithelial cells and subsequently the proliferation and differentiation of the adult epithelial cells (Ishizuya-Oka and Shimozawa, 1992b).

As described above, the epithelium is separated from the connective tissue by a special ECM, the basal lamina. Several lines of evidence implicate that the basal lamina plays a role during intestinal epithelial development in higher vertebrates. First of all, the composition and spatial distribution of ECM molecules change during intestinal development (Louvard et al., 1992; Simon-Assmann and Kedinger, 1993; Simon-Assmann et al., 1994). For example, the laminin genes are activated during epithelial morphogenesis, and the genes encoding the two subunits of laminin are differentially regulated (Simo et al., 1991). In addition, the composition of the glycosaminoglycans in ECM changes during intestinal morphogenesis and/or cell differentiation (Bouziges et al., 1991). Spatially, different ECM molecules, such as tenascin and fibronectin, are distributed differently along the crypt–villus axis and their expression profile changes during intestinal development (Simon-Assmann and Kedinger, 1993). Because epithelial cells are intimately associated with the basal lamina and their differentiation accompanies their migration from the crypt to the villus, the temporal and spatial changes in ECM composition will very likely affect epithelial development.

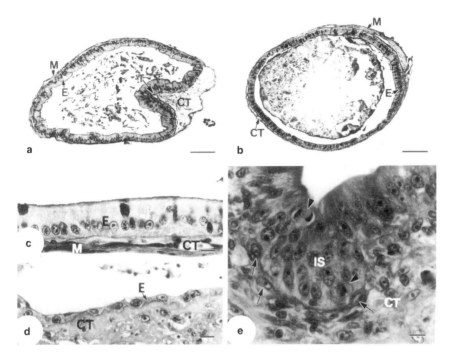

Figure 9.1 Region-specific responses to TH of the small intestine in organ cultures in vitro. (a) Cross-section of anterior small intestine of Xenopus laevis tadpoles at stage 57/58 before cultivation. The connective tissue (CT) is localized in a typhlosole (T). E, Epithelium; M, muscle. Scale bar: 100 μm. (b) Cross-section of posterior small intestine before cultivation. The connective tissue is very thin. Scale bar: 100 μm. (c) Cross-section of control posterior small intestine fragment cultured in CTS medium for 5 days in the absence of TH. A simple columnar structure of the larval epithelium remains. Connective tissue is undeveloped. Scale bar: 10 μm. (d) Cross-section of fragment of posterior small intestine cultured for 5 days in CTS medium in the presence of TH. No islets of adult epithelial cells are observed. The simple squamous epithelium consists of a small number of cells. Connective tissue cells just beneath the epithelium are few. Scale bar: 10 μm. (e) Cross-section of fragment of anterior small intestine cultured for 5 days with TH. An islet (IS) containing mitotic cells (arrowheads) develops among the larval epithelium. Connective tissue cells (arrows) surround the islet. Scale bar: 10 μm. Reprinted from Ishizuya-Oka and Shimozawa (1992b) with permission from Springer-Verlag.

More direct evidence for a role of ECM in intestinal development is obtained from experiments involving organ cultures (Altmann and Quaroni, 1990) and co-cultures of epithelial and fibroblast cells (Louvard et al., 1992; Simon-Assmann and Kedinger, 1993). The differentiation of epithelial cells is preceded by the formation of a basement membrane with ECM components synthesized by both cell types (Simon-Assmann et al., 1988; Kedinger et al., 1989). Furthermore, antibodies to laminin can inhibit epithelial differentiation when added to the co-culture (Simo et al., 1992). Finally, the inclusion of different basement membrane materials in intestinal organ cultures in vitro can directly affect epithelial cell differentiation (Altmann and Quaroni, 1990).

The frog intestinal basal lamina also undergoes extensive remodeling during metamorphosis (Ishizuya-Oka and Shimozawa, 1987b; Murata and Merker, 1991). In premetamorphic tadpoles, the intestinal basal lamina is a thin, flat, single-layered structure (Fig. 9.2a). Around stage 60 for *Xenopus laevis*, when adult epithelial cells begin to proliferate and larval epithelial cells begin to undergo apoptosis (McAvoy and Dixon, 1977; Ishizuya-Oka and Shimozawa, 1992b; Ishizuya-Oka and Ueda, 1996), the basal lamina lining both types of epithelia develops into a much thicker, multilayered structure through exten-

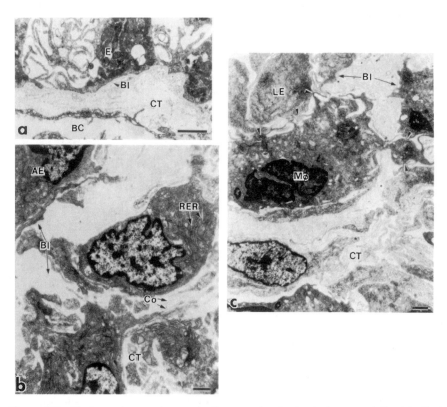

Figure 9.2 Electron microscopic examination reveals structural changes at the intestinal epithelial–connective tissue interface during metamorphosis. (a) A non-typhlosole region at stage 58. The basal laminal (Bl) is thin and lines the basal surface of the larval epithelium (E)(X12400). CT, connective tissue; BC, blood capillary. (b) At an early period (stage 61) of metamorphic climax. The basal lamina is thick due to vigorous folding. Fibroblasts just beneath the primordia of adult epithelium (AE) possess relatively developed rough endoplasmic reticulum (RER), and are surrounded by collagen fibrils (Co) (X7400). (c) Macrophage-like cells (Mϕ) migrate across the thick basal lamina into the degenerating larval epithelium at stage 62 (X6850). Bars: 1 μm. Reprinted from Shi and Ishizuya-Oka (1997b) with permission from Plenum Press.

sive folding (Fig. 9.2b,c). The basal lamina lining the larval epithelium remains thick until the larval epithelium finally disappears, that is, along with massive epithelial apoptosis (Ishizuya-Oka and Shimozawa, 1987b; Murata and Merker, 1991). Interestingly, the basal lamina appears to be much more permeable at the climax of metamorphosis (stages 60–63) in spite of the increased thickness. This permeability is reflected by the frequently observed migration of macrophages across the ECM into the degenerating epithelium, where they participate in the removal of degenerated epithelial cells (Fig. 9.1c). In addition, extensive contacts are present between the proliferating adult epithelial cells and the fibroblasts on the other side of the basal lamina (Ishizuya-Oka and Shimozawa, 1987b, 1992b).

Larval epithelial cell removal is essentially complete around stages 62/63. After stage 63, with the progress of intestinal morphogenesis, that is, intestinal fold formation, the adult epithelial cells differentiate. Concurrently, the basal lamina becomes thin and flat again, underlining the differentiating adult epithelium.

In addition to the above changes in the ECM, extensive ECM degradation and synthesis also occur. In particular, the ECM underlining the apoptotic epithelial cells is eventually removed and new basal lamina synthesis is required as the multilayered adult epithelial cells (cell islets) differentiate into a monolayer epithelium.

The intimate association of the basal lamina with surrounding cells, especially the epithelial cells, together with the close correlation of ECM remodeling with intestinal transformation, argues that ECM may play a critical role during metamorphosis. In particular, proper regulation of ECM remodeling is likely to be an important factor in epithelial cell fate determination, that is, death versus proliferation and differentiation.

9.3 REGULATION ON CELL–CELL INTERACTIONS THROUGH SECRETED MOLECULES IN THE INTESTINE

9.3.1 Sonic Hedgehog as a Signaling Molecule During Intestinal Remodeling

As described earlier, a subtractive differential screen has resulted in the isolation of many thyroid hormone regulated genes in *Xenopus laevis* intestine (Shi and Brown, 1993). One of the genes encodes a putative extracellular morphogen, the *Xenopus* homolog of the *Drosophila* hedgehog protein, that is, *Xenopus* sonic hedgehog (Stolow and Shi, 1995).

The *Drosophila* hedgehog was originally cloned as a segment polarity gene responsible for marking the boundaries of developing parasegments in the fly embryo (Lee et al., 1992). It has since been shown to be necessary for patterning of the adult eye, wing, and leg (Perrimon, 1995).

The morphogenic role of hedgehog is apparently conserved during evolution. Several different vertebrate homologs of the fly hedgehog gene have been cloned and shown to be involved in patterning and cell fate specification in a number of different developmental processes (Smith, 1994; Bumcrot and McMahon, 1995; Perrimon, 1995; Parisi and Lin, 1998). The most studied of the vertebrate genes is the sonic hedgehog. It functions as a signaling protein in a number of processes including establishment of polarity in the central nervous system, limb morphogenesis, and somite differentiation.

Four different hedgehog genes have been cloned in *Xenopus laevis* (Ekker et al., 1995; Stolow and Shi, 1995). These genes have distinct expression patterns during embryogenesis (Ekker et al., 1995), suggesting potentially different roles in development. The sonic hedgehog gene is the only one that has been analyzed in postembryonic tadpoles (Stolow and Shi, 1995). Although this gene is highly expressed from neurula to the feeding stages, its mRNA levels are considerably lower in older tadpoles. However, the gene is upregulated by T_3 in the intestine during metamorphosis. Its mRNA level peaks at the climax of metamorphosis (Stage 62, Fig. 9.3A). This corresponds to the period when larval epithelial cell death is essentially complete and adult epithelial morphogenesis (fold formation) is taking place (Fig. 3.1), suggesting that the sonic hedgehog may play a morphogenic role during adult epithelium development.

Interestingly, immunohistochemical analysis shows that the sonic hedgehog protein is synthesized in the epithelial cells (M. Stolow, A. Ishizuya-Oka, and Y.-B. Shi, unpublished data), similar to that observed in *Drosophila*, chicken, and mouse (Hoch and Pankratz, 1996; Bitgood and McMahon, 1995; Roberts et al., 1995). The protein levels appear to be the highest in the proliferating cell islets of adult epithelium but considerably lower in the larval or differentiated adult epithelial cells. These results suggest that sonic hedgehog may function in an autocrine fashion in the proliferation and/or the initiation of differentiation of adult epithelial cells. In this regard, it is worth noting that currently the origin of adult epithelial cells is unknown. The presence of high levels of hedgehog in or near the proliferating epithelial cells may also imply a role of sonic hedgehog in cell fate determination, that is, the generation of adult epithelial stem cells from existing cell types, especially from the larval epithelial cells that are mitotically active (McAvoy and Dixon, 1977).

Alternatively, the hedgehog protein can exert its effect on intestinal development by acting on the mesenchymal tissues. The best evidence for this comes from studies of gut development in other animal species. Hedgehog is known to influence the expression of several secreted signaling molecules including the *Drosophila* gene decapentaplegic and vertebrate bone morphogenetic protein (BMP) genes, all of which are members of the TGF-β superfamily (Perrimon, 1995). Interestingly, these genes are expressed in the mesenchyme of the developing gut of *Drosophila*, chick, and mouse (Bitgood and McMahon, 1995; Capovilla et al., 1994; Roberts et al., 1995). The coexpression of these genes with the hedgehog genes has led to the suggestion that the hedgehog protein influences mesenchymal tissue development.

The conservation in evolution of the hedgehog signaling pathway in *Drosophila* and mammals (Perrimon, 1995; Bumcrot and McMahon, 1995;

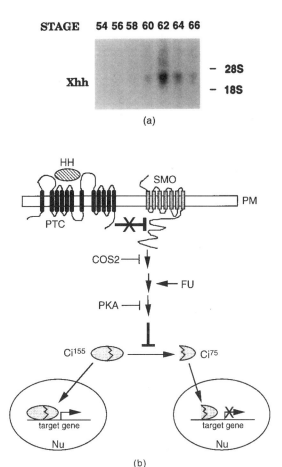

Figure 9.3 (A) Developmental expression of Xenopus sonic hedgehog (Xhh) in the intestine. RNA was isolated from tadpole intestine at different stages and analyzed by Northern blot hybridization (Stolow and Shi, 1995). (B) A model of hedgehog (HH) signaling pathway in Drosophila. HH binds to the transmembrane receptor Patched (PTC). This binding releases the repression of the activity of the transmembrane protein Smoothened (SMO). SMO in turn activates the transcription factor Ci by preventing its proteolysis and possibly enhancing the nuclear import of full-length Ci, which then activates target gene transcription. Two proteins, COS2 and protein kinase A (PKA), negatively influence Ci activation, whereas FU facilitates Ci activation. See text for details.

Parisi and Lin, 1998) argues for a similar potential role of hedgehog in amphibians. Consistently, at least the expression of one frog BMP gene has been shown to be induced by TH during intestinal metamorphosis (Amano, 1998), although its cell type specificity is unknown and the role of hedgehog, if any, in its expression remains to be determined. In addition, as mentioned above, adult epithelial formation in organ cultures requires the presence of connective tissue. Co-cultures of the larval intestinal connective tissue with larval epithelium in the presence of T_3 results in first the apoptosis of larval epithelial cells and then the development of adult epithelium (Ishizuya-Oka and Shimozawa, 1992b). Apparently, this requirement of the connective tissue is mediated by a soluble secreted factor(s) because adult epithelial development occurs even when the epithelium and the connective tissue are separated by a permeable membrane. It remains to be determined whether the frog hedgehog and BMPs are the factors mediating this epithelial-mesenchyme interaction. Nonetheless, these studies indicate that both cell–cell interaction and ECM remodeling are important for tissue transformation during metamorphosis.

9.3.2 Mechanism of Hedgehog Signaling

The *Drosophila* and vertebrate hedgehog proteins are secreted extracellularly. Interestingly, they are autocleaved into an *N*-terminal active fragment and a *C*-terminal nonfunctional fragment (Lee et al., 1994*a*; Stolow and Shi, 1995). The N-terminal fragment has all the biological activities attributed to hedgehog, whereas the *C*-terminal end is essential for the cleavage and post-translational modification (Porter et al., 1995; 1996a,b). The crystal structure of this *N*-terminal fragment reveals a surprising existence of a tetrahedrally coordinated zinc binding site that is structurally similar to that in zinc hydrolases such as thermolysin and carboxypeptidase A (Tanaka-Hall et al., 1995). It remains to be determined whether this putative protease domain can cleave any substrates and whether such cleavage has any roles in hedgehog signaling.

It is now assumed that hedgehog functions through the binding of the *N*-terminal active domain to a cell surface receptor on target cells (Fig. 9.3*B*; Ingham, 1998; Johnson and Scott, 1998). Genetic and biochemical studies have shown that products of the *Drosophila Patched* and *Smoothened* genes and their vertebrate homologs function as the receptor complex to mediate hedgehog signal (Alcedo et al., 1996; Marigo et al., 1996; Stone et al., 1996; Van den Heuvel and Ingham, 1996). Hedgehog binds directly to *Patched*, as integral plasma membrane protein with 12 transmembrane domains (Marigo et al., 1996; Stone et al., 1996). *Patched*, however, is not typical of receptors for secreted polypeptides and functions indirectly by blocking the function of *Smoothened*, a 7-transmembrane domain protein (Fig. 9.3*B*; Alcedo et al., 1996; Van den Heuvel and Ingham, 1996). *Smoothened* is similar to the family of G protein-coupled receptors and bears sequence homology to members of the *Frizzled* family of serpentine proteins, some of which act as receptors for Wnt proteins (Bhanot et al., 1996; Ingham, 1998). The binding of hedgehog to *Patched* releases the repression of *Smoothened* by *Patched* to allow the activation of the transcription factor *Cubitus interruptus* (Ci, a homolog of the mammalian Gli genes) (Parisi and Lin, 1998; Ingham, 1998; Johnson and Scott, 1998).

How *Smoothened* activates Ci is not entirely clear. Two proteins, protein kinase A and the product of the gene *Costal-2* (COS2), act negatively, whereas the products of the gene *Fused* (FU) acts positively in this activation process (Ruiz i Altaba, 1997; Ingham, 1998; Johnson and Scott, 1998). In the absence of hedgehog, Ci is cleaved into a smaller, 75 kDa protein that acts as a transcription suppressor (Aza-Blanc et al., 1997). In the presence of hedgehog, this cleavage is inhibited. This allows the accumulation of nuclear, 155 kDa full-length Ci to activate target genes, thus completing the hedgehog signaling cascade.

9.4 MATRIX METALLOPROTEINASES AS REGULATORS OF EXTRACELLULAR MATRIX

Matrix metalloproteinases (MMPs) are extracellular enzymes that are capable of degrading various components of the ECM (Alexander and Werb, 1991; Woessner, 1991; Matrisian, 1992; Birkedal-Hansen et al., 1993; Werb, 1997; Parks and Mecham, 1998; Barrett et al., 1998). This growing family of enzymes include collagenases, gelatinases, and stromelysins, each of which has different but often overlapping substrate specificities (Table 9.1; Sang and Douglas, 1996; Uria and Werb, 1998). They are secreted into the ECM as proenzymes with the exceptions of stromelysin-3 (ST3), which appears to be secreted in the active form (Pei and Weiss, 1995), and membrane-type MMPs, which exist as membrane-bound active enzymes. The proenzymes are enzymatically inactive owing to the formation of the fourth coordination bond with the catalytic Zn^{2+} ion by the conserved cysteine residue in the propeptide (Fig. 9.4; Van Wart and Birkedal-Hansen, 1990; Nagase et al., 1992; Kleiner and Stetler-Stevenson, 1993; Murphy et al., 1994; Birdedal-Hansen et al., 1993). The proenzymes can be activated in the ECM or on the cell surface through the proteolytic removal of the propeptide (Nagase et al., 1992; Murphy et al., 1994; Barrett et al., 1998; Nagase, 1998). The mature enzyme has a catalytic domain at the N-terminal half of the protein, which contains a conserved Zn^{2+} binding site. Once activated, these MMPs can degrade components of ECM (Table 9.1). Thus differential expression and activation of various MMPs can result in specific remodeling of the ECM.

Many MMPs have been implicated to play a role in the metastasis of cancerous cells owing to the upregulation of their expression in this process and the fact that metastasis requires extensive ECM degradation/modification (Matrisian, 1992, Stetler-Stevenson et al., 1993; Tryggvason et al., 1987; Sato and Seiki, 1996). In addition, the developmental expression profiles of MMP genes, for example, the activation of gelatinase A (GelA) and stromelysin-1 during mammary gland involution (Talhouk et al., 1992; Lund et al., 1996), suggest that MMPs are also critical players in a number of developmental processes (Matrisian and Hogan, 1990; Salamonsen, 1996; Powell and Matrisian, 1997; Sang, 1998). The importance of MMPs in development has been

TABLE 9.1 The Matrix Metalloproteins Family[a]

Enzyme	Alternative Names	Extracellular Matrix Substrates
Collagenases		
Interstitial collagenase (MMP-1)	Fibroblast collagenase, tissue collagenase, type I collagenase, vertebrate collagenase	Collagen I, II, III, VII, X, gelatin, entactin, tenascin, aggrecan, proMMP 2, 9
Neutrophil collagenase (MMP-8)	—	Collagen I, II, III, gelatin, aggrecan
Collagenase-3 (MMP-13)	—	Collagen, I, II, III, gelatin, aggrecan
Collagenase-4 (MMP-18)	—	Collagen I, gelatin
Gelatinases		
Gelatines A (MMP-2)	72-kDa type IV collagenase	Collagen I, IV, VII, X; gelatin, fibronectin, laminin, aggrecan, elastin, proMMP 9
Gelatine B (MMP-9)	92-kDa type IV collagenase	Collagen IV, V, XI, gelatin, elastin entactin, aggrecan
Stromelysins		
Stromelysin 1 (MMP-3)	Collagenase activating protein, procollagenase activator, proteoglycanase, transin-1, pTR1 protein (rat)	Collagen II, IV, V, IX, X, XI, gelatin laminin fibronectin, elastin, tenascin, aggrecan, vitronectin, proteoglycans, proMMPs 1, 8, 9, 13
Stromelysin 2 (MMP-10)	Transin-2	Collagen III, IV, V, IX, laminin, fibronectin aggrecan, elastin, proteoglycans
Stromelysin 3 (MMP-11)	—	Collagen IV, fibronectin, laminin, aggrecan
Membrane-type MMPs		
MT1-MMP (MMP14)	—	ProMMP 2, 13, collagen I, II, III, fibronectin, laminin, vitronectin, gelatin, proteoglycans
MT2-MMP (MMP-15)	—	Unknown
MT3-MMP (MMP-16)	—	ProMMP2
MT4-MMP (MMP-17)	—	Unknown
Other MMPs		
Matrilysin (MMP-7)	Matrin, putative metallopreteinase-1 (PUMP-1), uterine metalloproteinase	Collagen IV, gelatin, laminin, fibronectin, entactin, elastin, aggrecan, tenascin, proteoglycans, proMMP1, 2, 9,
Macrophage metalloelastase (MMP-12)	—	Elastin, fibronectin, laminin
Enamelysin (MMP19, 20)	—	Amelogenin

[a] Based on Sang and Douglas (1996) and Uria and Werb (1998).

(a)

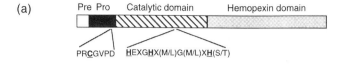

(b)

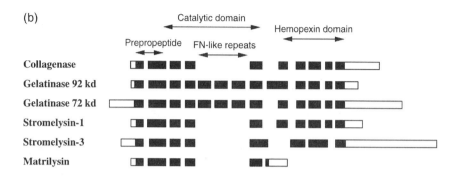

Figure 9.4 *Structure of MMP genes. (A) MMPs in general contain four domains. These are the pre- and propeptides, and catalytic and hemopexin domains from N- to C-terminus, respectively. The exceptions are that matrilysin does not have a hemopexin domain and membrane-type MMPs contain a transmembrane domain at the carboxyl end. A conserved peptide sequence is present in the propeptide where a C residue (underlined) is involved in coordination with the catalytic Zn atom in the inactive proenzyme. In the catalytic domain, a conserved region contains three H residues (underlined) that coordinate with the catalytic Zn atom. (B) Gene organizations of MMPs. MMPs have similar intron/exon organizations with the exception of stromelysin-3. Stromelysin-3 (Anglard et al., 1995; Li et al., 1998b) has only eight exons instead of 10 exons as in the interstitial collagenase (Fini et al., 1987; Collier et al., 1988) and stromelysin-1 (Breathnach et al., 1987) genes, or 13 exons as in gelatinases (Huhtala et al., 1990, 1991; Collier et al., 1991), owing to the presence of the fibronectin-like repeats, or only six exons as in matrilysin (Gaire et al., 1994), owing to the lack of the carboxyl hemopexin domain. The individual exons are shown as bars. The dotted bars indicate the coding region and the open bars show the 5′- and 3′-UTRs. The individual domains of the coding region are indicated on the top.*

substantiated by gene knockout studies in mice (Shapiro, 1998). For example, homozygous deletion of gelatinase B gene leads to delayed apoptosis of hypertrophic chondrocytes and abnormal skeletal development (Vu et al., 1998).

The participation of MMPs in amphibian metamorphosis was first implicated more than 30 years ago by the drastic increase in collagen degradation activity in the resorbing tadpole tail (Gross, 1966). The first frog MMP genes cloned were two *Xenopus* MMP genes isolated as thyroid hormone-inducible genes from subtractive differential screens (Wang and Brown, 1993; Shi and Brown, 1993). The full-length proteins encoded by these two genes share strong homology ($\geqslant 64\%$ identity) to mammalian collagenase-3 and stromelysin-3, respectively (Brown et al., 1996; Patterton et al., 1995) and are of similar sizes as their mammalian homologs, suggesting a functional conservation as well. Another frog MMP, the *Rana catesbeiana* collagenase-1, was cloned by screening an expression cDNA library with an antiserum against purified *Rana* tail collagenase and found to be regulated by TH during metamorphosis (Oofusa et al., 1994; Oofusa and Yoshizato, 1996). Although the *Rana* collagenase-1 is much smaller than the mammalian collagenase-1 (384 vs. 469 amino acids), it shares more than 81% identity with the human collagenase-1, suggesting that it is likely to be the homolog of the human collagenase. Expression profiles of these and other MMPs have provided evidence for their participation in various processes of metamorphosis.

9.4.1 Correlation of *Xenopus* Stromelysin-3 Expression with Cell Death During Metamorphosis

Among the known amphibian MMPs, the *Xenopus* stromelysin-3 (ST3) gene is of particular interest. This is in part because its human homolog is expressed in most, if not all, human carcinomas (Basset et al., 1990; Muller et al., 1993). Furthermore, both the human and mouse ST_3 are expressed during development in tissues where cell death takes place (Basset et al., 1990; Lefbvre et al., 1992). These results suggest that ST3 is involved in both apoptosis and cell migration, the processes that also occur during frog intestinal remodeling.

As mentioned above, the *Xenopus* ST3 gene was isolated as a thyroid hormone response gene. It can be precociously activated in all organs/regions of premetamorphic tadpoles if exogenous TH is added to the rearing water (Fig. 9.5A, Wang and Brown, 1993; Shi and Brown, 1993; Patterton et al., 1995). Furthermore, this regulation is direct and at the transcriptional level, for it occurs even if new protein synthesis is blocked by protein synthesis inhibitors.

The developmental expression of ST3 mRNA correlates strongly with organ-specific metamorphosis (Fig. 9.5B; Patterton et al., 1995). In the tail, the ST3 expression is low until around stage 62, when it is drastically upregulated, coinciding with the onset of massive tail resorption through programmed cell death (Nieuwkoop and Fabor, 1956; Kerr et al., 1974). In contrast, the ST3 mRNA levels are much lower in the hind limb throughout limb development (Fig. 9.5B). However, even in this case, higher levels of the mRNA are present during limb morphogenesis when interdigital cells degenerate at stages 54–56, considerably earlier than tail resorption. Finally, high levels of ST3 mRNA are present during larval intestinal epithelial degeneration and adult cell proliferation (stages 60-62; Fig. 9.5B).

Thus ST3 expression is temporally correlated with the stages when cell death occurs in all these organs, and the levels of its mRNA appear to correlate with the extents of cell death in these organs (Fig. 9.5B; Patterton et al., 1995). More importantly, the activation of the ST3 gene occurs prior to cell death. Thus, in the intestine, high levels of ST3 mRNA are already present at stage 60 when larval apoptosis is first detected by TUNEL (Chapter 4; Ishizuya-Oka

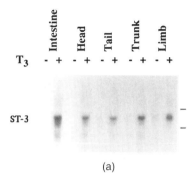

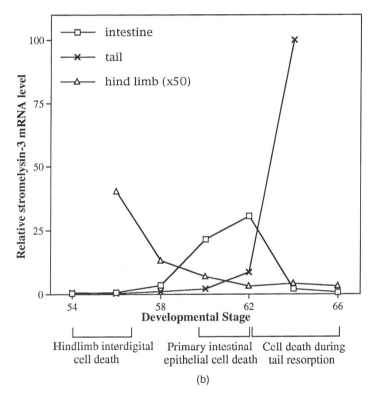

Figure 9.5 Stromelysin-3 (ST3) is regulated by thyroid hormone (T_3) and its expression correlates with cell death during metamorphosis. (A) ST3 can be activated by exogenous T_3 in different organs/regions of premetamorphic Xenopus laevis tadpoles (stage 52–54). Tadpoles were treated with 5nM T_3 and RNA from different organs was analyzed by Northern blot hybridization with a Xenopus ST3 cDNA probe. Each lane had 10 µg total RNA except the hind limb lanes (Limb), which had only 1.5 µg. The bars indicate the positions of 28S and 18S rRNA. (B) ST3 is expressed only during the period of cell death in the hind limb, tail, and intestine. The relative mRNA levels were quantified by Northern blot hybridization (based on the data of Patterton et al., 1995). Note that ST3 mRNA levels in the hind limb are about 50 times lower than those in the intestine and tail, probably corresponding to the much smaller fraction of the cells undergoing apoptosis in the limb. Reprinted from Shi and Ishizuya-Oka (1997b) with permission from Plenum Press.

and Ueda, 1996). Similarly, ST3 mRNA reaches high levels by stage 62 in the tail when massive tail resorption (length reduction) has yet to occur. These results suggest that ST3 is involved in regulating larval cell death in both intestine and tail. However, because adult epithelial proliferation in the intestine is rapid around stages 60–62 and differentiation also starts around stage 62, it is possible that ST3 may also be involved for adult epithelial development.

In situ hybridization analysis has revealed that ST3 expression is spatially correlated with apoptosis in different organs during *Xenopus* metamorphosis. In the tail, ST3 expression is localized to the earliest resorbed structures, that is, the dorsal and ventral fins (Berry et al., 1998a). Similarly, within the tadpole head, ST3 is highly expressed in the cells of the resorbing epithelial pharyngobranchial tract (Berry et al., 1998b). In the intestine, ST3 mRNA is localized in the fibroblastic cells adjacent to the epithelium, but not actually in the apoptotic cells of the intestine (Patterton et al., 1995). Thus, as a putative MMP, ST3 influences epithelial apoptosis by modifying ECM, in particular, the basal lamina that separates the ST3-expressing fibroblasts and the dying epithelial cells. Consistent with this, ST3 expression is not only temporally but also spatially correlated with basal lamina (the ECM that separates the epithelium and connective tissue) modification (Fig. 9.6; Ishizuya-Oka et al., 1996). In both pre- and postmetamorphic intestine, fibroblasts just beneath the thin and flat basal lamina do not express ST3. However, during metamorphosis, the intestinal basal lamina adjacent to ST3-expressing fibroblasts is thick, multiply folded, but more permeable as described above. It is, therefore, possible that ST3 causes specific degradation/cleavage of certain ECM components that helps the folding of the ECM and results in the increased permeability. Such modifications may facilitate larval epithelial apoptosis and adult epithelial development. Although such a function for ST3 remains to be proven, it is interesting to note that overexpression of the MMP stromelysin-1 in the mammary gland of transgenic mice leads to similar changes in basal lamina, alters mammary gland morphogenesis, and induces apoptosis (Witty et al., 1995a,b; Sympson et al., 1994; Alexander et al., 1996). Furthermore, homologous deletion of ST3 results in mice with reduced incidence of carcinogen-induced tumors whereas overexpression of ST3 in cultured cells leads to increased tumor formation when the ST3-expressing cells are injected into mice (Noel et al., 1996; Masson et al., 1998), again supporting a role of ST3 in ECM remodeling to influence cell behavior.

9.4.2 Differential Regulation of Different Matrix Metalloproteinases During Metamorphosis

The multicomponent nature of the ECM suggests the participation of different MMPs in its remodeling and degradation during metamorphosis. One way to investigate the involvement of MMPs is to assay their expression by using gelatin zymography. This approach makes use of the fact that many MMPs,

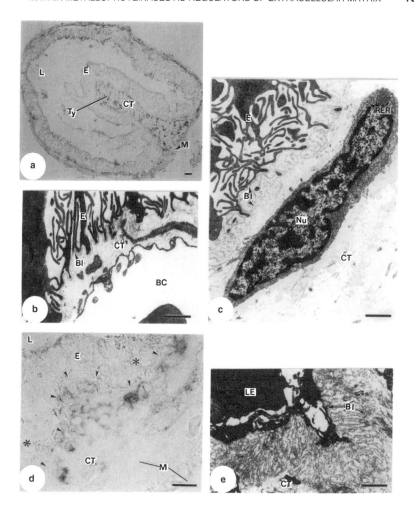

Figure 9.6 Remodeling of the basal lamina in the intestine is associated with ST3 expression during metamorphosis. (A), In situ hybridization using antisense ST3 probe on cross-sections of the anterior region of the small intestine. (A) At stage 59 the layer of connective tissue (CT) is thin except for the typhlosole (Ty). Hybridization signals are observed in some cells of the connective tissue near the muscular layer (M), but are weaker in the upper region of the typholosole. X150. (D) Small intestine at stage 61. The basal surface of the epithelium (E) is rugged because of the growth of adult epithelial primordia (asterisks) into the connective tissue. Most of the connective tissue cells just beneath the epithelium are positive, that is, ST3 expressing. X610. Bars: 20 μm. L, lumen. (B, C, E) Electron micrographs of epithelial-connective tissue interface of the small intestine. (B) Upper region of the typhlosole at stage 59 where ST3 expression is weak. The basal lamina (Bl) remains thin. X12400. (C) Bottom region of the typhlosole at stage 59 where ST3 expression is strong. The basal lamina is beginning to fold. A fibroblast-like cell near the epithelium is slender and possesses a large nucleus (Nu) and slightly developed rough endoplasmic reticulum (RER). X12400. (E) Basal lamina lining the basal surface of the larval epithelium (LE) at stage 61 when ST3 is highly expressed. The basal lamina is vigorously folding into accordionlike pleats. X12400. Bars: 1 μm. BC, blood capillary. From Stolow et al. (1997) with permission from Am Zool.

especially gelatinases, can be renatured, after separating on a denaturing polyacrylamide gel, to become active toward a gelatin substrate preembedded in the gel (the gelatin zymography method, McKerrow et al., 1985). Thus protein extracts from different organs at various metamorphic stages can be assayed by gelatin zymography to determine the relative levels of MMPs capable of degrading gelatin.

Using this method, three species (74 kDa, 61 kDa and 56 kDa) have been detected in the intestine, tail, and limb (Fig. 9.7A). In the intestine, like ST3, the levels of the 74 kDa and 56 kDa peptides are more abundant at stage 62 than at stage 64 and hence temporally correlated with the process of intestinal remodeling.

In the tail, both the 56 and 61 kDa enzymes increased markedly at stage 64, whereas the 74 kDa enzyme was replaced by an extremely high level of 67 kDa gelatin-degrading enzyme. Again like ST3, the levels of these MMPs were highly correlated with the process of tail resorption. In contrast, but similar to the lower levels of ST3 mRNA in the developing hind limb during metamorphosis, much lower levels of all three putative MMPs were detected enzymatically at stage 56 and further reduced after stage 58. Though it is difficult to identify the MMPs producing these patterns of gelatinase activity, it is very unlikely that any of the three bands detected represent ST3. This is because they are all larger than the expected size of the frog ST3. Furthermore, the mouse ST3 has been shown to have only a weak gelatinase activity (Murphy et al., 1993). It is, however, tempting to speculate that the 67 kDa species represents the activated or truncated form of the 74 kDa enzyme and both are encoded by frog gelatinase A (GelA) based on the molecular weight of the mammalian GelA and the putative *Xenopus* GelA mRNA regulation patterns in the tail and intestine (Fig. 9.7B, below).

A second method to determine MMP expression is to use homologous or heterologous cDNA probes to determine the mRNA levels for the corresponding MMPs. Overall, five genes have been analyzed in *Xenopus laevis* owing to the availability of appropriate cDNA probes. Three are homologous probes, encoding *Xenopus* ST3 (Patterton et al., 1995); collagenase-3 (Col3, Wang and Brown, 1993), and collagenase-4 (Col4, Stolow et al., 1996), respectively. The other two probes are the human cDNA clones for stromelysin-1 (ST1) and gelatinase A (GelA) (Patterton et al., 1995). The human GelA probe recognizes a *Xenopus* mRNA of similar size as the mammalian GelA mRNA whereas the ST1 probe hybridizes to two major RNAs of larger sizes than the mammalian ST1 mRNA (Patterton et al., 1995). However, both hybridization signals appear to be specific under the conditions used (Patterton et al., 1995), suggesting that the mRNAs detected likely encode the corresponding frog MMPs.

Unique but overlapping expression profiles have been observed for the five genes in the intestine and tail during metamorphosis (Fig. 9.7B, Patterton et al., 1995; Wang and Brown, 1993; Stolow et al., 1996). ST3 is the most upregulated gene in both the intestine and tail during metamorphosis whereas

ST1 has relatively constant mRNA levels throughout development in both organs. The two collagenase genes are upregulated during tail resorption (stages 62–64, Fig. 9.7B). However, their mRNA levels are only a fewfold higher in the metamorphosing intestine around stage 62 compared with those at other stages. The putative *Xenopus* GelA has an expression profile that is most similar to that of ST3 (Fig. 9.7B). Its mRNA is at very low levels prior to metamorphosis but drastically upregulated by stage 62 in both the intestine and tail. Subsequently, like ST3, the *Xenopus* GelA gene is repressed as intestinal remodeling is completed by the end of metamorphosis but remains high in the tail up to the end of tail resorption.

In addition to these *Xenopus* genes, the *Rana* collagenase-1 (Col1) has also been found to be induced during tail resorption (Oofusa et al., 1994). Furthermore, kinetic studies suggest that the *Rana* Col1 is an early TH response gene (Oofusa et al., 1994), in contrast to *Xenopus* Col3 and Col4 (Wang and Brown, 1993; Brown et al., 1996; Stolow et al., 1996). Promoter analysis has provided evidence that *Rana* Col1 is a direct TH response gene, containing a TRE (Oofusa and Yoshizato, 1996; Yoshizata, 1996). Such distinct properties compared to known *Xenopus* MMP genes suggest that it is very likely that a homolog of *Rana* Col1 is present in *Xenopus laevis*.

In situ hybridization analysis of *Xenopus* Col3 and ST3 has shown that they are expressed in distinct regions of the resorbing tail and remodeling head (Berry et al., 1998a,b). Both genes, however, are expressed in regions where tissue resorption takes place. For example, in the tail, ST3 is expressed in the dorsal and ventral fins, which are the first structures to resorb, whereas Col3 is highly expressed in the cells that line the notochord and within the notochord sheath, both of which resorb as well. Similarly, in the head, ST3 is highly expressed in cells of the resorbing epithelial pharyngobranchial tract and Col3 is expressed in branchial arches and anterior floor of the brain case. Again, all these structures resorb during metamorphosis.

In spite of the similar destiny of the ST3- and Col3-expressing structures, the difference in their temporal expression profiles suggest different functions for various MMPs. In particular, it is worth noting that high levels of GelA mRNA in the intestine are reached later (at stage 62) than those of ST3 (at stage 60, Fig. 9.7B). This difference is also observed when premetamorphic tadpoles are treated with T_3 to induce precocious intestinal remodeling (Patterton et al., 1995). Although the *Xenopus* ST3 gene is upregulated very quickly (within a few hours) by the T_3 treatment, the GelA gene upregulation is detectable only after a treatment of 3 days or longer (Fig. 9.8). Similarly, in the tail, ST3 gene also responds to exogenous T_3 treatment directly, within a few hours and much faster than the GelA and collagenase genes (Patterton et al., 1995; Wang and Brown, 1993; Stolow et al., 1996). As described above, stage 62 is the time when intestinal epithelial cell death is mostly complete and adult epithelial differentiation is taking place. The activation of GelA and to a lesser extent, Col4, at this stage in the intestine suggests that these MMPs are involved in the removal of the ECM associated with the degenerated larval

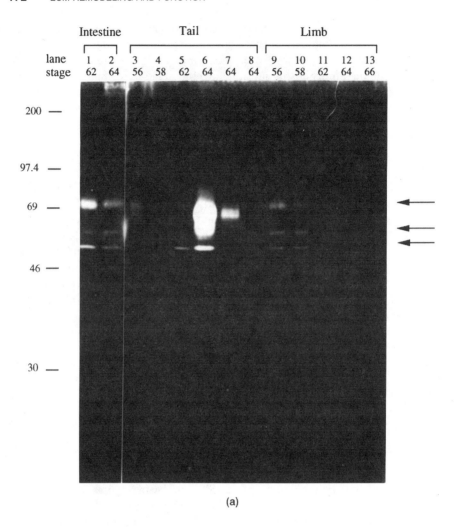

Figure 9.7 Multiple MMPs are differentially involved in metamorphosis. (A) *Gelatin zymography assay (McKerrow et al., 1985) for gelatin-degrading activity showing differential regulation of MMPs in different organs during metamorphosis. Each lane (except lanes 7 and 8) has 8 μg total protein extracted from the intestine, tail, or hind limb at the indicated developmental stage. Lanes 7 and 8 contained 1 and 0.25 μg stage 64 tail proteins, respectively. Protein size markers are indicated. Arrows point to the 74 kDa, 61 kDa, and 56 kDa gelatin degrading peptides. (B) Northern blot analyses demonstrate differential, organ-specific regulation of different MMP genes during* Xenopus *metamorphosis. Each lane had 10 g total RNA except the tail lane at stage 64, which had only 5 μg. The probes used were human stromelysin-1 (ST1, Wilhelm et al., 1987),* Xenopus *collagenase-3 (Col3, Wang and Brown, 1993),* Xenopus *collagenase-4 (Col4, Stolow et al., 1996), human gelatinase A (GelA, Collier et al., 1988), and* Xenopus *ST3 (Patterton et al., 1995). Reprinted from Shi and Ishizuya-Oka (1997b) with permission from Plenum Press.*

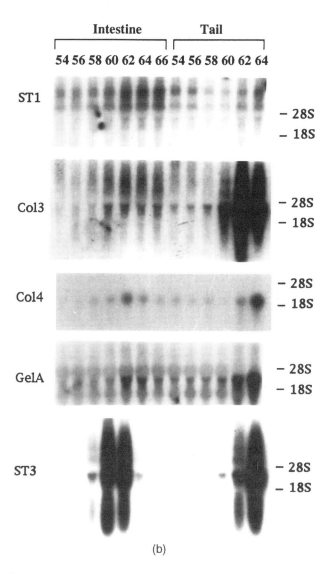

Figure 9.7 Continued.

epithelium and remodeling of the ECM for adult epithelium differentiation. It also reinforces the idea that the earlier activation of the ST3 gene in the intestine is important for the ECM remodeling that facilitates larval intestinal epithelial apoptosis.

How ST3 influences tissue remodeling is yet unknown. Interestingly, mammalian ST3 is secreted in its enzymatically active form when overexpressed in tissue culture cells (Pei and Weiss, 1995). This has allowed the identification of

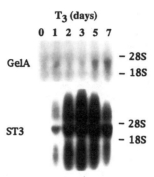

Figure 9.8 Differential kinetic responses to T_3 of Xenopus ST3 and GelA genes. Premetamorphic tadpoles were treated with T_3 for indicated periods of time and intestinal RNA was isolated for Northern blot analyses. Based on data of Patterton et al. (1995) with permission from Academic Press.

one potential physiological substrate, the α1-proteinase inhibitor, a non-ECM derived serine proteinase inhibitor (Pei et al., 1994). Although no natural ECM substrates for ST3 have been identified, such substrates may exist. On the other hand, the ability of ST3 to cleave a non-ECM substrate raises the possibility that ST3 may affect cell behavior through both ECM and non-ECM mediated pathways.

The other frog MMPs, that is, ST1, GelA, Col3, and Col4, are expected to digest specific ECM substrates just like their mammalian counterparts, thus leading to the remodeling and degradation of the ECM during metamorphosis. In particular, the upregulation of Col4 and GelA at stage 62 in the intestine when cell death is mostly completed implies that they may participate in ECM removal after cell death. It should be pointed out that direct demonstration of the ECM degradation activity for these frog MMPs is lacking with the exception of Col4, which has been shown to be capable of cleaving native type I collagen similarly as human collagenase-1 (Stolow et al., 1996). However, the sequence conservation of the frog MMPs with the mammalian homologs argues for a conservation at the function level.

Organ culture experiments have provided some direct evidence for the involvement of MMPs in tissue transformation. When tail fins are cultured in vitro in the presence of T_3, they undergo resorption just like in vivo. This resorption can be quantitated by measuring the remaining area of the tail fins. Using such an assay, Oofusa and Yoshizato (1991) demonstrated that inclusion of TIMP(s) purified from bovine dental pulp blocked the resorption of the dorsal tail fin of *Rana catesbeiana*. Likewise, intestinal fragments isolated from premetamorphic tadpoles of *Xenopus laevis* undergoes metamorphosis in vitro in the presence of physiological concentrations (5–10 nM) T_3 (Ishizuya-Oka and Shimozawa, 1991; Ishizuya-Oka et al., 1997c). Addition of a polyclonal antibody against the catalytic domain of ST3 inhibits (delays) the apoptotic degeneration of the larval epithelium and prevents the invagination of the proliferating adult epithelial cells into the connective tissue (Ishizuya-Oka and Shi, unpublished observation). These observations are consistent with the suggestions based on expression studies that ST3 is involved directly or indirectly in ECM remodeling that facilitates larval cell death and adult epithelial development.

9.5 MULTIPLE FACTORS CONTRIBUTE TO THE REGULATION OF CELL–CELL AND CELL–EXTRACELLULAR MATRIX INTERACTIONS IN DIFFERENT ORGANS

Analyses of the genes regulated by thyroid hormone also provide evidence for a role of cell–cell and cell–ECM interactions in the transformation of diverse organs. First of all, several matrix metalloproteinase genes, for example, stromelysin-3 and collagenase-3, are activated by T_3 in different organs, both resorbing ones like the tail and developing ones such as the hind limb (see above; also Patterton et al., 1995; Wang and Brown, 1993; Oofusa et al., 1994; Brown et al., 1996; Stolow et al., 1996). The activation of these MMP genes during tissue transformation are expected to modify/degrade ECM components, thus altering the nature of the ECM and its interactions with the surrounding cells. Secondly, at least in the resorbing tail, two early TH response genes have been found to encode the ECM component fibronectin and an ECM receptor subunit integrin α-1, respectively. Thus a second level of ECM remodeling takes place through the synthesis of new ECM components. Furthermore, ECM–cell interactions are regulated by T_3 not only by changing the nature of the ECM but also by altering ECM receptors (integrins), which directly link the extracellular matrices to intracellular processes (Brown and Yamada, 1995; Damsky and Werb, 1992; Roskelley et al., 1995). Finally, several early TH response genes encode extracellular or membrane-bound proteins (Table 6.2; Brown et al., 1996; Wang and Brown, 1993; Stolow and Shi , 1995). These proteins may participate directly in regulating cell–cell interactions. For example, the activation of sonic hedgehog by TH in the hind limb (Stolow and Shi, 1995) implicates a role in cell–cell interaction just like that during avian and mammalian limb development, although there, the hedgehog gene is activated by retinoic acid, the regulator of avian and mammalian limb morphogenesis.

The isolation of these TH response genes provides at least a partial molecular explanation for the well-documented transformations in different organs. For example, during limb development, sonic hedgehog may play a role in cell proliferation and morphogenesis by influencing cell–cell interactions. On the other hand, the activation of different MMPs is likely to be crucial during digit formation when interdigital cells undergo programmed cell death and interdigital ECM degradation and remodeling are actively taking place. Similarly, ECM degradation is an important and major step in the systematic resorption of the tail tissues. The coordination of the sequential resorption of different tail tissues, with the tail fin first and the muscle and other tissues later, will undoubtedly require complex interactions among different cell types and between cells and ECM. Thus it may not be surprising that even in this resorbing organ, selective ECM synthesis and ECM receptor expression are actually upregulated in spite of the concurrent increase in ECM-degrading MMP expression. Thus it is important to know the spatial temporal regulation of the expression and functions of these different classes of molecules.

9.6 EXTRACELLULAR MATRIX AS A REGULATOR OF CELL FATE

The remodeling of the ECM during metamorphosis together with the correlation of the expression of MMPs, especially ST3, with apoptosis suggests that ECM plays a role in TH-induced apoptosis. Direct support for a role of ECM on cell fate has come from a number of studies with mammalian systems. One of the best-studied systems is the involution of the mammary gland. As the epithelial cells undergo post-lactation apoptosis, a number of MMP genes are activated (Lund et al., 1996; Talhouk et al., 1992). More importantly, by culturing the epithelial cells on different ECM matrices, it has been shown that the ECM can directly influence cell differentiation and survival (Boudreau et al., 1995; Roskelley et al., 1995). Similarly, ECM has been shown to be essential for the survival of several other types of cells and blocking the function of ECM receptor integrins can induce cell death (Bates et al., 1994; Brooks et al., 1994; Montgomery et al., 1994; Ruoslahti and Reed, 1994; Frisch and Ruoslahti, 1997; Meredith and Schwartz, 1997; Shi et al., 1997).

As described in Chapter 4, tadpole intestinal epithelial cells can be cultured in vitro and induced to undergo apoptosis by TH just as in vivo (Su et al., 1997a,b). When the plastic culture dishes are coated with various ECM proteins, the cells become resistant to TH-induced cell death (Fig. 9.9; Su et al., 1997b). On the other hand, the ECM has no effect on TH stimulation of cell proliferation based on ^{3}H-thymidine incorporation or the downregulation of an epithelial specific gene, indicating that the apoptosis-inhibiting effect of the ECM is specific (Su et al., 1997b).

The mechanism by which ECM influences cellular function is still unknown. Clearly, one way to transduce the ECM signal into the cells is through cell surface ECM receptors, especially integrins (Damsky and Werb, 1992; Schmidt et al., 1993; Ruoslahti and Reed, 1994; Brown and Yamada, 1995; Stromblad et al., 1997). In the case of mammary gland development, it has been proposed that the interaction of ECM with its integrin receptors leads to the activation of focal adhesion tyrosine kinase (FAK), which in turn transduces the signal through the MAP kinase pathway to the nucleus (Roskelley et al., 1995; Schlaepfer and Hunter, 1998). This or similar mechanisms may be responsible for ECM-dependent gene transcription. It is presumably this ECM-regulated

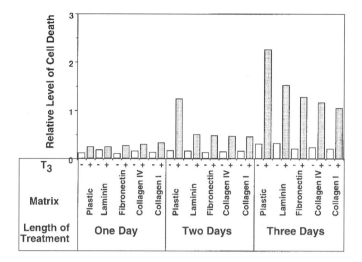

Figure 9.9 *Epithelial cells from stage 57/58 intestine cultured on matrix-coated dishes are more resistant to T_3-induced apoptosis. Intestinal epithelial cells were culture on plastic dishes with or without indicated coatings for 1–3 days in the presence or absence of 100 nM T_3. Cell death was quantified by measuring the extent of DNA fragmentation with an ELISA method. Based on Su et al. (1997b).*

gene expression that is ultimately responsible for mediating the biological effects of MMPs in influencing cell fate in processes such as intestinal remodeling and tail resorption during metamorphosis.

9.7 SUMMARY

Amphibian metamorphosis encompasses most if not all of the intra- and extracellular processes that are involved in vertebrate embryogenesis and organogenesis. The discovery that the putative morphogen hedgehog and MMP genes are TH response genes is consistent with the previously observed involvement of cell–cell and cell–ECM interactions during metamorphosis, especially in adult intestinal development. The modifications of the ECM, and the action of morphogens and other signaling molecules, are expected to alter cell–cell and cell–ECM interactions (Fig. 9.10), thus influencing cellular gene expression. The consequences of these changes are downstream cellular changes including apoptosis, cell proliferation, migration, and differentiation. The question is how those extracellular signals are transduced into the nucleus.

The hedgehog signaling pathway has been largely worked out through biochemical and genetic studies in *Drosophila*. The *N*-terminal, active domain of hedgehog binds to its cell surface receptor *Patched* to activate a signaling cascade, leading to the activation of the transcription factor *Cubitus interruptus*

178 ECM REMODELING AND FUNCTION

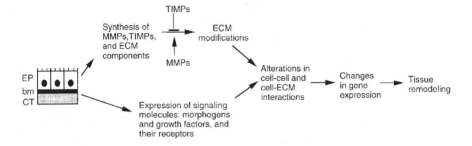

Figure 9.10 *Regulation of cell–cell and cell–ECM interaction during metamorphosis. The intestine is used as an example but the principles applies to other organ systems as well. The intestinal epithelial cells (EP) are separated from the connective tissue (CT) by a special ECM, the basement membrane (bm) or basal lamina. The EP and CT can contribute to ECM remodeling through the synthesis of MMPs, tissue inhibitors of MMPs (TIMPs), and ECM components. In addition, both cell types can express secreted signaling molecules and/or their receptors. ECM remodeling may also alter the availability of signaling molecules stored within ECM (Vukicevic et al., 1992; Werb et al., 1996). These changes thus will alter both cell–cell and cell–ECM interactions.*

(Ci) (Fig. 9.3B). It is very likely that a similar pathway exists in amphibians. The proof of such a pathway will require the cloning and characterization of the corresponding homologous genes in frogs.

The activation of MMP genes during metamorphosis is consistent with the idea that ECM remodeling is an important step in tissue remodeling. However, the examination and cloning have been limited to only a few MMPs. Other MMPs are likely to be involved. Furthermore, the activity of each MMP is regulated both through the control of the cleavage of the propeptide and the levels of tissue inhibitors of MMPs (TIMPs) (Alexander and Werb, 1991; Birkedal-Hansen et al., 1993; Nagase et al., 1992; Matrisian, 1992; Kleiner and Stetler-Stevenson, 1993). Essentially nothing is known about TIMPs during metamorphosis. Clearly, understanding of ECM remodeling would require the knowledge on the expression and function of different MMPs and TIMPs (Fig. 9.10).

One of most interesting players appears to be ST3. Its expression profiles clearly suggest a role in both larval epithelial apoptosis and proliferation and differentiation of adult intestinal epithelial cells. Such a role is also consistent with the observations in mammals, where ST3 is expressed in tissues undergoing apoptotic regression during development and in invasive human carcinomas (Basset et al., 1990; Lefebvre et al., 1992). However, in contrast to all other known MMPs, the ECM substrates have yet to be identified for mammalian or frog ST3, although mammalian ST3 can digest weakly some ECM proteins (Table 9.1). In addition, unlike most other MMPs, ST3 is secreted in its enzymatically active form when overexpressed in tissue culture cells (Pei and Weiss, 1995). This has allowed the identification of one potential physiological substrate, the α1-proteinase inhibitor, a non-ECM derived serine proteinase inhibitor (Pei et al., 1994). If the α1-proteinase inhibitor is indeed a physiological substrate of ST3 during metamorphosis, it would suggest that ST3 can regulate tissue remodeling through a novel mechanism unrelated to ECM remodeling.

10

Model Systems and Approaches to Study the Molecular Mechanisms of Amphibian Metamorphosis

10.1 INTRODUCTION

The advances in molecular biology have led to the cloning of thyroid hormone receptors in amphibians and a fairly detailed understanding of the biochemical and molecular properties of TRs. Furthermore, many genes that are regulated by TH directly or indirectly in different organs have been identified and their developmental expression profiles have provided strong support for their participation in tissue transformations. Thus the future challenges will be to investigate the molecular mechanisms governing their developmental- and tissue-specific regulation in various organs, and to determine the exact functions of these genes in different metamorphic processes such as larval cell apoptosis and adult tissue morphogenesis. Fortunately, several model systems and newly developed methodologies will facilitate such studies.

10.2 THYROID HORMONE REGULATION OF CELL FATE IN CELL CULTURES

The remodeling of various tadpole organs during metamorphosis involves an intricate control of cell proliferation and elimination (Chapter 4; Dodd and Dodd, 1976; Gilbert and Frieden, 1981). The development of adult organs requires first the proliferation and then differentiation of adult cells. This is especially true for adult-specific organs such as the limb. Even in such cases, specific cell death, for example, in the interdigital region of the limb, is likely to play important roles for proper morphogenesis. On the other hand, cell proliferation may also be an important factor even in organs undergoing complete resorption. This is in part because different cell types of the resorbing organs, such as the tail, are resorbed at distinct stages to coordinate effective resorption of the organs, while at the same time, the organs have to maintain certain physiological functions that are required before the completion of metamorphosis. It is commonly accepted that cell proliferation and differentiation are genetically controlled events critical for adult tissue development. However, extensive recent evidence, especially in the past decade or so, has shown that larval cell removal is an active, well-controlled cellular event. Such evidence has clearly shown that larval cell removal is through programmed cell death with apoptotic morphological changes (Chapter 4).

The observation of apoptotic cells in vivo suggests that metamorphic cell death is an active cellular response, directly or indirectly, to TH. As described in Chapter 4, this apoptotic response to TH is organ autonomous. Furthermore, a number of studies show that cell autonomous response to TH may be the primary event responsible for apoptosis during metamorphosis. First, tail epidermal cells isolated from *Rana catesbeiana* tadpoles are induced to die upon TH-treatment in vitro (Nishikawa and Yoshizato, 1986; Nishikawa et al., 1989). Similarly, Yaoita and Nakajima (1997) have established a stable cell line from *Xenopus laevis* tail muscle cells. These cells also can undergo apoptosis in response to TH. In addition, Ishizuya-Oka and Shimozawa (1992b) have shown that although the intestinal connective tissue is required for the development of adult intestinal epithelium in vitro, the tadpole intestinal epithelium undergoes apoptosis even when cultured alone in the presence of TH. More recently, the intestinal epithelial cells have been isolated from premetamorphic tadpoles of *Xenopus laevis* and cultured in vitro (Chapter 4; Su et al., 1997a, b). When treated with TH, these cells are stimulated to proliferate as reflected by the increased DNA synthesis. However, they are concurrently induced to undergo cell death with apoptotic morphology just like in vivo.

The availability of the cell line derived from tadpole tail and the ability to culture primary cells to respond to exogenous TH has allowed some initial investigations on the mechanisms of TH-induced cell death. The TH-dependent intestinal epithelial cell death can be inhibited by many known inhibitors, such as inhibitors of ICE-like proteases and nucleases, of mammalian apoptosis (Su et al., 1997b). Thus the presence of immunosuppressants cyclosporin A (CsA), a known inhibitor of activation-induced T cell death (Shi et al., 1989), during the TH treatment of these epithelial cells blocks the formation of the nucleosomal-sized DNA ladder (Fig. 10.1*A*; Su et al., 1997a). Furthermore, flow cytometry analysis has revealed that cells at different stages of cell cycle (i.e., with different DNA contents) can all undergo apoptosis in response to TH and CsA-inhibition of this TH-dependent apoptosis is independent cell cycle (Fig.

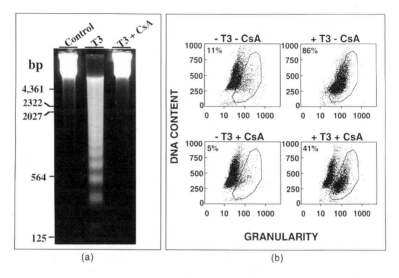

Figure 10.1 Tadpole intestinal epithelial cells undergo apoptosis when cultured in vitro in the presence of T_3. (A) T_3-treatment results in the formation of a nucleosomal-sized DNA ladder which can be inhibited by cyclosporin A (CsA), a known inhibitor of activation-induced T cell death (Shi et al., 1989). The epithelial cells were treated with 0 or 100 nM T_3 and/or 600 ng/ml CsA for 1 day. The genomic DNA was then isolated and analyzed on an agarose gel. (B) Flow cytometry analysis indicates that epithelial cells at different stages of cell cycle undergo apoptosis in response to T_3. The epithelial cells were cultured in the presence or absence of 100 nM T_3 and/or 600 ng/ml CsA for 3 days. The cell were then analyzed by flow cytometry. Although the boundary between live and apoptotic cells (encircled) was fixed with difficulty, the results clearly showed that cells with all different DNA contents or at different cell cycle stages (G2 at the top and G1 at the bottom) were present in the apoptotic region (as reflected by the increased cellular granularity). The percentage of the cells in the apoptotic region is indicated for each culturing condition. Note that CsA inhibits apoptosis independently of cell cycle. Reprinted from Shi et al. (1998) with permission from the Universidad del País Vasco.

10.1*B*). Finally, gene expression studies have shown that at least one TH-regulated gene, the intestinal fatty acid binding protein gene, can be downregulated by TH in this in vitro cell culture system, mimicking their downregulation in the larval epithelium in vivo during metamorphosis (Su et al., 1997b). Although CsA can block cell death in vitro, it fails to influence the downregulation of the intestinal fatty acid binding protein gene and the stimulation of epithelial cell proliferation by TH in this cell culture system. Thus the apoptosis of intestinal epithelial cells is a direct cellular response to TH and involves similar cell death effectors such as ICE-like proteases as in mammalian cell death (White, 1996). Furthermore, TH appears to induce multiple independent pathways that lead to apoptosis, DNA synthesis, and specific regulation of certain cell-type-specific genes that is apparently unrelated to either cell death or proliferation.

Like the intestinal epithelial cells, the stable cell line derived from tadpole tail muscle also undergo programmed cell death upon TH treatment (Yaoita and Nakajima, 1997). This TH-induced apoptosis is accompanied by the activation of the TRβ genes as well as at least one death effector gene, the CPP32 (caspase 3, see Chapter 4), an ICE-like protease whose homologs in mammals are known to be involved in apoptosis. Consistently, the cell death process can also be inhibited by inhibitors of ICE-like proteases. Thus different larval cells in tadpoles appear to employ apoptosis for their removal during metamorphosis.

With the increasing number of molecular probes and methodologies it will become easier to dissect the molecular basis underlying the regulation of different pathways induced by TH. For example, as described in Chapter 9, ECM inhibits TH-induced intestinal epithelial cell death in vitro. It is unclear how ECM influences TH-induced cell death. Are integrins or related cell surface ECM receptors involved in tranducing the ECM signal? Also, which TR subtype is involved in mediating the cell death versus proliferation signal? What are the roles of the TH-induced genes described in earlier chapters? In principle, one can introduce genes of interest to these cell cultures through transfection or alter gene function through other means and investigate the consequences of such manipulations on TH-induced larval apoptosis.

10.3 ORGAN CULTURE SYSTEMS

The first evidence showing that TH exerts it metamorphic effects directly on individual tissues came from organ culture experiments. Thus when tadpole tail and hind limb bud are cultured in vitro, they remain alive for days to weeks without significant changes (Chapter 2; Dodd and Dodd, 1976; Tata et al., 1991). Upon addition of physiological levels of TH to the culture medium, these organs undergo the same changes as those normally take place in vivo during natural metamorphosis. That is, the tail is gradually absorbed through apoptosis while the limb bud grows and differentiates. These studies thus not only demonstrate the direct effects of TH on target tissues but also indicate that individual tissues are genetically programmed to undergo pre-specified changes when TH is present.

Although TH is the causative factor for amphibian metamorphosis, other hormones can modulate this process. In particular, prolactin has been shown to inhibit the inductive affects of TH when added to tadpole rearing water both during TH-induced and natural metamorphosis. On the other hand, glucocorticoid can enhance the effects of TH. Again, by using organ culture systems, the effects of these hormones have been shown to be exerted directly on the target tissues. More importantly, by analyzing the regulation of known TH response genes, it has been shown that these hormones act by influencing TH-dependent gene expression. For example, the TH induction of the TRβ

genes in *Xenopus laevis* is directly at the transcriptional level by TRs themselves (Ranjan et al., 1994; Machuca et al., 1995). When prolactin is added to organ culture medium, this auto-induction of the TRβ genes is blocked (Tata et al., 1991), suggesting that prolactin directly interferes with TR function through yet unknown signal transduction pathways (see Chapter 2). Conceiveably, further studies with these organ culture systems will allow detailed dissections on how these various hormones interact with each other to effect tissue metamorphosis.

Another important application of organ culture systems is to study how cell–cell and cell–ECM interaction participate in tissue remodeling. As summarized in Chapter 9, tadpole intestinal fragments can be cultured in vitro and induced by TH to undergo the same changes as in intact tadpoles, that is, the apoptotic degeneration of larval epithelial cells and development of the adult epithelium. Although the larval epithelial cell death can be induced when the epithelium alone is cultured with TH, the connective tissue is required to be present together with the epithelium in order to allow adult epithelium development. Interestingly, adult epithelium can develop even when the larval epithelium is separated from the connective tissue by a permeable membrane (Ishizuya-Oka and Shimozawa, 1992b). Thus these experiments indicate not only that cell–cell interactions are important for adult intestinal epithelial development but also that a soluble factor(s) is involved in mediating these interactions.

Such recombinant organ culture systems offer an excellent opportunity to investigate factors that can influence cell–cell and cell–ECM interactions. For example, a number of ECM component or modifying (MMPs) genes and signal molecules like hedgehog are activated during metamorphosis (Chapter 6; Stolow and Shi, 1995; Stolow et al., 1996; Shi, 1996; Brown et al., 1996). These proteins function extracellularly and thus their function can be manipulated relatively easily in organ cultures. Thus functional or dominant negative proteins can be made and added to the organ culture to alter the levels and/or function of endogenous proteins. Similarly, function-blocking antibodies or inhibitors, for example, tissue inhibitors of MMPs or synthetic inhibitors, can be employed. Because these factors are likely to function in the interface between the epithelium and mesenchyme in the case of the intestine, reconstitution experiments using isolated epithelium and connective tissue in medium containing the overexpressed proteins or inhibiting factors can avoid any possible problems that may be associated with the ability of these factors to penetrate into the tissue. In fact, as reviewed in Chapter 9 (9.4.2), tail and intestinal organ cultures have been used to demonstrate the involvement of MMPs in tail resorption and intestinal remodeling by employing TIMPs and antibodies. These and similar experiments will determine whether extracellular factors and ECM participate in tissue remodeling.

The extracellular signals are most likely transduced through ECM receptors, such as integrins, and transmembrane receptors for growth factors. The roles of these receptors can be investigated by using various active or inactive forms of their ligands, and/or antibodies that recognize the extracellular domains of these receptor proteins, to the organ culture medium. The resulting changes in cellular response to TH will then reveal the function of these receptors in TH-dependent metamorphosis.

10.4 FROG EMBRYOS AS MODELS FOR STUDYING GENE REGULATION AND FUNCTION

The lack of proper genetic manipulation, for example, transgenesis or gene-knockout methodologies, has hampered the analysis of gene function in metamorphosing tadpoles. However, it is easy to introduce exogenous proteins into developing amphibian embryos. In particular, *Xenopus laevis* has been widely used for studying embryogenesis, especially mesoderm induction, through microinjection of various factors into fertilized eggs (Dawid, 1991; Phillips, 1991). This can be easily achieved by simply injecting mRNAs encoding the proteins of interest (Vize et al., 1991; Seidman and Soreq, 1997). Thus, given an appropriate molecular or phenotypic marker(s), it is possible to investigate the in vivo functions of genes that are normally expressed during metamorphosis.

As briefly described in Chapter 7, Puzianowska-Kuznicka et al. (1997) have employed this methodology to address the question whether TR and RXR are both required to mediate the effects of TH in vivo. They injected TR and RXR mRNAs into fertilized egg. These RNAs were efficiently translated in the embryos and the overexpressed proteins (TR and RXR) were present for at least 2 days, at which time normal embryos developed into tail bud stages. By phenotypic analyses, they showed that RXR is required for maximal function of TRs both in the absence or presence of TH. More importantly, the overexpressed receptors properly regulated the expression of endogenous genes that are normally regulated by TH during metamorphosis, that is, downregulation of these genes in the absence of TH and activation of these genes when TH is present. These experiments thus provide in vivo evidence for the role of TR and RXR during metamorphosis and in premetamorphic tadpoles (see Chapter 7 for more details on receptor function).

A problem associated with mRNA injection is that the mRNAs are short-lived and little overexpressed protein is left 2–3 days post-fertilization. Thus such an approach will not be useful for functional studies on proteins whose action targets are absent in early embryos, for example, an MMP. For these genes, an alternative is to microinject plasmid DNA bearing an appropriate promoter to direct their expression in embryos. The injected plasmid DNA can be maintained episomally for a long period and often is integrated into the genome, surviving at least till postmetamorphosis (Sargent and Mathers, 1991; Vize et al., 1991). A drawback of this approach is that the genes are often transmitted mosaically. In addition, their expression gradually decreases as the animals develop and only low levels of expression or no expression remains in postmetamorphic frogs. These shortcomings make such an approach not very useful to study genes that function intracellularly. However, for genes encoding extracellular proteins, such as MMPs and ECM proteins, mosaic expression is sufficient as long as high levels of the proteins can be synthesized by the transgenes. Thus this method, especially when a tissue-specific, developmentally regulated promoter is used, will be valuable to dissect the function of extracellular proteins.

Another use of embryo injection is for promoter studies in somatic tissues during development. A promoter of interest can be fused to a reporter and microinjected into fertilized eggs. Analysis of the levels of the reporter will allow the determination of DNA sequences necessary for proper temporal and spatial expression of the promoter. Furthermore, coinjection of an expression plasmid for a candidate factor should allow the analysis of the role of this factor in the regulation of this promoter in somatic cells during development, avoiding potential artifacts associated with promoter studies by using transfection into transformed cell lines or other ex vivo systems.

A recent improvement will make the DNA injection more useful for studies in embryos and maybe even tadpoles (Fu et al., 1998). This involves flanking the expression cassettes of interests with inverted terminal repeat sequences from an adeno-associated virus. Such an arrangement has been shown to increase the expression of both ubiquitous and tissue-specific transgenes injected in the plasmid DNA form into embryos. The plasmid DNA also segregates more efficiently throughout the embryos. The increased expression and reduced mosaicism will make the approach more suitable for studying transgene function.

The DNA injection experiments can also be carried out directly in tadpoles to study the gene/promoter of interest during metamorphosis (Sachs et al., 1998). This involves direct injection of plasmid DNA carrying a gene or promoter of interest into tadpole tissues. The DNA will be taken up by the cells and the effect of the gene or the activity of the promoter can be studied. This approach was first used to demonstrate that endogenous TRs can activate an exogenous TRE-containing promoter injected into tadpole tail in the presence of TH (De Luze et al., 1993). Subsequently, the same approach was employed to demonstrate that dominant negative TRs can block the ability of endogenous TR in the tail muscle cells to activate an exogenously introduced TRE-containing promoter (Ulisse et al., 1996). Furthermore, the inclusion of cationic lipids such as polyethyleninine in the DNA solution prior to injection dramatically enhances the uptake and activity of the injected DNA (Abdallah et al., 1995; Ouatas et al., 1998). Thus it is likely such as approach can be used to study gene function in tissue transformation during metamorphosis.

10.5 TRANSGENIC *XENOPUS LAEVIS* ANIMALS

The most physiological assay to study gene function is through gene knockout and transgenesis. So far, no methods have been developed for gene knockout in amphibians. However, two transgenic approaches have yielded varying degrees of success. These methods employ somatic nuclear transplantation and sperm-mediated transgenesis, respectively.

The first method is based on earlier observations that somatic nuclei of *Xenopus laevis* transplanted into the fertilized egg of *Xenopus laevis*, after removing the egg nucleus, can dedifferentiate and be used as the genetic material for tadpole development. This was first done more than 30 years ago by transplanting nuclei isolated from tadpole tissues into fertilized eggs (Gurdon, 1991). These somatic nuclei underwent dedifferentiation and allowed the development of feeding tadpoles. Furthermore, it was shown that many different cells could support the transplanted embryonic development. Interestingly, as tadpoles grow older, their tissue nuclei became less able to support animal development to tadpoles, and nuclei from younger tadpoles or embryos could support the development of animals to older stages, suggesting that dedifferentiation of somatic nuclei becomes more limited with more differentiated cells. In any case, these earlier studies pointed to a possible use of nuclear transplantation to introduce exogenous genes into developing embryos.

In 1994, Kroll and Gerhart reported the use of somatic nucleus transplantation to obtain transgenic *Xenopus laevis* embryos. This method involves first stably transfecting a gene of interest, under the control of an appropriate promoter, into a *Xenopus laevis* tissue culture cell line. Nuclei are then isolated from the stably transfected cell line after proper characterizations to ensure the gene is integrated into the genome with a desired number of copies and/or expression levels. These nuclei are then microinjected at a level of 1 nucleus/egg into fertilized *Xenopus* eggs. A significant fraction of the embryos developed from the injected eggs is from the transgenic somatic nuclei. These embryos often develop at least to gastrula stages, thus offering a chance to study gene function during early embryogenesis in a homogeneous environment. Unfortunately, they normally do not develop to tadpole stages, thus are of limited value for studying the function of genes participating in metamorphosis.

The inability of tissue culture cell nuclei to support development of tadpoles or frogs and the knowledge that more differentiated nuclei seem to be less able to support animal development prompted Kroll and Amaya (1996) to develop the second transgenic methodology by using undifferentiated nuclei, that is, sperm nuclei. Briefly, this method involves first placing the desired gene under the control of an appropriate promoter in a plasmid (Fig. 10.2). The plasmid is then linearized with a restriction enzyme. *Xenopus laevis* sperm nuclei are prepared and mixed with the linearized plasmid DNA in a high-speed extract made from *Xenopus laevis* eggs in the presence of low levels of the restriction enzyme used to linearize the plasmid. A short incubation of the mixture allows partial decondensation of the nuclei and partial digestion of the sperm

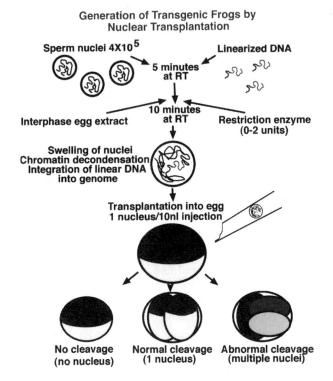

Figure 10.2 *Schematic flow chart of the transgenic procedure of Kroll and Amaya (1996). Sperm nuclei are mixed with linearized plasmid DNA carrying a gene of interest and incubated at room temperature (RT) in interphase egg extract in the presence of the same restriction endonuclease used to linearize the plasmid. This treatment results in swelling of the sperm nuclei and the incorporation of the plasmid DNA into sperm chromatin. The mixture is then injected at about one sperm nucleus/egg into* Xenopus *eggs to fertilize the eggs. An egg generally develops properly if receiving one sperm but no development or an abnormal embryo is the consequence if no or more than one sperm, respectively, is received by an egg.*

chromatin. This allows the insertion of the plasmid DNA into the chromatin through the cohesive ends generated by the restriction enzyme. The nuclei are then diluted and microinjected into *Xenopus laevis* eggs at one sperm nucleus/egg. Those eggs receiving one nucleus/egg develop normally and often carry the transgenes.

Although its efficiency is still relatively low, the technique compares quite favorably with those developed for mouse or zebrafish (Kroll and Amaya, 1996). More importantly, such transgenic animals can develop into frogs, thus allowing the study of gene function throughout amphibian development. In fact, by using this approach, Huang et al. (1999) demonstrated that overexpression of 5-deiodinase in *Xenopus laevis* tadpoles inhibited (delayed) metamorphosis and led to a resistance to TH treatment, in agreement with its proposed role based on its ability to inactivate TH and its developmental expression profiles (Chapter 8.4). The availability of many genes likely involved in metamorphosis together with the use of tissue-specific promoters will likely make this method highly valuable for functional studies on tissue remodeling during metamorphosis. The adaptation of this method to *Xenopus tropicalis* (Amaya et al., 1998), which is a true diploid organism with a shorter generation time, will make genetic studies of amphibian metamorphosis even more feasible.

11

Comparison with Insect Metamorphosis

11.1 INTRODUCTION

Like amphibians in vertebrates, the invertebrate insects also undergo varying degree of metamorphosis during their life cycle. There are three major patterns of postembryonic development in insects (Highnam, 1981; Gilbert and Frieden, 1981; Granger and Bollenbacher, 1981; Sehnal et al., 1996). The most primitive types of insects are wingless and undergo little change as the juvenile develops into the adult. The major differences separating adults from the juveniles are the presence of developed gonads and genitalia, in the adult. The transition from the juvenile to adult for this class of insect is referred as "ametabolous."

The second pattern of insect development is called "gradual" or "hemimetabolous." For this class of insects, the change from last-instar larva to adult at the final molt is more dramatic than that for the ametabolous insects. The larval form is distinguished from the adult form by the presence of vestigial wings and incomplete genitalia. However, different species within this class of insects vary considerably in the extent of changes taking place during metamorphosis, as reflected both in similarities between the larval and adult forms and in the degree of larval tissue replacement.

The most highly evolved insects undergo a third pattern of postembryonic development termed "holometabolous." This class of winged insects undergoes three distinct stages of development, that is, larval, pupal, and adult stages. Although the degree of larval tissue replacement still varies among different species, the metamorphosis of this class of insects is the most dramatic and often referred to as complete metemorphosis. Many of the larval tissues/organs degenerate and much of the adult body develops from imaginal buds or discs, which are nests of cells without any function in the larva but serving to produce adult organs/tissues during metamorphosis. This class of insects, including

Manduca sexta and *Drosophila melanogaster*, resemble anurans in vertebrates and are the focus of this chapter.

11.2 HORMONAL REGULATION OF INSECT METAMORPHOSIS

Metamorphosis in holometabolous insects like *Drosophila* affects most, if not all, organs/tissues in the animal (Bodenstein, 1950). Similar to amphibian metamorphosis, three major transformations take place as the larva is changed into the adult. These are remodeling, de novo development, and resorption. The first one is exemplified by the partial but drastic remodeling of the central nervous system (Truman, 1990; Truman et al., 1993). Many of the larval neurons, such as those in the brain and thoracic regions of the CNS, persist throughout metamorphosis into adult stages. However, they lose their larval-specific components and develop adult elements, for example, the loss of larval neurites followed by the outgrowth of adult-specific processes in the mesothoracic motoneuron MN5. The other larval neurons, on the other hand, undergo apoptosis concurrent with the degeneration of their target tissues/organs. Finally, adult-specific neurons develop de novo during metamorphosis.

The second major transformation is the de novo development of adult organs, such as the wings and legs. These organs are derived through proliferation and differentiation of cells preserved in the form of imaginal buds or discs in the larvae (Fristrom, 1981; Fristrom and Fristrom, 1993).

The last form of change during metamorphosis is the complete resorption of larval-specific organs/tissues. For example, the gastric caeca of the alimentary canal in *Drosophila* is completely resorbed through programmed cell death during metamorphosis (Skaer, 1993; Jiang et al., 1997). Similarly, the larval salivary gland undergos complete histolysis, although an adult salivary gland is subsequently developed de novo from imaginal discs (Skaer, 1993).

Like amphibian metamorphosis, insect metamorphosis is also under complex neuroendocrine control (Granger and Bollenbacher, 1981; Gilbert and Goodman, 1981; Riddiford, 1993a; Gilbert et al., 1996). In particular, the hormone ecdysone plays a causative role similar to that played by thyroid hormone in amphibians. There are multiple forms of ecdysone in insects and the physiologically active form is most likely 20-hydroxyecdysone (Ec) (Fig. 11.1; Gilbert and Goodman, 1981; Richard, 1981a). In *Drosophila melanogaster*, Ec titer peaks during metamorphosis (Fig. 11.2; Richards 1981b; Riddiford, 1993a). Interestingly, there are at least six pulses of Ec during *Drosophila* development (Fig. 11.2), three of which occur during embryonic and larval developmental stages and three during metamorphosis (from the end of the third instar stage to adulthood). Similar observations have also been made in another holometabolous insect, *Manduca sexta* (Bollenbacher et al., 1975; Riddiford and Truman, 1978; Riddiford, 1993b). The early Ec pulses are likely to be critical for preparation of the larvae for metamorphosis and the late pulses control the actual metamorphic process.

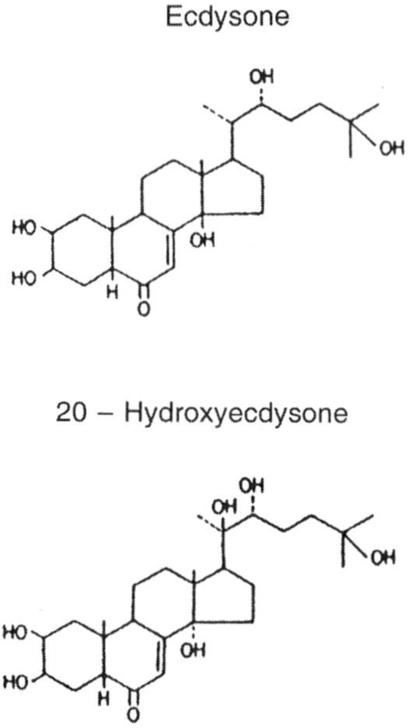

Ecdysone

20 – Hydroxyecdysone

Figure 11.1 Two common forms of ecdysone. There are a number of other derivatives but the physiologically active one is most likely 20-hydroxyecdysone (Gilbert and Goodman, 1981; Richards, 1981a).

Ec acts organ autonomously to induce metamorphosis. Imaginal discs cultured in vitro can be induced to produce the same transformations as those in vivo, including the formation of not only the basic structures such as the legs and wings but also musculature, neural tissue, and the fine sculpturing of bristles and hairs (Ashburner, 1972; Fristrom et al., 1973, 1977; Mandaron et al., 1977; Milner, 1977a,b; Fristrom, 1981). Similarly, larval-specific organs such as the salivary gland undergo degeneration through programmed cell death when cultured in vitro in the presence of Ec (Andres and Thummel, 1994; Jiang et al., 1997). Such organ- and/or cell-autonomy of Ec action is consistent with the transcriptional regulation mechanism of Ec through its nuclear receptors (see below).

While Ec initiates and controls metamorphosis, the juvenile hormone inhibits metamorphosis (Riddiford, 1993b; 1996). This effect is similar to that of prolactin on TH-dependent amphibian metamorphosis. Unlike prolactin, which has little expression in premetamorphic tadpoles and is upregulated during metamorphosis (see Chapter 2), juvenile hormone is present as several pulses in insect larvae (Fig. 11.2C; Riddiford and Truman, 1978; Riddiford, 1993a,b). Although some overlaps exist between the Ec and juvenile hormone pulses, little juvenile hormone is present at the pupal stages when a broad Ec pulse exists and drastic metamorphic transformations take place, supporting a

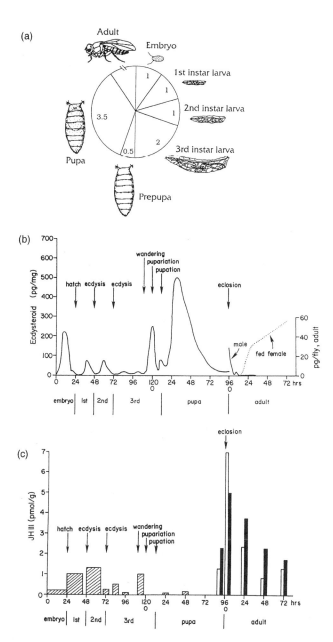

Figure 11.2 (A) Drosophila *life cycle. The six developmental stages in the life cycle are shown: embryonic, three larval instars, prepupal, and pupal. Each of these stages is characterized by a pulse of ecdysone. The numbers (in days) represent the duration of each stage.* (B) *Composite ecdysteroid titers in* D. melanogaster *in 20-HE (20-hydroxyecdysone) equivalents in whole-body homogenates.* (C) *Composite JH III (juvenile hormone) levels in whole-body homogenates of* D. melanogaster. *Reprinted from Andres and Thummel (1992) and Riddiford (1993a) with permission from Elsevier Science and Cold Spring Harbor Laboratory Press, respectively.*

role of juvenile hormone in preventing precocious metamorphosis and maintaining larval form during development. Unfortunately, the mechanism of juvenile hormone action is yet unclear and its receptor(s) has not been cloned.

11.3 ECDYSONE RECEPTORS

There are three isoforms of nuclear ecdysone receptors (EcRs), which are generated through alternative promoter usage and alternative splicing from a single gene (Koelle et al., 1991; Talbot et al., 1993). These receptors are members of the steroid/thyroid hormone receptor superfamily (Mangelsdorf et al., 1995). Interestingly, although Ec is a steroid, EcRs belong to the retinoid/thyroid hormone receptor subfamily but not to the steroid receptor subfamily (Mangelsdorf et al., 1995). Like thyroid hormone receptors, EcRs function as a heterodimer with another member of this subfamily, the ultraspiracle (UPS) in *Drosophila* (Yao et al., 1992; Thummel, 1995; Cherbas and Cherbas, 1996). At the amino acid sequence level, the *Drosophila* EcR shares extensive sequence homology with *Xenopus* TRs, and *Drosophila* USP is highly homologous to *Xenopus* RXRs (Table 11.1). In fact, *Drosophila* USP can functionally substitute for mammalian RXR in stimulating the function of TR and other members of the retinoid/thyroid hormone receptor subfamily (Yao et al., 1992).

In *Drosophila*, UPS gene is expressed throughout development at fairly constant levels (Fig. 11.3; Andres et al., 1993). The EcR mRNA is also

TABLE 11.1 Amino Acid Sequence Comparison between *Drosophila* and *Xenopus* Receptors[a]

Similarity/Identity Length, aa)	xTrαA	xTRβA1	EcR	xRXRα	xRXRβ	xRXRγ	USP
xTRα	100/100 (419)						
xTRβA1	90/86 (376)	100/100 (370)					
EcR	44/33 (382)	47/35 (382)	100/100 (878)				
xRXRα				100/100 (489)			
xRXRβ				94/79 (421)	100/100 (452)		
xRXRγ				77/71 (458)	81/75 (417)	100/100 (471)	
USP				67/56 (408)	69/57 (412)	65/54 (417)	100/100 (508)

[a]The sequences of *Xenopus* TR and RXRs are from Yaoita et al. (1990), Blumberg et al. (1992), and Marklew et al. (1994); those of *Drosophila* EcR and USP are from Koelle et al. (1991) and Oro and Evans (1990), respectively.

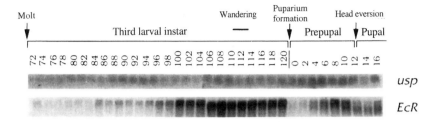

Figure 11.3 Developmental expression of EcR and USP in Drosophila. Samples of total RNA isolated at 2 hr intervals from staged third instar larvae and prepupae were fractionated by formaldehyde agarose gel electrophoresis, transferrred to nylon membranes, and analyzed by Northern blot hybridization. Shown at the top are the time points of third instar larval development in hours after egg laying and prepupal development in hours after puparium formation. Also shown are the developmental stages and the major ecdysteroid-regulated events that occur at the onset of metamorphosis: second to third instar larval molt, wandering, puparium formation, and head eversion. Modified after Andres et al., 1993 with permission from Academic Press.

detectable at embryonic to metamorphic stages and EcR proteins are present throughout the organism (Fig. 11.3; Koelle et al., 1991; Andres et al., 1993). Like TR genes in anurans, the EcR genes is also upregulated by its ligand Ec (Karim and Thummel, 1992). This upregulation may at least in part contribute to the rise in EcR expression from mid-third instar to puparium formation and the rise during the prepupal stages (after the rapid repression of EcR expression at puparation) (Fig. 11.3; Andres et al., 1993). These increases produce high levels of EcRs during the period when several Ec pulses, including the broadest pulse during the pupal stages, are present to induce the larval to adult transformation. Furthermore, the EcR mRNA levels correlate well with the expression of several classes of Ec response genes with different temporal expression profiles (Andres et al., 1993), arguing for a role of EcRs in the regulation of such genes.

Detailed analyses of the expression of different EcR isoforms have revealed that they are regulated differently during *Drosophila* development (Robinow et al., 1993; Talbot et al., 1993; Truman et al., 1994). Although essentially all organs/tissues change during metamorphosis in *Drosophila*, they undergo distinct transformations and/or metamorphose at different developmental stages (Bodenstein, 1950). Interestingly, the expression patterns of the EcR-A and EcR-B1 isoforms at the onset of metamorphosis are similar in those tissues belonging to the same metamorphic class (Talbot et al., 1993). For example, different imaginal discs uniformly express high ratios of EcR-A to EcR-B1. On the other hand, the imaginal cells of the midgut islands and histoblast nests have the highest ratios of EcR-B1 to EcR-A. These different EcR profiles correlate with the distinct metamorphic changes of the two tissue types. The imaginal disc cells proliferate and differentiate to form the adult organs de novo, while the imaginal cells of the midgut islands and histoblast nests proliferate and migrate extensively to generate the adult midgut and abdominal epidermis, respectively (Bodenstein, 1950; Madhavan and Schneiderman, 1977; Roseland and Schneiderman, 1979). Finally, larva-specific tissues that are destined to die exhibit high ratios of EcR-B1 to EcR-A based on antibody staining. Thus different EcRs may have distinct functions during metamorphosis. That is, high ratios of EcR-A to EcR-B1 mediate de novo adult organ development whereas high ratios of EcR-B1 to EcR-A facilitate organ remodeling and tissue resorptions. This same conclusion was also reached from studies of EcR expression in the central nervous system (Robinow et al., 1993; Truman et al., 1994; Truman, 1996).

11.4 ECDYSONE-INDUCED GENE REGULATION CASCADE

11.4.1 The Ashburner Model

The histolysis of the larval salivary gland has served as an excellent model for studying the molecular genetics of insect metamorphosis. In response to Ec, many loci of the polytene chromosome in the salivary gland begin to be highly transcribed and can be observed as puffs by electron microscopy (Fig. 11.4; Ashburner, 1967, 1972; Russell and Ashburner, 1996). These polytene puffs are also induced when the salivary gland is cultured in vitro in the presence of Ec (Ashburner, 1972). There are at least two classes of Ec-inducible polytene puffs. The early puffs are relatively few (six major ones in the salivary gland) and are induced within minutes after Ec addition (Ashburner et al., 1974). After reaching maximal sizes within about 4 hr, these early puffs regress. Concurrently, a large number of late puffs appear. This sequential appearance of the puffs is identical to the one that normally occurs at the end of larval development (Ashburner et al., 1974).

Interestingly, the addition of protein synthesis inhibitors to the organ culture blocks the induction of late puffs but not the early ones, indicating that the early puffs are the direct consequence of Ec action whereas the late ones require new protein synthesis to form (Ashburner et al., 1974). Furthermore, the regression of the early puffs is also inhibited by the protein synthesis inhibitors, arguing that new protein synthesis is necessary for this regression.

Based on these and other studies such as dose dependence and time course of Ec treatment, Ashburner et al. (1974) proposed a gene regulation cascade model to explain this sequential puffing pattern (Fig. 11.5). According to the model, the hormone ecdysone binds to EcR and receptor–hormone complex, then activates the genes located at the early puffs and at the same time represses the genes at the late puffs. The products of the early genes, which are direct response genes, in turn activate the genes at the late puffs (late response genes) and repress their own expression. Such an autoregulatory loop ensures that the early genes are expressed only transiently to initiate the downstream metamorphic events through the action of late response genes.

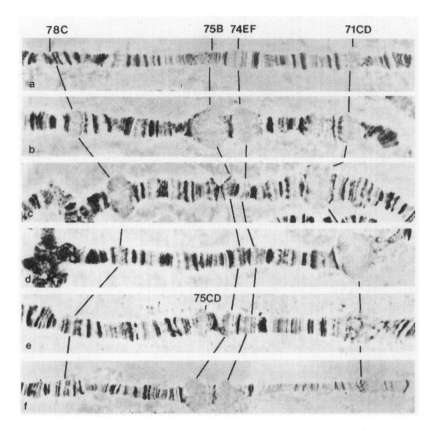

Figure 11.4 The sequential induction and regression of puffs on the left arm of chromosome 3 in salivary glands dissected at various puff stages (PS) from larvae and prepupae: (a) PS1; (b) PS6; (c) PS10-11 (d) PS12; (e) PS18; and (f) PS19. The 74EF and 75B early puffs, the 78C early-late puff, and the 71CD and 75CD late puffs are indicated. Reprinted from Russell and Ashburner (1996) with permission from Academic Press.

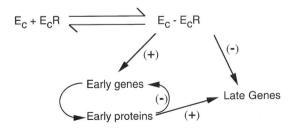

Figure 11.5 The Ashburner model for ecdysone-induced gene regulation cascade in the salivary gland. (+) and (−) indicate stimulatory and inhibitory effects, respectively.

11.4.2 Genes Regulated by Ecdysone

Direct support for the Ashburner model has come from molecular cloning and functional studies of the genes encoded at the puff loci (Thummel, 1990; Andres and Thummel, 1992; Russell and Ashburner, 1996). As predicted from the model, the genes BR-C (Broad-Complex), E74, and E75, located at the early puff loci 2B5, 74EF, and 75B, respectively, encode transcription factors and are primary or direct response genes of ecdysone (Table 11.2). In addition, their mRNAs are only transiently expressed in response to Ec pulses and precede those of the late genes like 4F (Table 11.2), just like the sequential appearance of their respective puffs (Fig. 11.6; Andres et al., 1993; Huet et al., 1993). Furthermore, molecular and genetic studies have provided evidence for involvement of BR-C and E74 genes in the regulation of late Ec-response genes during *Drosophila* metamorphosis (Karim et al., 1993; Fletcher and Thummel, 1995). Finally, at least in the case of E74A protein encoded by the E74 gene, it has been shown that it is bound to both early and late Ec-inducible puffs in the salivary gland and that it directly regulates the transcription of the late gene L71-6 (Urness and Thummel, 1990, 1995), supporting its dual role in early gene repression and late gene activation.

The early Ec-response genes such as BR-C and E74 are not restricted to the salivary gland. In fact, most, if not all, early genes are expressed in multiple

TABLE 11.2 Some Known Ecdysone Response Genes in *Drosophila*[a]

Gene	Nature of the Protein/Function	Response to Ec
BR-C	Four zinc-finger transcription factors	Direct
E74A/B	EST transcription factors	Direct
E93	Putative transcription factor	Direct
E75A/B/C	Nuclear hormone receptors of unknown ligand	Direct
FTZ-F1	Nuclear hormone receptors of unknown ligand	Direct
DHR3	Nuclear hormone receptors of unknown ligand	Direct
E63-1	Ca^{2+}-binding protein	Direct
IMP-E1	Secreted protein with three EGF-like repeats followed by a mucin-like domain	Direct
IMP-E2	Apically secreted protein with 16 EIK-like repeats	Direct
IMP-E3	Apically secreted protein	Direct
IMP-L1	Secreted protein	Indirect
EDG-78E	Member of cutin cuticle protein family	Indirect
EDG-84A	Similar to *Locusta* and *Hyalophora* cuticle proteins	Indirect
4F	Unknown	Indirect
L71	Small secreted polypeptide resembling defensins and venom toxins	Indirect

[a] For references, see Andres and Thummel (1992, 1995); Andres et al. (1993); Baehrecke and Thummel (1995); Thummel (1995, 1997); Bayer et al. (1996b); Wright et al. (1996).

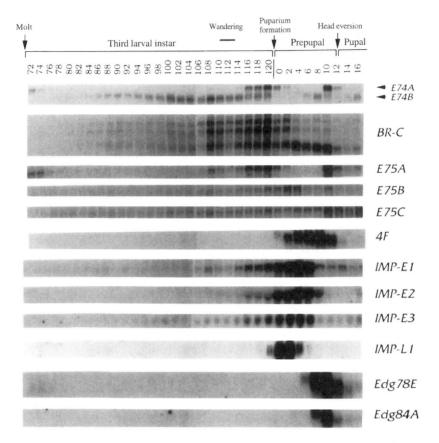

Figure 11.6 Expression of ecdysone-response genes during Drosophila development. E74A, E74B, BR-C, E75A, E75B, E75C, IMP-EI, IMP-E2, and IMP-E3 are early genes, whereas 4F, IMP-L1, Edg 78E, and Edg 84A are late genes. For other descriptions, see Figure 11.3. Modified after Andres et al. (1993) with permission from Academic Press.

tissues/organs (Huet et al., 1993; Bayer et al., 1996a, b), although the E63-1 gene encoded by the 63F early puff is regulated by Ec only in the salivary gland (Andres and Thummel, 1995). Thus the early genes are likely involved in the metamorphosis of many different organs/tissues and the tissue specificity of metamorphic transformations may be determined through the action of downstream, tissue-specific, late Ec-response genes (Thummel, 1990).

In addition to the genes encoded by the early puffs, there are many early or direct Ec-response genes that are not located at polytene puffs (Natzle et al., 1986; Natzle, 1993; Thummel, 1990; Andres and Thummel, 1992; Hurban and Thummel, 1993). Many of these early response genes are also expressed in multiple organs (Andres and Thummel, 1992; Bayer et al., 1996b). Furthermore, their developmental expression profiles also resemble those of the genes at early puffs (Andres et al., 1993; Hurban and Thummel, 1993). For example, the early genes IMP-E1, IMP-E2, and IMP-E3, which were isolated from the imaginal discs, are expressed earlier than the late genes like IMP-L1, Edg78E, and Edg84A, also isolated from the imaginal discs, during *Drosophila* metamorphosis (Fig. 11.6; Table 11.2; Natzle et al., 1986; Andres et al., 1993). Thus the early genes are likely to play a role early in the metamorphic transition. Interestingly, the early genes such as IMP-E1, IMP-E2, IMP-E3, and E63-1, do not encode transcription factors (Table 11.2; Natzle et al., 1988; Moore et al., 1990; Paine-Saunders et al., 1990; Andres and Thummel, 1992; Bayer et al., 1996b). Their involvement in the regulation of downstream, late Ec-response genes, if any, is likely to be indirect, in contrast to the predictions based on the simple Ashburner model.

The identification of early Ec-response genes in multiple organs and the ubiquitous expression of many of these genes argue that the basic principle, that is, the transcriptional regulation cascade, of the Ashburner model is true for most, if not all, organ transformations during metamorphosis, although the model itself is clearly oversimplified. The understanding of the exact gene regulation cascades leading to the metamorphic transitions in individual organs will require the identification of late Ec response genes and determination of how the early genes influence downstream gene expression. Unfortunately, relatively little is known in these areas. Although some late genes have been identified in the salivary gland and imaginal discs, the mechanisms by which the early genes regulate the late genes largely remain to be elucidated (Table 11.2; Thummel, 1990; 1996; Bayer et al., 1996b).

11.4.3 A Gene Regulation Cascade for Tissue Resorption

The best studied metamorphic transformation is the histolysis of the salivary gland through programmed cell death. This degeneration process involves multiple genes induced by two pulses of Ec between the third instar and early pupal stages. Some of the genes that most likely participate in this process are listed in a gene regulation cascade model in Fig. 11.7 (Baehrecke, 1996; Thummel, 1996; Jiang et al., 1997; Lam et al., 1997; White et al., 1997). During the early third-instar stages, both EcR and USP are expressed and low levels of Ec are also present (Figs. 11.2 and 11.3). They may be responsible for the low levels of expression of the early genes E74B and BR-C. When the Ec titer rises to high levels at the late third instar stages, the expression of the early genes E74A, E75A, and BR-C are upregulated to high levels in a process that requires EcR as well as BR-C (Fig. 11.7). The roles of these three genes on the expression of downstream genes in the cascade are yet unclear. However, because they all encode transcription factors, they can regulate late genes such as L71-6 (Urness and Thummel, 1990, 1995) and presumably participate either directly or indirectly to ensure the proper response of downstream genes in the cascade to the changing concentration of Ec. The high titer of Ec at the late third instar stages also leads to the repression of βFTZ-F1 and activation of

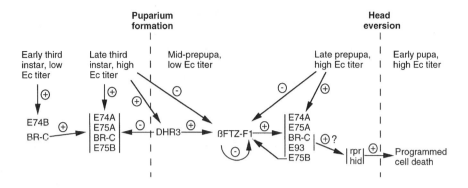

Figure 11.7 A gene regulation cascade model for salivary gland histolysis. Based on Thummel (1997), Jiang et al. (1997), Lam et al. (1997), and White et al. (1997). See text for details.

DHR3 genes, both of which encode proteins of the nuclear hormone receptor superfamily and have no known ligands at the present time (Table 11.2; Lavorgna et al., 1991, 1993; Koelle et al., 1992; Thummel, 1995).

As development proceeds into the mid-prepupal stages, the Ec titer is low again. The DHR3 then activates the βFTZ-F1 gene, presumably directly as several DHR3 binding sites are present in the βFTZ-F1 gene (Lam et al., 1997). In addition, DHR3 expression is correlated with the repression of the early genes, and overexpression of DHR3 in *Drosophila* larva leads to the reduction of the expression of early genes and increase of βFTZ-F1 expression (Lam et al., 1997; White et al., 1997). Thus DHR3 is also involved in the down-regulation of early genes at the mid-prepupal stages when Ec titer is low. This repression may be through direct inhibition of EcR function as DHR3 and EcR interact with each other in a yeast-two-hybrid assay (White et al., 1997).

The expression of βFTZ-F1 facilitates the re-expression of the early genes at the late prepupal stages when Ec titer arises again (Fig. 11.7; Woodard et al., 1994; Broadus et al., 1999). The expression of βFTZ-F1 is transient and several mechanisms appear to contribute to its downregulation at late prepupal stages (Andres et al., 1993; Woodard et al., 1994). First, the high titer of Ec at late prepupal stages represses its expression. Second, βFTZ-F1 inhibits its own transcription. Finally, E75B, which is upregulated again at the late prepupal stages and is also an orphan nuclear receptor like DHR3 and βFTZ-1 (Table 11.2; Segraves and Hogness, 1990; Thummel, 1995), can interact with DHR3 to block the induction of βFTZ-F1.

Although βFTZ-F1 expression is only transient, it is necessary and sufficient to direct the Ec induction of the early gene E93 at the late prepupal stages, possibly through direct recognition of its binding sites present in the E93 gene (Woodard et al., 1994; Broadus et al., 1999). The E93 gene is expressed in a stage- and tissue-specific manner that correlates with tissue resorption through programmed cell death (Baehrecke and Thummel, 1995). Although the exact function of E93 is unclear, it is likely to encode a transcription factor (Baehrecke and Thummel, 1995). Based on genetic and molecular studies, it has been proposed that E93 functions together with BR-C at the late prepupal stages to induce the expression of the *Drosophila* death genes rpr and hid, leading to the histolysis of the salivary gland (Jiang et al., 1997).

11.5 SIMILARITIES AND DIFFERENCES BETWEEN AMPHIBIAN AND INSECT METAMORPHOSIS

The molecular studies of amphibian and insect metamorphosis have largely been done on *Xenopus laevis* and *Drosophila melanogaster*, respectively. The general conclusions from such studies are, however, believed to be applicable to anuran and holometabolous insect metamorphosis, respectively, and are indeed supported by studies, wherever available, on other species such as *Rana catesbeiana* for anurans and *Manduca sexta* for insects.

Both insect and amphibian metamorphosis involve three contrasting types of transformations, resorption, remodeling, and de novo development. Despite these diverse changes, all are controlled essentially by a single hormone, thyroid hormone for amphibians and ecdysone for insects. The hormonal signals are mediated by heterodimers of two members of the nuclear receptor superfamily and extensive similarities exist between the respective amphibian and insect receptors, especially between RXR and USP, which can functionally interchange for each other.

The hormones clearly induce a cascade of gene regulation in respective species to effect the metamorphic changes. In the case of *Drosophila*, strong evidence supports a critical role of a cascade involving sequential activation and repression of a series of transcription factors that lead to the degeneration of larval tissues through programmed cell death. However, the participation of non-transcription factors can not be ruled out. As in amphibians, not all early Ec-response genes in insects encode transcription factors. Many early Ec-response genes in *Drosophila* encode extracellular or membrane-bound proteins (Table 11.2; Andres and Thummel, 1992; Bayer et al., 1996b), just like those in *Xenopus laevis*. In addition, the downregulation of gene expression by hormone-bound EcR or TR complex is also likely to play a role during metamorphosis but little information on this is available. Thus the actual gene regulation program is likely to be much more complex. It involves genes both activated and repressed by the hormone-receptor complexes (Fig. 11.8). Those genes encoding transcription factors are likely to regulate directly the transcription of downstream or late response genes, whereas those encoding other proteins may indirectly influence downstream gene expression through regulation of cell–cell and cell–extracullar matrix interactions and other signal transduction pathways. Such coordinated gene regulation involving simultaneous induction of both intra- and extra-cellular processes by the hormone

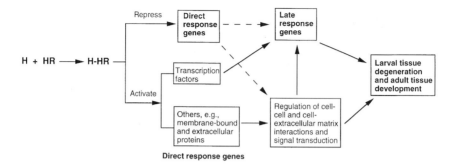

Figure 11.8 *A simplified gene regulation cascade model for the regulation of insect and amphibian metamorphosis by Ec or TH (H) through their nuclear receptors (HR). Currently, there is little information on the downregulated, direct hormone response genes. Their roles in the downstream events are uncertain and thus are indicated by dashed arrows. The known early upregulated genes include both transcription factors and others, such as potential signaling molecules. Thus they can affect downstream events through direct control of gene transcription or indirect effects on gene expression by means of various signal transduction pathways (Andres and Thummel, 1992; von Kalm et al., 1995; Shi, 1996; 1997).*

may be essential for the spatial and temporal regulation of larval tissue degeneration and adult organ development.

In spite of the impressive similarities between anuran and holometabolous insect metamorphosis, some distinct differences exist in their underlying molecular programs. First of all, amphibian metamorphosis requires the continuous presence of TH, whereas in insects, several Ec pulses are required to coordinate proper development (Fig. 11.2; Riddiford and Truman, 1978; Richards, 1981b; Gilbert et al., 1996). Although TR/RXR and EcR/USP heterodimers can mediate the transcriptional effects of TH and Ec, respectively, recent evidence has revealed that another member of the nuclear receptor family, DHR78, is required for Ec signaling at the onset of *Drosophila* metamorphosis (Fisk and Thummel, 1998). TRs, however, appear to be the only factor required for TH to regulate properly the expression of its response genes (Puzianowska-Kuznicka et al., 1997).

The expression of the direct Ec-response genes is only transient in response to each hormone pulse in *Drosophila* (Paine-Saunders et al., 1990; Karim and Thummel, 1992; Andres et al., 1993). In *Xenopus laevis*, the direct response genes are expressed for much longer periods, if not through the entire metamorphic process (Buckbinder and Brown, 1992; Shi and Brown, 1993; Wang and Brown, 1993; Shi, 1996; Denver et al., 1997). Although several early genes in the tadpole intestine are eventually repressed, it is unclear whether this repression involves the products of the early genes themselves as in *Drosophila*.

In addition, the direct Ec-response genes change their expression very quickly (within 1 hr) upon Ec addition to organ cultures of *Drosophila* imaginal discs (Paine-Saunders et al., 1990) or larvae (Karim and Thummel,

1992). In *Xenopus*, a lag of several hours exists between the addition of TH and the increase in the mRNA levels of the early TH response genes (Buckbinder and Brown, 1992; Kanamori and Brown, 1992; Shi and Brown, 1993; Wang and Brown, 1993; Denver et al., 1997). The mechanisms behind such different kinetics are unclear. It is unlikely that the *Xenopus* genes are not true direct TH response genes because their regulation by TH does not require new protein synthesis. Furthermore, at least one of the direct TH-response genes, the *Xenopus* TRβA genes, has been shown to be regulated by TR/RXR through direct recognition of at least one thyroid hormone response element located near the transcription start site (Ranjan et al., 1994; Machuca et al., 1995; Wong et al., 1995, 1998a).

A common problem in the study of molecular mechanisms governing metamorphosis is the lack of sufficient downstream or late response genes. The feasibility for genetic approaches in *Drosophila* has enabled a greater progress in dissecting the gene regulation cascade induced by Ec, especially with regard to salivary gland histolysis. Using a genetic screen with criteria defined for metamorphosis, a gene that is important for leg development during metamorphosis has been isolated (D'Avino and Thummel, 1998). Similar screens and other genetic studies may allow the isolation of additional genes in the cascade and provide insights into the molecular pathways linking the early genes to late genes and eventually to the tissue specific metamorphosis. However, the final proof of the molecular pathways in both insect and amphibian metamorphosis will require the characterization of the proteins, especially those encoded by the upstream genes, and the promoter and enhancer elements governing the expression of hormone response genes at various steps in the gene regulation cascade.

12

Roles of Thyroid Hormone and Its Receptors in Mammalian Development and Human Diseases

12.1 INTRODUCTION

The absolute dependence of frog development on thyroid hormone is perhaps the most extreme role that the hormone plays in any vertebrates, including mammals. On the other hand, extensive conservations exist across vertebrate species. Not only are the basic body plan and the functions of various organs highly similar or identical among different species such as amphibians and mammals, but often the developmental processes for various organs are also conserved and involve homologous genes. For example, the adult intestines of both frogs and mammals have similar structure organizations with extensive epithelial folding to increase the luminal surface for nutrient absorption (see Chapter 3). The anuran and mammalian intestines also share similar developmental pathway. They both initially develop into a simple tubular structure consisting of mostly endodermally derived epithelial cells. Although this initial simple organ is more developed in anuran tadpoles for it to be functional because the tadpoles are independent living beings, both the amphibian and mammalian larval/embryonic intestines subsequently undergo morphogenetic transformations to form the complex adult organ (Chapter 3; Glass, 1968; Dauca et al., 1990; Shi and Ishuzuya-Oka, 1996).

In addition to intestinal development, many other remodeling processes that occur during amphibian metamorphosis also take place during post embryonic

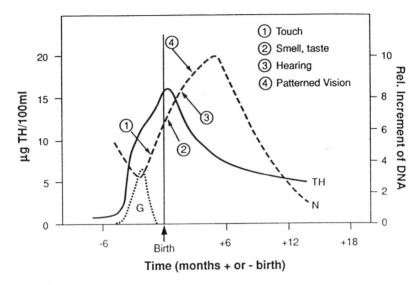

Figure 12.1 *Idealized curves showing the close association between the appearance of thyroid hormones (TH) in human fetal plasma, the proliferation of neuronal (N) and glial (G) cells in the brain, and the onset of acquisition of sensory perceptions (arrows) during the perinatal period and after birth. The proliferation of neural cells is represented as relative increments of DNA at different times during development. From Tata (1993).*

development in mammals (Tata,1993; Atkinson, 1994). For example, both amphibians and mammals undergo a postembryonic transition of the expression of the hemoglobin genes, from larval (tadpole) or fetal types to frog or adult types, respectively. Likewise, the rise in the serum albumin level during metamorphosis resembles the increase in the albumin level in mammals during late fetal development. Other similarities in postembryonic development include skin keratinization and induction of urea cycle enzymes as well as the developmental progression and restructuring of the central and peripheral nervous systems. Such conservations or similarities argue for a role of TH during postembryonic development in mammals. Consistently, high levels of TH are present in the human fetal plasma during the several months around birth when extensive organ development and maturation such as brain development take place (Fig. 12.1; Tata, 1993).

12.2 THYROID HORMONE IN HUMAN DEVELOPMENT AND DISEASES

The critical effects of TH on human development have been known for centuries. The most obvious and earliest known abnormalities of human body and behavior associated with TH deficiency are goiter (a lump in the neck due to thyroid gland enlargement caused by iodine deficiency) and cretinism (a form of mental deficiency together with defects in skeletal growth caused by the lack of TH during fetal development) (Hetzel, 1989). One of the oldest records of goiter was from the Chinese Emperor Shen-Nung (2838–2698 BC). Similar records of goiter and cretinism can be found in other ancient cultures, that is, Hindu, Greece, and Roman cultures. These developmental abnormalities can treated or prevented with exogenous TH, again known for centuries (Hetzel, 1989; Mandel et al., 1993).

Goiter and cretinism are the consequences of TH deficiency in humans. The development of goiter appears to be simply due to the reduced negative feedback by TH on the thyroid gland through the pituitary (Hetzel, 1989). The reduction of circulating TH leads to an increase in TSH output from the pituitary (see Chapter 2), consequently overstimulating the thyroid gland and causing its overgrowth. Cretinism, on the other hand, is a serious human disease caused by abnormal development of a number of organs due to the lack of sufficient TH.

TH deficiency can arise from the lack of iodine, removal of the thyroid gland, or absence of the gland due to diseases or congenital defects, although human cretinism was best recognized in its endemic form, which results from lack of iodine in the general population and can be easily prevented. Congenital hypothyroidism, which results from the loss of thyroid gland function in utero or in the neonate and affects about one in 4000 newborn infants (DeGroot et al., 1984), requires early neonatal diagnosis and treatment with TH replacement therapy to prevent mental and growth retardation.

TH appears to begin to function very early during human fetal development (Hetzel, 1989). The thyroid gland is developed by the end of the first 12 weeks of gestation. The pituitary develops around the same period and thus TSH is available to stimulate the release of TH from the thyroid gland (see Chapter 2). TSH becomes detectable in the blood between 18 and 22 weeks. Shortly after, T_4 is also detectable in the blood although T_3 levels remain low until just before birth when 5'-deiodinase activity increases, leading to a rapid increase in T_3 concentration.

Even before the maturation of the fetal neuroendocrine–thyroid axis, maternal TH, which can cross the placental barrier whereas TSH cannot, may influence fetal growth and development (Hetzel, 1989). Maternal TH, although possibly at only low levels in the fetus, may be responsible for the apparently normal development of congenital hypothyroid infants if TH replacement is initiated shortly after birth (Larsen et al., 1989). Such observations also suggest that defects caused by TH deficiency during fetal development can be reversed by TH treatment.

Upon birth, the human neonate changes from intra-uterine to extra-uterine life and is accompanied by an increased demand for oxygen and a higher rate of metabolism (Hetzel, 1989). Thyroid hormone is known to stimulate metabolic rate both in vitro and in vivo in humans and animals (Guernsey and Edelman, 1983; Freake and Oppenheimer, 1995; Silva, 1995). T_4 is inactive or less active than T_3. Thus the rise of T_3 concentration around birth is likely to be critical to prepare the neonate for extra-uterine living habitat.

In addition to increased metabolism after birth, the period of perinatal development is also associated with increased proliferation of glial and neuronal cells and acquisition of several brain functions, as exemplified by the sensory processes (Fig. 12.1; Tata, 1993). Again, the high levels of T_3 at this period is critical for these and other neural development processes. Although the exact processes affected by TH in the developing human brain are not clear. Animal studies have shown than TH influences many aspects of the brain development (Dussault and Ruel, 1987). At anatomical and histological levels, both the forebrain and cerebellum require TH for normal maturation. For example, thyroidectomy of newborn rats results in severe damage in the pyramidal cells in the forebrain, as indicated by the marked reduction in the number of spines present along the apical shaft of the pyramidals. Similarly, the maturation of the Purkinje cells in the cerebellum is permanently interrupted by TH deficiency as reflected by their arborization and the number of dendritic spines. Hypothyroidism also delays the appearance of the external germinal layer and decreases the number and density of synaptic contacts with the already defective Purkinje cells, leading to a permanent impairment of neuronal connectivity. These anatomical alterations are accompanied by extensive biochemical changes in the brain, among them the changes in oxygen consumption and the metabolism of glucose and polyamines. In addition, the content and/or composition of myelin, neurotransmitters and neuropeptides, and microtubules in the brain are also altered by hypothyroidism.

In humans, TH deficiency during neonatal period leads to irreversible, profound neurological deficit and mental retardation (Hetzel et al., 1989; Porterfield and Hendrich, 1993; Hsu and Brent, 1998). Such defects can be reduced significantly if maternal supplementation of iodine is provided in the first trimester (Xue-Yi et al., 1994), again supporting the importance of TH in brain development during this period.

Thyroid hormone continues to play important roles during human postnatal development well past birth (Hetzel, 1989). It is essential for normal growth and development and TH deficiency is associated with severe retardation of growth and maturation of almost all organ system. Body weight does not increase and bone growth is also retarded if TH is deficient. Most dramatic effects are to those tissues that are rapidly proliferating. The severe consequence of insufficient or lack of TH during fetal and postnatal development is cretinism, which affects many aspects of the patient. In addition to mental retardation and short stature, the skin is coarse and dry, and the voice becomes husky (Table 12.1; Schwartz, 1983; Rivkees et al., 1988; Hetzel, 1989; Raz and Kelley, 1997). Delayed skeletal maturation is also present, including epiphyxed dysgenesis in the limbs. Hearing loss also occurs in many cases of congenital hypothyroidism (Debruyne et al., 1983; Anand et al., 1989) and is more prevalent if hypothothyroidism is present during fetal development, for example, in the case of endenic cretinism or congenital iodine deficiency (Delong et al., 1985).

TABLE 12.1 Clinical Features of Hypothyroidism and Resistance to Thyroid Hormone Syndromes (RTH) in Human[a]

Feature	Hypothyroidism	RTH
Brain	Mental retardation, low IQ	More severe mental retardation, low IQ
Stature	Dwarfism due to growth retardation	Usually normal but short stature in some cases
Hearing	Hearing loss in many but not all cases	Deaf-mutism
Skeletal	Retarded growth and delayed maturation, e.g., epiphyseal dysgenesis	Usually normal but delayed bone maturation, and dysmorphic facies in some cases
Reflexes	Delayed relaxation	Excessively brisk
Skin	Coarse, dry	Usually normal
Voice	Husky	Usually normal
Goiter	Present in endemic but not sporadic (congenital) cases	Usually present
TH	Low levels or absent	Elevated levels
TH treatment	Improvement	No effect

[a] Not all features are present in all cases. Based on Schwartz (1983); Hetzel (1989); Refetoff et al. (1993); Hauser (1994); Chatterjee and Beck-Peccoz (1994); Brucker-Davis et al., 1995; Raz and Kelley (1997).

In addition to the developmental effects above, TH is also important for normal function of various organs. TH deficiency leads to reduced metabolic rate (Guernsey and Edelman, 1983; Freake and Oppenheimer, 1995; Silva, 1995). Normal cardiovascular function also requires the presence of proper levels of TH. Abnormal TH levels are associated with a number of cardiovascular symptoms and signs (Table 12.2, Franklyn and Gammage, 1996). TH deficiency results in low heart rate and cardiac outputs as well as hyperlipidemia whereas hyperthyroidism causes tachycardia that can lead to arrhythmia, elevated cardiac outputs and hypolipidemia (Dillman, 1996; Franklyn and Gammage, 1996). Thus, although the role of TH in human development is not as dominating as in anurans, proper regulation of TH levels is critical for normal development and function throughout human life.

12.3 ROLES OF THYROID HORMONE RECEPTORS IN THYROID HORMONE RESISTANCE SYNDROMES

12.3.1 Thyroid Hormone Resistance Syndromes

Another major human abnormality associated with TH action is the syndrome of resistance to thyroid hormone (RTH). This was first described in 1967 by Refetoff et al. in two siblings who were euthyroid. Despite their high levels of

TABLE 12.2 Clinical Features of Hyperthyroidism and Hypothyroidism[a]

Hyperthyroidism		Hypothyroidism	
Symptom	Sign	Symptom	Sign
Palpitation	Sinus tachycardia	Shortness of breath	Sinus bradycardia
Shortness of breath	Atrial premature beats	Ankle swelling	Atrioventricular block
Ankle swelling	Atrial fibrillation	Chest pain	Heart failure
			Diastolic hypertension
			Percardial effusion
Chest pain	Systolic hypertension		
	Hyperdynamic circulation		
	Cardiac systolic murmurs		

[a] From Franklyn and Gammage (1996).

circulating TH, they had a number of abnormalities, including deaf-mutism, stippled femoral epiphyses with delayed bone maturation and short stature, dysmorphic facies, and winging of the scapulae and pectus carinatum (Refetoff et al., 1967, 1972, 1993). Some patients with RTH have clincal features commonly associated with hypothyroidism, including short stature and goiter; the majority of them are clinically euthyroid with higher than normal levels of circulating TH, reflecting a compensatory physiological response of the thyroid-pituitary axis to tissue resistance (Table 12.1; Usala and Weintraub, 1991; Refetoff et al., 1993; Hauser, 1994; Chatterjee and Beck-Peccoz, 1994; Brucker-Davis et al., 1995).

There are three forms of the disorder (Refetoff et al., 1993). The predominant one is generalized resistance to TH (GRTH). This form is characterized by high levels of circulating TH maintained by the secretion of pituitary TSH in response to hypothalamic TRH (see Chapter 2, except that in mammals TRH instead of CRF as in anurans functions to stimulate TSH release). There is a global insensitivity to TH. Thus, despite higher than normal levels of free TH, the plasma TSH levels are often inappropriately normal or elevated, reflecting the insensitivity of the pituitary to the negative feedback of TH (see Chapter 2).

Pituitary resistance to TH (PRTH), is also known as nonneoplastic inappropriate secretion of TSH or nonneoplastic central hyperthyroidism. It differs from GRTH by the apparent manifestations of TH excess at the level of peripheral tissues. It is derived from selective pituitary defect to respond to TH. Patients with PRTH have elevated levels of free TH but have nonsuppressed, basal serum TSH, similar to GRTH patients. However, patients have signs of thyrotoxicosis or hyperthyroidism, for example, hypermetabolism.

The last form has been reported in detail for only a single patient (Kaplan et al., 1981). In this case, the resistance to TH is confined to peripheral tissues (PTRTH or peripheral tissue resistance to TH, Refetoff et al., 1993). The patient has normal serum TSH and TH levels but requires high doses of TH to maintain a normal basal metabolic rate. Without TH treatment, PTRTH patients are expected to be hypometabolic and manifest symptoms and signs of hypothyroidism.

Of all known RTH cases, the vast majority of the patients have GRTH (Refetoff et al., 1993). This is dominantly inherited with highly variable clinical features (Refetoff et al., 1993; Chatterjee and Beck-Peccoz, 1994). Both GRTH and PRTH are caused by mutations in the TRβ receptor gene.

12.3.2 Human Thyroid Hormone Receptors

As in anurans, two types of nuclear TH receptors exist in humans. These are TRα and TRβ, located on chromosomes 17 and 3, respectively (Naylor and Bishop, 1989; Solomon and Barker, 1989; Hsu and Brent, 1998). Each gene produces at least two alternative spliced transcripts, encoding TRα1 and TRα2 or TRβ1 and TRβ2, respectively (Fig. 12.2; Benbrook and Pfahl, 1987; Nakai et al., 1998a,b; Weinberger et al., 1986, 1988; Miyajima et al., 1989; Lazar, 1993; Sakurai et al., 1989, 1990; Laudet et al., 1991; Takeda et al., 1992a,b; Yen et al., 1992; Refetoff et al., 1993; Hsu and Brent, 1998). TRα1 and TRα2 differ only at the very C-terminus (Fig. 12.2). Technically, TRα2 is not a thyroid hormone receptor because the ligand binding domain is no longer capable of binding TH owing to the replacement of the important C-terminus of the TRα1 by an unrelated exon (exon 10). No TRα2 form has been cloned in amphibians, although it is present in other mammals (Lazar, 1993). All the other three TR forms can bind to TH with similar high affinities. TRβ2 has a longer N-terminus with different sequences than TRβ1. In addition, TRβ2 has much more limited expression profiles whereas TRα1, TRα2, and TRβ1 are expressed widely in different tissues (Lazar, 1993).

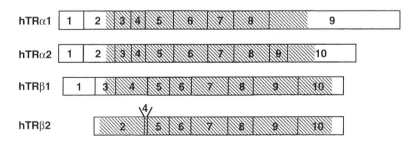

Figure 12.2 Organizations of human TRα and TRβ isoforms. The shaded boxes indicate coding regions and the open ones are untranslated regions. Each gene has at least two protein products due to alternative splicing. TRα2 is technically not a thyroid hormone receptor because it does not bind TH owing to the replacement of the critical hormone-binding region encoded by part of exon 9 at the carboxyl terminal end of TRα1 by an unrelated sequence encoded by exon 10. TRβ1 and TRβ2 differ only at the N-terminal end and both function as normal TH receptors.

TRs are highly conserved during evolution. For example, the *Xenopus* TRs and human TRs share more than 84% identity and 90% similarity among them (Table 12.3). Higher levels of sequence identity exist between the same TR subtype in the two species than between the two subtypes of the same species. For example, human and *Xenopus* TRβ are 93% identical but *Xenopus* TRα and TRβ share only 86% identity (Table 12.3). The DNA binding (in the N-terminal half) and hormone binding (C-terminal half) domains are essentially identical between the *Xenopus* and human receptors, although the very N-terminus is highly diverged among different receptors (Fig. 12.3A). These conservations again support a conserved role of TH in development.

In addition to the sequence conservation, the expression patterns of TR genes during development are also quite consistent in different species, including anurans, birds, and mammals (Yaoita et al., 1990; Mellstrom et al., 1991; Porterfield and Hendrich, 1993; Wong and Shi, 1995; Hsu and Brent, 1998). TRα1 is expressed first, usually ahead of thyroid gland maturation (e.g., in *Xenopus*; Yaoita and Brown, 1990; Kawahara et al., 1991; Wong and Shi, 1995). TRβ gene is not expressed or is expressed at low levels prior to TH synthesis but is upregulated by TH. In fact, both human and *Xenopus* TRβ genes have been shown to be upregulated by TH (Ranjan et al., 1994; Suzuki et al., 1994, 1995; Machuca et al., 1995; Nagasawa et al., 1997; Wong et al., 1998a). This autoregulation is mediated in both cases through direct binding by TRs to at least one thyroid hormone response element (TRE) in the respective promoter, even though the other promoter/enhancer elements are divergent between the human and frog genes.

The TRα and TRβ genes have different developmental and tissue-specific expression profiles (Brent, 1994; Porterfield and Hendrich, 1993; Hsu and Brent, 1998). Other than this, the functional significance of these receptor isoforms during development are largely unknown. However, there is some evidence for different affinities for TH analogs (Schueler et al., 1990) and isoform-specific gene regulation and antagonism by mutant TRs (Lezoualch et al., 1992; Zavacki et al., 1993; Safer et al., 1997). Thus it is likely that different TRs will have different functions even when expressed within a single tissue.

Human TRs are also believed to function mainly through their heterodimers formed with RXRs or 9-*cis*-retinoic acid receptors. As is the case for TRs, RXRs are also highly conserved throughout evolution. The *Xenopus* RXRs and human RXRs share extensively homology (Table 12.3), especially in the DNA and hormone binding domains, which are again essentially identical among the frog and human RXRs (Fig. 12.3B). However, very divergent sequences are present in the N-terminal domain, and the hinge region between the DNA binding and the hormone binding domains (Fig. 12.3B). Overall, the frog and human RXRs share 71–91% identity and 77–93% similarity among them (Table 12.3). Again, higher levels of homologies exist between the same RXR isotypes in *Xenopus* and human than those between two isotypes in a single species, for example, 91% identity between human and *Xenopus* RXR but only 71–79% identity between *Xenopus* RXRα and *Xenopus* RXRβ or RXRγ (Table

TABLE 12.3 Amino Acid Sequence Homology between *Xenopus laevis* and Human Nuclear Receptors[a]

Similar/Identity	xTRαA	xTRβA1	hTRα	hTRβ	xRXRα	xRXRβ	xRXRγ	hRXRα	hRXRβ	hRXRγ
xTRαA	100/100									
xTRβA1	90/86	100/100								
hTRα	**92/89**	91/85	100/100							
hTRβ	89/84	**95/93**	90/84	100/100						
xRXRα					100/100					
xRXRβ					84/79	100/100				
xRXRγ					77/71	81/75	100/100			
hRXRα					**93/91**	84/78	77/72	100/100		
hRXRβ					81/76	**83/80**	77/73	79/75	100/100	
hRXRγ					82/78	78/73	**81/77**	81/77	77/72	100/100

[a] The receptor sequences are from the following references: *Xenopus* TRs; Yaoita et al. (1990); *Xenopus* RXRα and RXRγ; Blumburg et al. (1992); *Xenopus* RXRβ; Marklew et al. (1994); human TRα, Nakai et al. (1988a); human TRβ, Weinberger et al. (1988), and Sakurai et al. (1990); human RXRα, Mangelsdorf et al. (1990); human RXRβ; Leid et al. (1992); human RXRγ; Fleischhauer et al. (1992); Mangelsdorf et al. (1992). The numbers in bold refer to the homologies between homologs of the two species.

```
xTRβA1
hTRβ    MTPNSMTENG LTAWDKPKHC PDREHDWKLV GMSEACLHRK SHSERRSTLK  50
hTRα                                                MEQKPS      6
xTRαA                                               MDQNLS      6

xTRβA1                                              MEGYIPSY    8
hTRβ    NEQSSPHLIQ TTWTSSIFHL DHDDVNDQSV SSAQTFQTEE KKCK~~~~~~ 100
hTRα    KVECGSDPEE NSARSPDGKR KRKNGQCSLK TSMS       ~~~~~~     46
xTRαA   GLDCLSEPDE K  RWPDGKR KRKNSQCMGK SGMSGDSLVS LPSA~~~~~~ 54

xTRβA1  LDKDELCVVC GDKATGYHYR CITCEGCKGF FRRTIQKNLH PSYSCKYEGK 58
hTRβ    ~~~~~~~~~~ ~~~~~~~~~~ ~~~~~~~~~~ ~~~~~~~~~~ ~~~~~~~~~~ 150
hTRα    ~~~~~Q~~~~ ~~~~~~~~~~ ~~~~~~~~~~ ~~~~~~~~~~ ~T~~~~~DSC 96
xTRαA   ~~~~~P~~~~ S~~~~~~~~~ ~~~~~~~~~~ ~~~~~~~~~~ ~~~~~~~D~C 104

xTRβA1  CVIDKVTRNQ CQECRFKKCI AVGMATDLVL DDNKRLAKRK LIEENREKRR 108
hTRβ    ~~~~~~~~~~ ~~~~~~~~~~ Y~~~~~~~~~ ~~S~~~~~~~ ~~~~~~~~~~ 200
hTRα    ~~~~~I~~~~ ~~L~~~~~~~ ~~~~~M~~~~ ~~S~~V~~~~ ~~~Q~~~~~~ 146
xTRαA   ~I~~~I~~~~ ~~L~~~~~~~ ~~~~~M~~~~ ~~G~~V~~~~ ~~~~~~QR~~ 154

xTRβA1  KDEIQKSLVQ KPEPTQEEWE LIQVVTEAHV ATNAQGSHWK QKRKFLPEDI 158
hTRβ    RE~L~~~~GH ~~~~~D~~~~ ~~KT~~~~~~ ~~~~~~~~~~ ~~~~~~~~~~ 250
hTRα    ~E~MIR~~Q~ R~~~~P~~~D ~~HIA~~~~R ~~~~~~~~~~ ~~~~~~~D~~ 196
xTRαA   ~E~MI~T~Q~ R~~~SS~~~~ ~~RI~~~~~R S~~~~~~~~~ ~~~~~~~~~~ 204

xTRβA1  GQAPIVNAPE GGKVDLEAFS QFTKIITPAI TRVVDFAKKL PMFCELPCED 208
hTRβ    ~~~~~~~~~~ ~~~~~~~~~~ H~~~~~~~~~ ~~~~~~~~~~ ~~~~~~~~~~ 300
hTRα    ~~S~~~SM~D ~~~~~~~~~~ E~~~~~~~~~ ~~~~~~~~~~ ~~~~~~~~~~ 246
xTRαA   ~~S~MASM~D ~D~~~~~~~~ E~~~~~~~~~ ~~~~~~~~~~ ~~~~~~~~~~ 254

xTRβA1  QIILLKGCCM EIMSLRAAVR YDPESETLTL NGEMAVTRGQ LKNGGLGVVS 258
hTRβ    ~~~~~~~~~~ ~~~~~~~~~~ ~~~~~~~~~~ ~~~~~~~~~~ ~~~~~~~~~~ 350
hTRα    ~~~~~~~~~~ ~~~~~~~~~~ ~~~~~D~~~~ S~~~~~~E~ ~~~~~~~~~~ 296
xTRαA   ~~~~~~~~~~ ~~~~~~~~~~ ~~~D~~~~~~ S~~~~~K~E~ ~~~~~~~~~~ 304

xTRβA1  DAIFDGVSL SSFSLDDTEV ALLQAVLLMS SDRPGLASVE RIEKCQEGFL 308
hTRβ    ~~~~~~~M~~ ~~~N~~~~~~ ~~~~~~~~~~ ~~~~~~~C~~ ~~~~Y~DS~~ 400
hTRα    ~~~~E~~K~~ ~~~N~~~~~~ ~~~~~~~~~~ ~~~~~~LC~D K~~~S~~AY~ 346
xTRαA   ~~~~~~~R~~ ~~~N~~~~~~ ~~~~~~~~~~ ~~~~~~ICTD K~~~~~ETY~ 354

xTRβA1  LAFEHYINYR KHNIAHFWPK LLMKVTDLRM IGACHASRFL HMKVECPTEL 358
hTRβ    ~~~~~~~~~~ ~~~VT~~~~~ ~~~~~~~~~~ ~~~~~~~~~~ ~~~~~~~~~~ 450
hTRα    ~~~~~~V~H~ ~~~IP~~~~~ ~~~~~~~~~~ ~~~~~~~~~~ ~~~~~~~~~~ 396
xTRαA   ~~~~~~~~H~ ~~~~P~~~~~ ~~~~~~~~~~ ~~~~~~~~~~ ~~~~~~~~~~ 404

xTRβA1  FPPLFLEVFE D                                            369
hTRβ    ~~~~~~~~~~ ~                                            461
hTRα    ~~~~~~~~~~ ~QEV                                         410
xTRαA   ~~~~~~~~~~ ~QEV                                         418
```

(a)

Figure 12.3 Amino acid sequence comparisons among Xenopus and human TRs and RXRs. The sequences are derived from the references in Table 12.3. (A) TRs. The residues similar to Xenopus TRβA1 (xTRβA1) are indicated by "~" for Xenopus TRαA (xTRαA) and for human TRα1 and TRβ1 (hTRα and hTRβ, respectively). The open spaces indicate that no residues are present at the locations. Note that the sequences are essentially identical in the DNA and

ROLES OF TRs IN TH RESISTANCE SYNDROMES

```
xRXRβ
hRXRβ    MSWAARPPFL PQRHAAGQCG PVGVRKEMHC GVASRWRRRR PWLDPAAAAA  50
xRXRα    MS SAAMDTKHPL                                           12
hRXRα
xRXRγ
hRXRγ

xRXRβ    MGDS RVC QSPDTSSLSP    PLGHSFSDT                         26
hRXRβ    AAVAGGEQQT PEPEPGEAGR DG~~~~G~DS R~~~S~~~PN~ L~Q~VPPPSP 100
xRXRα    PLGGRTCAD~ LRCTTSWTAG YDFSSQVNSS SLSSSGLRGS MTAPLLHPSL   62
hRXRα               MDTKHFLP  LDFSTQVNSS LTSP~G RGS MAAPSLHPSL    37
xRXRγ       MHLATETA ~SMATYSSTY FNSSLHAHST SVS S~N~AA MNSLDTHPGY  47
hRXRγ          MYGNY SHFMKFPA~Y GGSPGHTGST SMSPSAA~~T GKPMD~HPSY  45

xRXRβ    PPPPSAP LH PSMVGSAMTS SVNSPLGSIG SPFPVINCSV GSPGIPGTPS   75
hRXRβ    ~G~~LP~STA ~~LG~~GAPP PPPM~PPPL~ ~~~~~~SS~M ~S~~L~PPAP  150
xRXRα    GNSGLNNS~G SPTQLPSP    LS~~INGM~ P~~S~~SPPL ~~ SMAIPST  108
hRXRα    G       ~GIG SPGQLHSPI~ TLS~~INGM~ P~~S~~SSPM ~~HSMSVPTT 82
xRXRγ      MGNS~N GPRSMTTNMN ~MC~~GNN~~ L~YR~~AS~M ~~HS  LPSP    91
hRXRγ    TDTPVS APRT    L~ A~GT~~NAL~ ~~YR~~TSAM ~~~SGALAAP      87

xRXRβ    IGY GPVSSP QINSTVNLSG LHHVGSSEDV KPP   LGMR SMQ SHPNGG  120
hRXRβ    P~FS~~~~~P ~~~~~~S~P~ GGS~PP~~~ ~~~   V~~V~ GLHCPP~P~~ 197
xRXRα    P~LGYGTG~P ~~H~PM~    S~S~T~~I ~~~PGIN~IL KVPM   HPS~  151
hRXRα    PTLGFSTG~P ~LS~PM~    P~S~~~~I ~~~LGLN~VL KVPA  HPS~   125
xRXRγ    TILNY~GHES PPFNIL~    N~SC~~~I ~~~~GLSSLG ~PCMNNYSCN   136
hRXRγ    P~INLVAPPS SQLNV~~    S~S~~~~I ~~~~GLP~IG N  MNYPST    129

xRXRβ    TVSG  KRLC AICGDRSSGK HYGVHSCEGC KGFFKRTIRK DLTYTCRDSK  168
hRXRβ    PGA~  ~~~~  ~~~~~~~~~ ~~~~Y~~~~~ ~~~~~~~~~~ ~~~~~~~~N~ 245
xRXRα    AMASFT~HI~ ~~~~~~~~~~ ~~~~Y~~~~~ ~~~~~~~V~~ ~~~~~~~~~~ 201
hRXRα    NMAS~~~HI~ ~~~~~~~~~~ ~~~~Y~~~~~ ~~~~~~~V~~ ~~~~~~~~N~ 175
xRXRγ    SPGAL~~HI~ ~~~~~~~~~~ ~~~~Y~~~~~ ~~~~~~~~~~ ~~~~~~~~~~ 186
hRXRγ    SPGSLV~HI~ ~~~~~~~~~~ ~~~~Y~~~~~ ~~~~~~~~~~ ~~~~~~~~N~ 179

xRXRβ    DCIVDKRQRN RCQYCRYQKC LATGMKREAV QEERQRGRER DGEAELSGA   217
hRXRβ    ~~T~~~~~~~ ~~~~~~~~~~ ~~~~~~~~~~ ~~~~~~~~DK ~~DG~GA~G  294
xRXRα    ~~MI~~~~~~ ~~~~~~~~~~ ~~~M~~~~~~ ~~~~~~~K~~ NEN~V~S~NS 251
hRXRα    ~~LI~~~~~~ ~~~~~~~~~~ ~~~M~~~~~~ ~~~~~~~KD~ ~EN~V~STSS 225
xRXRγ    ~~LI~~~~~~ ~~~~~~~~~~ ~~~M~~~~~~ ~~~~~~~S~~~ S~T~~~STSS 236
hRXRγ    ~~LI~~~~~~ ~~~~~~~~~~ ~~~M~~~~~~ ~~~~~~~S~~~ AES~~~CATS 229

xRXRβ    INEEMPVEKI LEAELAVEQK SDQSLEG      GGSPSD PVTNICQDAD    260
hRXRβ    AP~~~~~~DR~ ~~~~~~~~~~ ~~~GVE~PGG TGGS~S~~N~ ~~~~~~~A~~ 344
xRXRα    A~~D~~~~~~ ~~~~~~~~P~ TETYTE   AN MGLAPN~~~~ ~~~~~~~A~~ 299
hRXRα    A~~D~~~~R~ ~~~~~~~~P~ TETYVE   AN MGLNPS~~~L~ ~~~~~~~~A~~ 273
xRXRγ    TS~~~~~~R~ ~~~~~~~~DP~ IEAFGD   AG L  PN~TN~ ~~~~~~HA~~ 281
hRXRγ    GH~D~~~~R~ ~~~~~~~~P~ TESYGD   MN M  EN~TN~ ~~~~~~HA~~ 274

xRXRβ    KQLFTLVEWA KRIPHFSELP ELPLDDQVIL LRAGWNELLI ASFSHRSISE  310
hRXRβ    ~~~~~~~~~~ ~~~~~~~~    S~~~~~~~~~ ~~~~~~~~~~ ~~~~~~~~DV 391
xRXRα    ~~~~~~~~~~ ~~~~~~~~    ~~~~~~~~~~ ~~~~~~~~~~ ~~~~~~~~AV 346
hRXRα    ~~~~~~~~~~ ~~~~~~~~    ~~~~~~~~~~ ~~~~~~~~~~ ~~~~~~~~AV 320
xRXRγ    ~~~~~~~~~~ ~~~~Y~~    D~~~E~~~~~ ~~~~~~~~~~ ~~~~~~~~VSV 328
hRXRγ    ~~~~~~~~~~ ~~~~~~~~    D~T~E~~~~~ ~~~~~~~~~~ ~~~~~~~VSV 321

xRXRβ    KDGILLATGL HVHRNSAHSA GVGAIFERVL TELVSKMRDM RMDKTELGCL  360
hRXRβ    R~~~~~~~~~ ~~~~~~~~~~ ~~~~~~~~~~ ~~~~~~~~~~ ~~~~~~~~~~ 441
xRXRα    ~~~~~~~~~~ ~~~~~~~~~~ ~~~~~~D~~~ ~~~~~~~~~~ Q~~~~~~~~~ 396
hRXRα    ~~~~~~~~~~ ~~~~~~~~~~ ~~~~~~D~~~ ~~~~~~~~~~ Q~~~~~~~~~ 370
xRXRγ    Q~~~~~~~~~ ~~~~~S~~~~~ ~~~~~~D~~~ ~~~~~~~~~~ D~~~S~~~~~ 378
hRXRγ    Q~~~~~~~~~ ~~~~~S~~~~~ ~~~S~~D~~~ ~~~~~~~~~~ Q~~~SL~~~~ 371

xRXRβ    RAIILFNPDA KGLSNPGDVE VLREKVYACL ESYCKQKYPD QQGRFAKLLL   410
hRXRβ    ~~~~~~~~~~ ~~~~~~~E~~ ~~~~~~~~S~ ~T~~~~~~~E ~~~~~~~~~~  491
xRXRα    ~~~V~~~~~S ~~~~~~LE~~ A~~~~~~~~S~ E~~~~~~~~~E ~P~~~~~~~~ 446
hRXRα    ~~~V~~~~~S ~~~~~~AE~~ A~~~~~~~~S~ ~A~~~H~~~ ~P~~~~~~~~  420
xRXRγ    ~~~V~~~~~~ ~~~~~AAE~~ A~~~~~~~~T~ ~~~T~~~~ ~P~~~~~~~~   428
hRXRγ    ~~~V~~~~~~ ~~~~~~SE~~ T~~~~~~~~T~ ~A~T~~~~~E ~P~~~~~~~~  421

xRXRβ    RLPALRSIGL KCLEHLFFFK LIGDTPIDTF LMEMLEAPHQ LS           452
hRXRβ    ~~~~~~~~~~ ~~~~~~~~~~ ~~~~~~~~~~ ~~~~~~~~~~ ~A           533
xRXRα    ~~~~~~~~~~ ~~~~~~~~~~ ~~~~~~~~~~ ~~~~~~~~~~ MT           488
hRXRα    ~~~~~~~~~~ ~~~~~~~~~~ ~~~~~~~~~~ ~~~~~~~~~~ MT           462
xRXRγ    ~~~~~~~~~~ ~~~~~~~~~~ ~~~~~~~~~~ ~~~~~~T~~~ I~           470
hRXRγ    ~~~~~~~~~~ ~~~~~~~~~~ ~~~~~~~~~~ ~~~~~~A~L~ IT           463
```

(b)

Figure 12.3 Continued. hormone binding domains (amino acids 4–111 and 124–369, respectively, in xTRβA1), whereas the amino terminal end is highly variable in both length and sequence. (B) RXRs. Again the DNA binding domain and the hormone binding domains are highly conserved but the hinge regions between these two domains and the amino terminal end are divergent.

12.3). Such cross-species conservations argue that different RXR isotypes are likely to have differential effects on the function of TH through their interactions with TRs.

12.3.3 Thyroid Hormone Receptor Mutations in Human Diseases

Molecular cloning of TRs have allowed the analyses of possible changes in TRs in diseases related to thyroid hormone action. These studies have demonstrated that TR mutations are responsible for generalized resistance to thyroid hormone (GRTH) and pituitary resistance to thyroid hormone (PRTH) (Usala and Weintraub, 1991; Parrilla et al., 1991; Takeda et al., 1992a,b; Refetoff et al., 1993; Chatterjee and Beck-Peccoz, 1994; Collingwood et al., 1994, 1998; Jameson, 1994). Surprisingly, all known cases are due to mutations in the TRβ gene and none in the TRα gene.

Mutations in the TRβ genes have been found in many known GRTH and PRTH patients. Three clusters of mutations are present. Two are in the ligand binding domain (amino acids 310–383 and 429–460) and one in the hinge region or the N-terminal end of the ligand binding domain (amino acids 234–282) (Fig. 12.4). Four types of mutations are present in these clusters. These are amino acid substitutions, deletions, frame shifts and a truncation due to the generation of a stop codon. These changes are located away from domains important for dimerization and thus mutant receptors are capable of binding TRE as homo- and heterodimers (Darling et al., 1991; Glass et al., 1989; Forman et al., 1989; Forman and Samuels, 1990; Spanjaard et al., 1991; O'Donnell et al., 1991; Refetoff et al., 1993; Jameson, 1994; Collingwood et al., 1994, 1998).

Not surprisingly, based on their locations in the ligand binding domain, most of the mutations affect TH binding to some degrees. On the other hand, only less than half of the mutant receptors have severe defects in TH binding ($<1/10$ of the binding affinity of the wild type TRβ, Fig. 12.4), even though all confer resistance to TH. In addition, the degree of reduction in TH binding affinity does not correlate with the clinical severity of hormone resistance as determined by the levels of plasma TH needed to maintain a normal serum TSH concentration (Refetoff et al., 1993). Thus there are likely multiple mechanisms contributing to RTH.

As described in Chapter 5, TRs most likely function as heterodimers with RXRs and mediate transcriptional repression or activation depending upon its association with transcriptional corepressors or coactivators, respectively. Thus a mutant TR can dominantly affect the function of a wild type TR through heterodimerization with the wild type TR, competing for a TRE as heterodimers with RXR against the wild type TR/RXR, and/or competing for coactivators and corepressors. If a mutant TR is functionally defective, such interferences may explain the dominant inheritance of RTH phenotypes, especially considering that the mutations are localized in regions away from those critical for TRE binding and dimerization (Fig. 12.4).

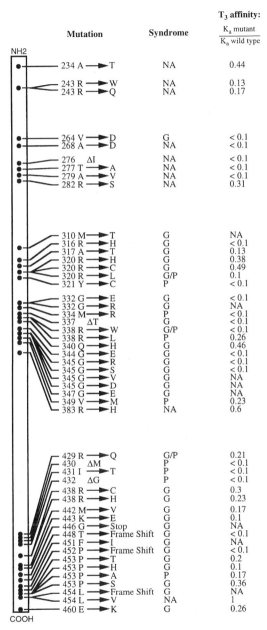

Figure 12.4 Some of the known mutations in the hTRβ gene of subjects with syndromes of GRTH (G) and PRTH (P), and their effects on TRβ's affinity for T_3 (K_a). On the left is a diagram representing the T_3-binding domain of TRβ spanning from amino acid 243 to 461 at the carboxyl terminus (COOH). The numbers refer to human TRβ1 sequence (Sakurai et al., 1990). NA: not available. For references see Parrilla et al. (1991), Takeda et al. (1992b), Behr and Loos (1992); Refetoff et al. (1993); Collingwood et al. (1994, 1998); Adams et al. (1994); Onigata et al. (1995); Yagi et al. (1997); Safer et al. (1997); Macchia et al. (1997); Miyoshi et al. (1998).

Given the presence of a variety of mutations, it can also be expected that different RTH mutations in TRβ can impair TR's response to TH through distinct mechanisms (Usala and Weintraub, 1991; Refetoff et al., 1993; Chatterjee and Beck-Peccoz, 1994; Jameson, 1994; Collingwood et al., 1994, 1997, 1998; Clifton-Bligh et al., 1998; Yoh et al., 1997). The most obvious one is for those mutant TRs defective in TH binding. They are expected to be constitutive repressors similar to unliganded wild type TRs (Fig. 12.5B; see Chapter 5; Baniahmad et al., 1992; Refetoff et al., 1993; Chatterjee and Beck-Peccoz, 1994; Jameson, 1994). It is more complicated for those mutations that have little effect on TH binding or affect both TH binding and other aspects of receptor function. Recent studies have shown that often these mutations can influence the interactions between TR and TR-corepressors and/or coactivators (Collingwood et al., 1997; Yoh et al., 1997; Clifton-Bligh et al., 1998). Thus another possible mechanism for the impaired TR response to TH may be the inability of TH-bound mutant TR to release corepressors and recruit coactivators (Fig. 12.5A). Finally, the mutant TR may bind to TH and this binding releases corepressors but liganded mutant TR can not recruit coactivators, and is thus unable to activate the target gene above basal level transcription (Fig. 12.5C). Clearly, other mechanisms may also exist. A careful dissection of the different steps involved in transcriptional regulation by mutant TRs and a detailed understanding of cofactor expression and function will be needed to pinpoint the exact mechanisms by which various mutant TRs cause RTH.

12.4 TR KNOCKOUT STUDIES IN MICE

The advances in mouse genetics have made it possible to investigate TR function during mammalian development directly. In particular, gene knockout approaches have been applied to remove TRα or TRβ genes selectively and examine the developmental consequences and effects on organ function in adult mice. Given the conserved nature of TH function, and TR sequences and expression profiles in different animal species, such studies are likely to help revealing the roles of TR in animal development and organ function that are applicable to human diseases.

12.4.1 TRα Gene

There have been two independent knockouts on the TRα gene. Given the fact that no human TRα mutations have been found in human RTH patients and the evolutionary conservations in its sequence and expression profiles, one would expect that TRα gene would be essential for mouse development and survival. Surprisingly, a carefully designed approach to remove selectively only the TRα1 transcript but not the alternatively spliced product encoding a non-TH binding version of the receptor, TRα2, produced mice that have apparently normal growth, fertility, and survival (Wikstrom et al., 1998).

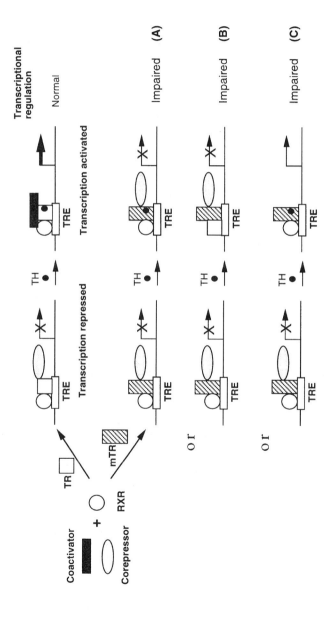

Figure 12.5 Possible mechanisms by which human TRβ mutants cause resistance to TH. Normally, TRs function as heterodimers with RXRs and regulate transcription by recruiting cofactors into the target promoters. In the absence of TH, corepressors are recruited, leading to target gene repression (Top). Upon binding of TH to TRs, the corepressors are released and coactivators are recruited to activate the genes. The mutant TRs (mTR) may have lost the ability to release corepressors and recruit coactivators even though they can bind to TH, thus leading to target gene repression even when TH is present (A). Alternatively, mTRs may fail to bind to TH, and thus are unable to respond to TH (B). Finally, mTRs may be able to bind to TH and release corepressors but fail to recruit coactivators, leading to impaired response to TH (C).

However, the mutant mice have reduced serum T_4 levels and heart rate is also lower than wild type mice (Table 12.4, Wikstrom et al., 1997; Hsu and Brent, 1998). These mice also have prolonged QT and QRS durations as determined based on averaged electrocardiograms. Furthermore, they have a lower body temperature even when their activity levels are the same as wild type mice, and they exhibit a mild hypothyroidism. These results are consistent with importance of TH in normal heart function (Table 12.2) and reveal the existence of specific roles for TRα1 in regulation of tightly controlled physiological functions, such as cardiac pacemaking, ventricular repolarization, and control of body temperature (Wikstrom et al., 1998).

In contrast to the mild phenotypes of the TRα1-specific knockout, removal of both TRα1 and TRα2 expression has a profound effect on mice development (Fraichard et al., 1997). Even though the heterozygous mice were identical to wild type ones, the homozygous knockout mice were born normally but had growth arrest after 2 weeks and died within 5 weeks. Although there were no consistent cellular or morphological abnormalities in the brain of the homozygous mice, significant defects were detected in several other tissues, including the intestine and bone (Table 12.4; Fraichard et al., 1997; Hsu and Brent, 1998). Among them, the intestine had a two- and threefold reduction in diameter and had a reduced number of epithelial cells per crypt–villus unit. The overall intestine appeared to be normally developed but immature. The bone had reduced longitudinal area and was demineralized compared to wild type animals. In addition, the thyroid was hypoplastic and serum TH levels were lower than normal. TSH levels in the serum were also reduced compared to control mice. Such contrasting phenotypes between TRα1-specific and TRα1-TRα2 knockouts argue that TRα2 has important roles and that different TRα isoforms have distinct functions. Thus it is important to consider TRα2 when addressing the roles of TH and TRs because TRα2 is functionally similar, if not identical, to unliganded TRα1.

These knockout studies also provide an alternative explanation for the lack of known TRα mutations in human TH-associated diseases. That is, patients with defective TRα genes may have subnormal levels of TH and TSH, which would contrast with the hallmarks of RTH, thus escaping the traditional diagnosis. However, because TRα1 and TRα2 share the vast majority of the coding regions, mutations affecting TRα1, other than those at the extreme carboxyl terminus, also affect TRα2. Given the postnatal lethal phenotype of the TRα1/TRα2 knockout, it is possible that some mutations in human TRα gene are indeed fatal and thus do not exist in any human patients.

12.4.2 TRβ Gene

The TRβ gene knockout has been done to inactivate both TRβ1 and TRβ2 isoforms in mice (Forrest et al., 1996a,b). As expected, TRβ expression was eliminated. No apparent compensating increase in TRα expression was detected. Again, the heterozygous mice are normal. The homozygous TRβ knockout

TABLE 12.4 Phenotypes of Thyroid Hormone Receptor Knockout Mice[a]

Tissue	TRα1	TRα1/α2	TRβ1/β2	TRα/TRβ
Thyroid	Normal	Hypoplastic	Hyperplastic	Hyperplastic
Serum T_4 concentration	Low	Low	Elevated	Highly elevated
TSH growth	Reduced	Reduced	Elevated	Highly elevated
	Normal	Normal up to two weeks postnatal but then severely retarded	Normal	Normal up to two weeks postnatal but then severely retarded
Survival	Normal	Neonatal death at 4–6 weeks	Normal	Neonatal death at 5 weeks
Bone	Normal	Delayed maturation	Normal	Delayed maturation
Brain	Normal	Normal	Normal	
Other[b]	Reduced heart rate, prolonged QT interval, lower body temperature	Delayed maturation of small intestine	Functional cochlear defect	Delay maturation of the small intestine

[a] Based on Forrest et al. (1996a,b); Fraichard et al., 1997; Wikstrom et al., 1998; Hsu and Brent (1988); and Gauthier et al., 1999.
[b] It is possible that the double-knockout mice may have other alterations such as changes in heart rate as observed in single knockout, although they are not reported due to limited analyses.

mice are viable and display normal growth rates and weight gain. They show no gross abnormalities in most organs and are also fertile. Interestingly, the mice exhibit many changes that are associated with human GRTH. These include the presence of a goiter due to a hyperplastic thyroid gland and elevated levels of serum TH (Table 12.4; Forrest et al., 1996a,b; Hsu and Brent, 1998). They also have a permanent hearing deficit across a wide range of frequencies due to a functional cochlear defect. More interestingly, mice expressing a transgene of an RTH-associated TRβ mutant also have elevated TH levels and poor weight gain and growth, similarly to the TRβ knockout (Wong et al., 1997c). Thus TRβ inactivation is similar in terms of mouse development to a dominant negative point mutation in TRβ, in agreement with the fact that GRTH can be due to a loss of TRβ gene or TRβ mutations in human (Refetoff et al., 1993).

Somewhat surprisingly, there are no gross abnormalities of brain development and several TH-dependent cerebellar genes are expressed at normal levels in the TRβ knockout mice (Sandhofer et al., 1996) in spite of the large amount of clinical and in vitro evidence pointing toward a critical role of TH on brain development. On the other hand, more detailed analyses are needed to investigate the possible subtle abnormalities and functional defects of the brain.

12.4.3 Double Knockout

By crossing mice with mutant TRα and TRβ genes, Gauthier et al. (1999) generated mice with inactivated TRα and TRβ genes. In agreement with the phenotypes of mice with individual knockout of the TRα or TRβ gene, the double knockout mice have defects that mainly due to the sum of the two individual gene knockouts (Table 12.4; Gauthier et al., 1999). Thus TRα and TRβ genes have distinct functions during development, in particular TRα (TRα1 and/or TRα2) appears to be more important for the development of the bone and intestine and the survival of the animal while TRβ appears to be more critical for proper regulation of TSH expression. On the other hand, both the TSH and TH levels are much more elevated in the double knockout than in TRβ knockout mice (Table 12.4; Gauthier et al., 1999), suggesting that TRα can also repress TSH expression in the absence of TRβ, although less effectively. Thus, in some tissues, TR functional redundancy does exist.

12.5 CONCLUSIONS

Thyroid hormone is clearly important for development and normal organ function in diverse vertebrate organisms. The absolute dependence of amphibian metamorphosis on TH is perhaps the most dramatic and most extensive regulatory effects of TH on animal development. However, many aspects of human organ function depend on TH as well and abnormal levels of TH cause defects in human development and organ function that bear similarities to anurans. The critical roles of TH in humans are further supported by the discovery of TRβ mutations associated with human RTH diseases. The direct involvement of TRβ mutations in RTH diseases is supported by TRβ knockout studies in mice that reproduce some of the phenotypic defects observed in humans. Furthermore, the TRα and TRβ knockout studies also reveal unexpected findings, for example, the lack of major effects of TRα knockout on mouse development, the heart defects caused by the lack of TRα, and the apparent normal development of the brain in TRβ null mice. Such findings may be due to species-specific functions of TRs and/or the lack of proper assays of organ functions. In addition, they also suggest the possible existence of human abnormalities related to TRs that are yet to be identified. Thus further studies, especially at the molecular level, in different animal model systems will be required to understand fully the various physiological consequences of TH action or mis-action and their underlying molecular mechanisms.

References

Abdallah, B., Sachs, L., and Demeneix, B. A. (1995). Non-viral gene transfer: Applications in developmental biology and gene therapy. *Biol. Cell* 85, 1–7.

Adams, J. M. and Cory, S. (1998). The Bcl-2 Protein Family: Arbiters of cell survival. *Science* 281, 1322–1326.

Adams, M., Matthews, C., Collingwood, T. N., Tone, Y., Beck-Peccoz, P., and Chatterjee, V. K. (1994). Genetic analysis of 29 kindreds with generalized and pituitary resistance to thyroid hormone. *J. Clin. Invest.* 94, 506–515.

Adler, L. (1914). Metamorphostudien an Batrachierlarven. I. Exstirpation endokriner Drisen. A. Exstirpation der Hypophyse. *Wilhelm Roux. Arch. Entwicklungsmech. Organismen* 39, 21–45.

Alcedo, J., Ayzenzon, M., Von Ohlen, T., Noll, M., and Hooper, J. E. (1996). The Drosophila smoothened gene encodes a seven-pass membrane protein, a putative receptor for the hedgehog signal. *Cell* 86, 221–232.

Aleschin, B. (1926). Die aktuelle reaktion des gewebssaftes bei normaler und beschleunigter metamorphose von Rana temporaria. *Biochem. Z.* 171, 79–82.

Alexander, C. M. and Werb, Z. (1991). Extracellular matrix degradation. In *Cell Biology of Extracellular Matrix*, 2nd ed. E. D. Hay, Plenum Press, New York, pp. 255–302.

Alexander, C. M., Howard, E. W., Bissell, M. J., and Werb, Z. (1996). Rescue of mammary epithelial cell apoptosis and entactin degradation by a tissue inhibitor of metalloproteinases-1 transgene. *J. Cell Biol.* 135:1669–1677.

Ali, S., Pelligrini, I., and Kelly, P. A. (1991). A prolactin-dependent immune cell line (Nb2) expresses a mutant form of prolactin receptor. *J. Biol. Chem.* 266, 20110–20117.

Allen, B. M. (1916). The results of extirpation of the anterior lobe of the hypophysis and of the thyroid of Rana pipiens larvae. *Science* 44, 755–758.

Allen, B. M. (1918). The results of thyroid removal in the larvae of Rana pipiens. *J. Exp. Zool.* 24, 499–519.

Allen, B. M. (1929). The influence of the thyroid gland and hypophysis upon growth and development of amphibian larvae. Quart. Rev. Biol. 4, 325–352.

Allfrey, V., Faulkner, R. M. and Mirsky, A. E. (1964). Acetylation and methylation of histones and their possible role in the regulation of RNA synthesis. Proc. Natl. Acad. Sci. 51, 786–794.

Almer, A., Rudolph, H., Hinnen, A., and Horz, W. (1986). Removal of positioned nucleosomes from the yeast PHO5 promoter upon PHO5 induction releases additional activating DNA elements. EMBO J. 5, 2689–2696.

Almouzni, G., Clark, D. J., Mechali, M., and Wolffe, A. P. (1990). Chromatin assembly on replicating DNA in vitro. Nucl. Acids Res. 18, 5767–5774.

Alnemri, E. S., Livingston, D. J., Nicholson, D. W., Salvesen, G., Thornberry, N. A., Wong, W. W., and Yuan, J. (1996). Human ICE/CED-3 protease nomenclature. Cell 87, 171.

Altmann, G. G. and Quaroni, A. (1990). Behavior of fetal intestinal organ culture explanted onto a collagen substratum. Development 110, 5353–5370.

Amano, T. (1998). Isolation of genes involved in intestinal remodeling during anuran metamorphosis. Wound Rep. Reg. 6, 302–313.

Amaya, E., Offield, M. F., and Grainger, R. M. (1998). Frog genetics: Xenopus tropicalis jumps into the future. Trends Genetics 14, 253–255.

Anand, V. T., Mann, S. B. S., Dash, R. J., and Mehra, Y. N. (1989). Auditory investigations in hypothyroidism. Acta Otolaryngol. 108, 83–87.

Andres, A. J. and Thummel, C. S. (1992). Hormones, puffs and flies: the molecular control of metamorphosis by ecdysone. Trends Genet. 8, 132–138.

Andres, A. J. and Thummel, C. S. (1994). Methods for quantitative analysis of transcription in larvae and prepupae. In *Methods in Cell Biology, Drosophila melanogaster: Practical Uses in Cell and Molecular Biology*, Vol. 44 (L. S. B. Goldstein and E. A. Fyrberg, Eds.) Academic Press, New York, pp. 565–573.

Andres, A. J. and Thummel, C. S. (1995). The Drosophila 63F early puff contains E63-1, an ecdysone-inducible gene that encodes a novel Ca^{2+}-binding protein. Development 121, 2667–2679.

Andres, A. J., Fletcher, J. C., Karim, F. D. and Thummel, C. S. (1993). Molecular analysis of the initiation of insect metamorphosis: A comparative study of Drosophila ecdysteroid-regulated transcription. Develop. Biol. 160, 388–404.

Anglard, P., Melot, T., Guerin, E., Thomas, G., and Basset, P. (1995). Structure and promoter characterization of the human stromelysin-3 Gene. J. Biol. Chem. 270, 20337–20344.

Apt, D., Liu, Y., and Bernard, H. U. (1994). Cloning and functional analysis of spliced isoforms of human nuclear factor I-X: interference with transcriptional activation by NFI/CTF in a cell-type specific manner. Nucl. Acids Res. 19, 3825–3833.

Archer, T. K., Cordingley, M. G., Wolford, R. G., and Hager, G. L. (1991). Transcription factor access is mediated by accurately positioned nucleosomes on the mouse mammary tumor virus promoter. Mol. Cell Biol. 11, 688–698.

Archer, T. K., Lefebvre, P., Wolford, R. G., and Hager, G. L. (1992). Transcription factor loading on the MMTV promoter: a bimodal mechanism for promoter activation. Science 255, 1573–1576.

Ashburner, M. (1967). Patterns of puffing activity in the salivary gland chromosomes of

Drosophila. I. Autosomal puffing patterns in a laboratory stock of Drosophila melanogaster. Chromosoma 21, 398–428.

Ashburner, M. (1972). Patterns of puffing activity in the salivary gland chromosomes of Drosophila. VI. Induction by ecdysone in salivary glands of D. Melanogaster cultured in vitro. Chromosoma 38, 255–281.

Ashburner, M., Chihara, C., Meltzer, P., and Richards, G. (1974). Temporal control of puffing activity in polytene chromosome. Cold Spring Harbor Symp. Quant. Biol. 38, 655–662.

Ashizawa, K. and Cheng, S.-Y. (1992). Regulation of thyroid hormone receptor-mediated transcription by a cytosol protein. Proc. Natl. Acad. Sci. USA 89, 9277–9281.

Ashley, H., Katti, P., and Frieden, E. (1968). Urea excretion in the bullfrog tadpole: effects of temperature, metamorphosis, and thyroid hormones. Develop. Biol. 17, 293–307.

Atkinson, B. G. (1981). Biological basis of tissue regression and synthesis. In *Metamorphosis: A Problem in Developmental Biology*, 2nd ed. (L. I. Gilbert and E. Frieden, eds.) Plenum Press, New York, pp 397–444.

Atkinson, B. G. (1994). Metamorphosis: Model systems for studying gene expression in postembryonic development. Develop Genet. 15, 313–319.

Atkinson, B. G. and Just, J. J. (1975). Biochemical and histological changes in the respiratory system of Rana catesbeiana larvae during normal and induced metamorphosis. Develop. Biol. 45, 151–165.

Atkinson, B. G., Helbing, C., and Chen, Y. (1994). Reprogramming of gene expression in the liver of Rana catesbeiana tadpoles during spontaneous and thyroid homrone induced metamorphosis. In Perspectives in Comparative Endocrinology. (D. G. Davey, S. S. Tobe, R. G. Peter, eds.), National Research Council of Canada, Ottawa, pp. 416–423.

Atkinson, B. G., Helbing, C., and Chen, Y. (1996). Reprogramming of genes expressed in amphibian liver during metamorphosis. In *Metamorphosis: Postembryonic Reprogramming of Gene Expression in Amphibian and Insect Cells*. (L. I. Gilbert, J. R. Tata, B. G. Atkinson, eds.), Academic Press, New York, pp. 539–566.

Atkinson, B. G., Workman, A. S., and Chen, Y. (1998). Thyroid hormone induces a reprogramming of gene expression in the liver of premetamorphic Rana catesbeiana tadpoles. Wound Rep. Reg. 6, 323–336.

Aza-Blanc, P., Ramirez-Weber, F.-A., Laget, M.-P., Schwartz, C., and Kornberg, T. B. (1997). Proteolysis that is inhibited by hedgehog targets cubitus interruptus protein to the nucleus and converts it to a repressor. Cell 89, 1043–1053.

Baehrecke, E. H. (1996). Ecdysone signaling cascade and regulation of Drosophila metamorphosis. Arch. Insect Biochem. Physiol. 33, 231–244.

Baehrecke, E. H. and Thummel, C. S. (1995). The Drosophila E93 gene from the 93F early puff displays stage- and tissue-specific regulation by 20-hydroxyecdysone. Develop. Biol. 171, 85–97.

Baker, B. S. and Tata, J. R. (1992). Prolactin prevents the autoinduction of thyroid hormone receptor mRNAs during amphibian metamorphosis. Develop. Biol. 149, 463–467.

Balinsky, J. B., Cragg, M. M., and Baldwin, E. (1961). The adaptation of amphibian waste nitrogen excretion to dehydration. Comp. Biochem. Physiol. 3, 236–244.

Balls, M. and Bownes, M. (1985). *Metamorphosis*. Clarendon Press, Oxford.

Balls, M., Clothier, R. H., Rowles, J. M., Kiteley, N. A., and Bennett, G. W. (1985). TRH distribution, levels, and significance during the development of Xenopus laevis. In *Metamorphosis* (B. Balls and M. Bownes, eds.), Clarendon, Oxford, pp. 260–272.

Baniahmad, A., Kohne, A. C., and Renkawitz, R. (1992). A transferable silencing domain is present in the thyroid hormone receptor, in the v-erbA oncogene product and in the retinoic acid receptor. EMBO J. 11, 1015–1023.

Baniahmad, C., Nawaz, Z., Baniahmad, A., Gleeson, M. A. G., Tsai, M.-J., and O'Malley, B. W. (1995). Enhancement of human estrogen receptor activity by SPT6: A potential coactivator. Mol. Endocrinol. 9, 34–43.

Barrett, J. A., Rawloings, N. D., and Woessner, J. F. (1998). *Handbook of Proteolytic Enzymes*. Academic Press, New York.

Barsano, C. P. and De Groot, L. (1983). Nuclear cytoplasmic interrelationships. In *Molecular basis of thyroid hormone action* (J. Oppenheimer and H. Samuels, eds.), Academic Press, New York, pp. 139–177.

Basset, P., Bellocq, J. P., Wolf, C., Stoll, I., Hutin, P., Limacher, J. M., Podhajcer, O. L., Chenard, M. P., Rio, M. C., and Chambon, P. (1990). A novel metalloproteinase gene specifically expressed in stromal cells of breast carcinomas. Nature 348, 699–704.

Bates, R. C., Buret, A., van Helden, D. F., Horton, M. A., and Burns, G. F. (1994). Apoptosis induced by inhibition of intercellular contact. J. Cell Biol. 125, 403–415.

Bayer, C. A., Holley, B., and Fristrom, J. W. (1996a). A switch in broad-complex zinc-finger isoform expression is regulated posttranscriptionally during the metamorphosis of Drosophila imaginal discs. Develop. Biol. 177, 1–14.

Bayer, C. . Von Koln, L., and Fristrom, I. W. (1996b). Gene regulation in imaginal disc and salivary gland development during Drosophilia metamorphosis. In Metamorphosis: Postembryonic reprogramming of gene expression in amphibian and insect cells. (L. I. Gilbert, J. R. Tata, B. G. Atkinson, eds.). Academic Press, New York, pp 322–362.

Becker, K. B., Schneider, M. J., Davey, J. C., and Galton, V. A. (1995). The type III 5-deiodinase in Rana catesbeiana tadpoles is encoded by a thyroid hormone-responsive gene. Endocrinology 136, 4424–4431.

Becker, K. B., Stephens, K. C., Davey, J. C., Schneider, M. J., and Galton, V. A. (1997). The type 2 and type 3 iodothyronine deiodinases play important roles in coordinating development in Rana catesbeiana tadpoles. Endocrinology 138, 2989–2997.

Behr, M. and Loos U. (1992). A point mutation (Ala 229 to Thr) in the hinge domain of the c-erbAβ thyroid hormone recep- for gene in a family with generalized thyroid hormone resistance. Mol. Endocrinol. 6, 1119–1126.

Benbrook D. and Pfahl, M. (1987). A novel thyroid hormone receptor encoded by a cDNA clone from a human testis library. Science 238, 788–791.

Bennett, T. P. and Frieden, E. (1962). Metamorphosis and biochemical adaptation in amphibia. In *Comparative Biochemistry* (M. Florkin and H. S. Mason, eds.), Vol. 4, Academic Press, New York, pp. 483–556.

Benvenga, S. and Robbins, J. (1993). Lipoprotein-thyroid hormone interactions. Trends Endocrinol. Metab. 4, 194–198.

Benvenga, S. and Robbins, J. (1998). Thyroid hormone efflux from monolayer cultures of human fibroblasts and hepatocytes. Effect of lipoproteins and other thyroxine transport proteins. Endocrinology 139, 4311–4318.

Bergeron, L., Perez, G. I., Macdonald, G., Shi, L., Sun, Y., Jurisicova, A., Varmuza, S., Latham, K. E., Flaws, J. A., Salter, J. C. M., et al. (1998). Defects in regulation of apoptosis in caspase-2-deficient mice. Genes Dev. 12, 1304–1314.

Berman, R., Bern, H. A., Nicoll, C. S., and Strohman, R. C. (1964). Growth promoting effects of mammalian prolactin and growth hormone in tadpoles of Rana catesbeiana. J. Exp. Zool. 156, 353–360.

Berry, D. L., Schwartzman, R. A., and Brown, D. D. (1998a). The expression pattern of thyroid hormone response genes in the tadpole tail identifies multiple resorption programs. Develop. Biol. 203, 12–23.

Berry, D. L., Rose, C. S., Remo, B. F., and Brown, D. D. (1998b). The expression pattern of thyroid hormone response genes in remodeling tadpole tissues defines distinct growth and resorption gene expression programs. Develop. Biol. 203, 24–35.

Berry, M. J., Banu, L., and Larsen, P. R. (1991). Type I iodothyronine deiodinase is a selenocysteine-containing enzyme. Nature 349, 438–440.

Bevins, C. L. and Zasloff, M. (1990). Peptides from frog skin. Ann. Rev. Biochem. 59, 395–414.

Bhanot, P., Brink, M., Samos, C. H., Hsieh, J.-C., Wang, Y., Macke, J. P., Andrew, D., Nathans, J., and Nusse, R. (1996). A new member of the *frizzled* family from *Drosophila* functions as a Wingless receptor. Nature 382, 225–230.

Bhat, M. K., Parkison, C., McPhie, P., Liang, C. M., and Cheng, S. Y. (1993). Conformational changes of human beta 1 thyroid hormone receptor induced by binding of 3, 3′, 5-triiodo-L-thyronine. Biochem. Biophys. Res. Commun. 195, 385–392.

Bignon, C., Binart, N., Ormandy, C., Schuler, L. A., Kelly, P. A., and Djiane, J. (1997). Long and short forms of the ovine prolactin receptor: cDNA cloning and genomic analysis reveal that the two forms arise by different alternative splicing mechanisms in ruminants and in rodents. J. Mol. Endocrinol. 19, 109–120.

Birkedal-Hansen, H., Moore, W. G. I., Bodden, M. K., Windsor, L. J., Birkedal-Hansen, B., DeCarlo, A., and Engler, J. A. (1993). Matrix metalloproteinases: A review. Crit. Rev. Oral Biol. Med. 4 197–250.

Bitgood, M. J. and McMahon, A. P. (1995). Hedgehog and Bmp genes are coexpressed at many diverse sites of cell-cell interaction in the mouse embryo. Develop. Biol. 172, 126–138.

Bjerknes, M. and Cheng, H. (1981). The stem-cell zone of the small intestinal epithelium. IV. Evidence for controls over orientation of boundaries between the stem-cell zone, proliferation zone, and the maturation zone. Am. J. Anat. 160, 105–115.

Blanco, J. C. G., Minucci, S., Lu, J., Yang, X.-J., Walker, K. K., Chen, H., Evans, R. M., Nakatani, Y. and Ozato, K. (1998) The histone acetylase PCAF is a nuclear receptor coactivator. Genes Develop. 12, 1638–1651.

Blondeau, J.-P., Osty, J. and Francon, J. (1988). Characterization of the thyroid hormone transport system of isolated hepatocytes. J. Biol. Chem. 263, 2685–2692.

Blumberg, B., Mangelsdorf, P. J., Dyck, J. A., Bittner, D. A., Evans, R. M., and De Robertis, E. M. (1992). Multiple retinoid-responsive receptors in a single cell:

Families of retinoid "X" receptors and retinoic acid receptors in the Xenopus egg. Proc. Natl. Acad. Sci. USA 89, 2321–2325.

Bodenstein, D. (1950). The postembryonic development of Drosophila. In *Biology of Drosophila* (M. Demerec, ed.), Hafner, New York, pp. 275–367.

Bollenbacher, W. E., Vedeckis, W. V., Gilbert, L. I., and O'Connor, J. D. (1975). Ecdysone titers and prothoracic gland activity during the larval-pupal development of Manduca sexta. Dev. Biol. 44, 46–53.

Bonneville, M. A. (1963) Fine structural changes in the intestinal epithelium of the bullfrog during metamorphosis. J. Cell Biol. 18, 579–597.

Bonneville, M. A. and Weinstock, M. (1970). Brush border development in the intestinal absorptive cells of Xenopus during metamorphosis. J. Cell Biol. 44, 151–171.

Botte, V. and Buonanno, C. (1962). Determinazione quantitativa delle fenilfosfatasi alcalina ed acid adell'intestino di Rana esculenta nel corso della metamorfosi in condizioni normali e sperimentali. Boll. Zool. 29, 471–478.

Boudreau, N., Sympson, C. J., Werb, Z., and Bissell, M. J. (1995). Suppression of ICE and apoptosis in mammary epithelial cells by extracellular matrix. Science 267, 891–893.

Bourguet, W., Ruff, M., Chambon, P., Gronemeyer, H., and Moras, D. (1995). Crystal structure of the ligand-binding domain of the human nuclear receptor RXR-α. Nature 375, 377–382.

Boutin, J. M., Jolicoeur, C., Okamura, H. Gagnon, J., Edery, M., Shirota, M., Banville, D., Dusanter-Fourt, I., Djiane, J., and Kelly, P. A. (1988). Cloning and expression of the rat prolactin receptor, a member of the growth hormone/prolactin receptor gene family. Cell 53, 69–77.

Bouziges, F., Simo, P., and Kedinger, M. (1991). Changes in glycosaminoglycan expression in the rat developing intestine. Cell. Biol. Int. Rep. 152, 97–106.

Brachet, J. (1960). *The Biochemistry of Development*, Pergamon, New York.

Bray, T. and Sicard, R. E. (1982). Correlation among the changes in the levels of thyroid hormones, thyrotropin and thyrotropin-releasing hormone during the development of Xenopus laevis. Exp. Cell Biol. 50, 101–107.

Breathnach, R., Matrisian, L. M., Gesnel, M.-C., Staub, A., and Leroy, P. (1987). Sequences coding for part of oncogene-induced transin are highly conserved in a related rate gene. Nucl. Acids Res. 15, 1139–1151.

Brent, G. A. (1994). The molecular basis of thyroid hormone action. New Engl. J. Med. 331, 847–853.

Brent, G. A., Dunn, M. K., Harney, J. W., Gulick, T., Larsen, P. R., and Moore, D. D. (1993). Thyroid hormone aporeceptor represses T_3-inducible promoters and blocks activity of the retinoid acid receptor. New Biol. 1, 329–336.

Broadus, J., McCabe, J. R., Endrizzi, B., Thummel, C. S., and Woodard, C. T. (1999). The Drosphilia βFTZ-F1 orphan nuclear receptor provides competence for stage-specific responses to the steroid hormone ecdysone Mol. Cell. 3, 143–149.

Brooks, A. R., Sweeney, G., and Old, R. W. (1989). Structure and functional expression of a cloned Xenopus thyroid hormone receptors. Nucl. Acids Res. 17, 9395–9405.

Brooks, P. C., Montgomery, A. M. P., Rosenfeld, M., Reisfeld, R. A., Hu, T., Klier, G., and Cheresh, D. A. (1994). Integrin $a_v\beta_3$ antagonists promote tumor regression by inducing apoptosis of angiogenic blood vessels. Cell 79, 1157–1164.

Brown, D. D. (1997). The role of thyroid hormone in zebrafish and axolotl development. Proc. Natl. Acad. Sci. USA 94, 13011–13016.

Brown, D. D., Wang, Z., Kanamori, A., Eliceiri, B., Furlow, J. D., and Schwartzman, R. (1995). Amphibian metamorphosis: A complex program of gene expression changes controlled by the thyroid hormone. Recent Prog. Hormone Res. 50, 309–315.

Brown, D. D., Wang, Z., Furlow, J. D., Kanamori, A., Schwartzman, R. A., Rmo, B. F., and Pinder, A. (1996). The thyroid hormone-induced tail resorption program during *Xenopus laevis* metamorphosis. Proc. Natl. Acad. Sci. USA 93, 1924–1929.

Brown, G. W., Jr. and Cohen, P. P. (1958). Biosynthesis of urea in metamorphosing tadpoles. In *The Chemical Basis of Development* (W. McElroy and B. Glass, eds.), Johns Hopkins Press, Baltimore, MD, pp. 495–513.

Brown, G. W., Jr., Brown, W. R., and Cohen, P. P. (1959). Comparative biochemistry of urea synthesis. J. Biol. Chem. 234, 1775–1780.

Brown, K. E. and Yamada, K. M. (1995). The role of integrins during vertebrate development. Develop. Biol. 6, 69–77.

Broyles, R. H. (1981). Changes in the blood during amphibian metamorphosis. In *Metamorphosis: A Problem in Developmental Biology*, 2nd ed. (L. I. Gilbert and E. Frieden, eds.) Plenum Press, New York, pp. 461–490.

Brucker-Davis, F., Skarulis, M. C., Grace, M. B., Benichou, J., Hauser, P., Wiggs, E., and Weintraub, B. D. (1995). Genetic and clinical features of 42 kindreds with resistance to thyroid hormone. Ann. Intern. Med. 123, 572–583.

Buckbinder, L. and Brown, D. D. (1992). Thyroid hormone-induced gene expression changes in the developing frog limb. J. Biol. Chem. 267, 25786–25791.

Buckbinder, L. and Brown, D. D. (1993). Expression of the Xenopus laevis prolactin and thyrotropin genes during metamorphosis. Proc. Natl. Acad. Sci. USA 90, 3820–3824.

Bumcrot, D. A. and McMahon, A. P. (1995). Sonic hedgehog, a secreted signalling molecule known to play a role in the patterning of the central nervous system and the limb in vertebrates, also controls differentiation of the somites. Curr. Biol. 5, 612–614.

Burris, T. P., Nawaz, Z., Tsai, M.-J., and O'malley, B. W. (1995). A nuclear hormone receptor-associated protein that inhibits transactivation by the thyroid hormone and retinoic acid receptors. Proc. Natl. Acad. Sci. USA 92, 9525–9529.

Campbell, G. S., Argetsinger, L. S., Ihle, J. N., Kelly, P. A., Rillema, J. A., and Carter-Su, C. (1994). Activation of JAK2 tyrosine kinase by prolactin receptors in Nb_2 cells and mouse mammary gland explants. Proc. Natl. Acad. Sci. USA 91, 5232–5236.

Cao, Z., Umek, R. M., and McKnight, S. L. (1991). Regulated expression of three C/EBP isoforms during adipose conversion of 3T3-LI cells. Genes Develop. 5, 1538–1552.

Capovilla, M., Brandt, M., and Botas, J. (1994). Direct regulation of decapentaglegic by ultrabithorax and its role in Drosophila midgut morphogenesis. Cell 76, 461–475.

Carr, J. A. and Norris, D. O. (1990). Immunohistochemical localization of corticotropin releasing factor- and arginine vasotocin-like immunoreactivities in the brain and pituitary of the American bullfrog (Rana catesbeiana) during development and metamorphosis. Gen. Comp. Endocrinol. 78, 180–188.

Carroll, R. L. (1977). Patterns of amphibian evolution: An extended example of the incompleteness of the fossil record. In *Patterns of Evolution, as Illustrated by the Fossil Record*, Vol. 5 (A. Hallam, ed.), Elsevier, Amsterdam, pp. 405–437.

Carstensen, H., Burgers, A. C. J., and Li, C. H. (1961). Demonstration of aldosterone and corticosterone as the principal steroids formed in incubates of adrenals of the American bullfrog (Rana catesbeiana) and stimulation of their production by mammalian adrenocorticotropin. Gen. and Comp. Endocrinol. 1, 37–50.

Cecconi, F., Alvarez-Bolado, G., Meyer, B. I., Roth, K. A., and Gruss, P. (1998). Apaf1 (CED-4 homolog) regulates programmed cell death in mammalian development. Cell 94, 727–737.

Cereghini, S., Raymondjean, M., Carranca, A. G., Herbomel, P., and Yaniv, M. (1987). Factor involved in control of tissue-specific expression of albumin gene. Cell 50, 627–638.

Chakravarti D., Lamorte V. J., Nelson M. C., Nakajima T., Schulman I. G., Juguilon H., Montminy M., and Evans R. M. (1996). Role of CBP P300 in nuclear receptor signalling. Nature 383, 99–103.

Chatterjee, V. K. K., and Beck-Peccoz, P. (1994). Thyroid hormone resistance. Bailliere's clinical endocrinology and metabolism. 8, 267–283.

Chen, H., Lin, R. J., Schiltz, R. L., Chakravarti, D., Nash, A., Nagy, L., Privalsky, M. L., Nakatani, Y., and Evans, R. M. (1997). Nuclear receptor coactivator ACTR is a novel histone acetyltransferase and forms a multimeric activation complex with P/CAF and CBP/p300. Cell 90, 569–580.

Chen, J. D. and Evans, R. M. (1995). A transcriptional co-repressor that interacts with nuclear hormone receptors. Nature 377, 454–457.

Chen, J. D. and Li, H. (1998). Coactivation and corepression in transcriptional regulation by steroid/nuclear hormone receptors. Crit. Rev. Eukaryotic Gene Express. 8, 169–190.

Chen, Y. and Atkinson, B. G. (1997). Role for the Rana catesbeiana homologue of C/EBPα in the reprogramming of gene expression in the liver of metamorphosing tadpoles. Develop. Genetics 20, 152–162.

Chen, Y., Hu, H., and Atkinson, B. G. (1994). Characterization and expression of C/EBP-like genes in the liver of Rana catesbeiana tadpoles during spontaneous and thyroid hormone-induced metamorphosis. Develop. Gene. 15, 366–377.

Cheng, H. and Leblond, C. P. (1974). Origin, differentiation and renewal of the four main epithelial cell types in the mouse small intestine. I. Columnar cell. Am. J. Anat. 141, 503–520.

Cheng, S.-Y. (1991). In *Thyroid Hormone Metabolism. Regulation and Clinical Implications* (S.-Y. Wu, ed.), Blackwell Scientific Publications, Boston, pp 145–166.

Cheng, S.-Y., Gong, Q-H., Parkison, C., Robinson, E. A., Appella, E., Merlino, G. T., and Pastan, I. (1987). The nucleotide sequence of a human cellular thyroid hormone binding protein present in endoplasmic reticulum. J. Biol. Chem. 262, 11221–11227.

Cherbas, P., and Cherbas, L. (1996). Molecular aspects of ecdysterolol hormone action. In Metamorphosis postembryonic reprogramming of gene expression in amphibian and insect cells. (L. I. Gilbert, J. R. Tata, and B. G. Atkinson, eds.) Academic Press, N.Y. pp. 175–221.

Chin, Y. E., Kitagawa, M. Kuida, K., Flavell, R. A., and Fu, X.-Y. (1997). Activation of the STAT signaling pathway can cause expression of caspase 1 and apoptosis. Mol. Cell Biol. 17, 5328–5337.

Chinnaiyan, A. M., O'Rourke, K., Lane, B. R., and Dixit, V. M. (1997) Interaction of CED-4 with CED-3 and CED-9: A molecular framework for cell death. Science 275, 1122–1126.

Clarke, D. L. and Linzer, D. I. H. (1993). Changes in prolactin receptor expression during pregnancy in the mouse ovary. Endocrinol. 133, 224–232.

Clemons, G. K. and Nicoll, C. S. (1997a). Development and preliminary application of a homologous radioimmunoassay for bullfrog prolactin. Gen. Comp. Endocrinol. 32, 531–535.

Clemons, G. K. and Nicoll, C. S. (1977b). Effects of antisera to bullfrog prolactin and growth hormone on metamorphosis of Rana catesbeiana tadpoles. Gen. Comp. Endocrinol. 31, 495–497.

Clifton-Bligh, R. J., de Zegher, F., Wagner, R. L., Collingwood, T. N., Francois, I., Van Helvoirt, M., Fletterick, R. J., and Chatterjee, V. K. K. (1998). A novel TRβ mutation (R383H) in resistance tothyroid hormone syndrome predominantly impairs corepressor release and negative transcriptional regulation. Mol. Endocrinol. 12, 609–621.

Cohen, P. P. (1970). Biochemical differentiation during amphibian metamorphosis. Science 168, 533–543.

Coleman, J. R. (1962). Deoxyribonuclease activity in the development of the leopard frog Rana pipiens. Develop. Biol. 5, 232–251.

Coleman, J. R. (1963). Acid deoxyribonuclease activity in amphibian metamorphosis. Biochim. Biophys. Acta 68, 141–143.

Collier, I. E., Smith, J., Kronberger, A., Bauer, E. A., Wihelm, S. M., Eisen, A. Z., and Goldberg, G. I. (1988). The structure of the human skin fibroblast collagenase gene. J. Biol. Chem. 263, 10711–10713.

Collier, I. E., Bruns, G. A. P., Goldberg, G. I., and Grehard, D. S. (1991). On the structure and chromosome location of the 72- and 92-kDa human type IV collagenase genes. Genomics 9, 429–434.

Collingwood, T. N., Adams, M., Tone, Y., and Chatterjee, V. K. K. (1994). Spectrum of transcriptional dimerization, and dominant negative properties of twenty different mutant thyroid hormone β-receptors in thyroid hormone resistance syndrome. Mol. Endocrinol. 8, 1262–1277.

Collingwood, T. N., Rajanayagam, O., Adams, M., Wagner, R., Cavailles, V., Kalkhoven, E., Matthews, C., Nystrom, E., Stenlof, K., Lindstedt, G., et al. (1997). A natural transactivation mutation in the thyroid hormone β receptor: Impaired interaction with putative transcriptional mediators. Proc. Natl. Acad. Sci. USA 94, 248–253.

Collingwood, T. N., Wagner, R., Matthews, C. H., Clifton-Bligh, R. J., Gurnell, M., Rajanayagam, Ol., Agostini, M., Fletterick, R. J., Beck-Peccoz, P., Reinhardt, W., et al. (1998). A role for helix 3 of the TRβ ligand-binding domain in coativator recruitment identified by characterization of a third cluster of mutations in resistance to thyroid hormone. EMBO J. 17, 4760–4770.

Coté, J., Quinn, J., Workman, J. L., and Peterson, C. L. (1994). Stimulation of GAL4 derivative binding to nucleosomal DNA by the yeast SWI/SNF complex. Science 265, 53–60.

Cowell, I. G. and Hurst, H. C. (1994) Transcriptional repression by the human bZIP factor E4BP4: definition of a minimal repression domain. Nucl. Acids Res. 22, 59–65.

Cowell, I. G., Skinner, A., and Hurst, H. C. (1992). Transcriptional repression by a novel member of the bZIP family of transcription factors. Mol. Cell. Biol. 12, 3070–3077.

Cruz-Reyes, J., and Tata, J. R. (1995). Cloning, characterization and expression of two Xenopus bcl-2-like cell-survival genes. Gene, 158, 171–179.

Cryns, V. and Yuan, J. (1998). Proteases to die for. Genes Develop. 12, 1551–1570.

Damm, K., Thompson, C. C., and Evans, R. M. (1989). Protein encoded by v-erbA functions as athyroid-hormone receptor antagonist. Nature 339, 593–597.

Damsky, C. H. and Werb, Z. (1992). Signal transduction by integrin receptors for extracellular matrix: cooperative processing of extracellular information. Curr. Biol. 4, 772–781.

Darimont, B. D., Wagner, R. L., Apriletti, J. W., Stallcup, M. R., Kushner, P. J., Baxter, J. D., Fletterick, R. J., and Yamamoto, K. R. (1998). Structure and specificity of nuclear receptor–coactivator interactions. Genes Develop. 12, 3343–3356.

Darling, D. S., Beebe, J. S., Burnside, J., Winslow, E. R., and Chin, W. W. (1991). 3, 5, 3′-triiodothyronine (T_3) receptor-auxiliary protein (TRAP) binds DNA and forms heterodimers with the T_3 receptor. Mol. Endocrinol. 5, 73–84.

Darnell Jr., J. E. (1997). STATs and gene regulation. Science 277, 1630–1635.

Darnell Jr., J. E., Kerr, I. M., and Stark, G. R. (1994). Jak-STAT pathways and transcriptional activation in response to IFNs and other extracellular signaling proteins. Science 264, 1415–1421.

Dauca, M. and Hourdry, J. (1985). Transformations in the intestinal epithelium during anurametamorphosis. In *Metamorphosis* (M. Balls and M. Bownes, eds), Clarendon Press, Oxford, UK, pp. 36–58.

Dauca, M., Hourdry, J., Hugon, J. S., and Menard, D. (1980). Amphibian intestinal brush border enzymes during thyroxine-induced metamorphosis: A biochemical and cytochemical study. Histochemistry 70, 33–42.

Dauca, M., Hourdry, J., Hugon, J. S., and Menard, D. (1981). Amphibian intestinal brush border membranes. III. Comparison during metamorphosis of the protein, glycoprotein and enzyme patterns after gel electrophoretic separation of SDS-solubilized membranes. Comp. Biochem. Physiol. 69B, 15–22.

Dauca, M., Bouziges, F., Colin, S., Kedinger, M., Keller, J.-M., Schilt, J., Simon-Assmann, P., and Haffen, K. (1990). Development of the vertebrate small intestine and mechanisms of cell differentiation. Int. J. Dev. Biol. 34, 205–218.

Davey, J. C., Schneidr, M. J., and Galton, V. A. (1994). Cloning of a thyroid hormone-responsive Rana catesbeiana c-erbA-β gene. Develop. Gene. 15, 339–346.

Davey, J. C., Becker, K. B., Schneider, M. J., St. Germain, D. L., and Galton, V. A. (1995). Cloning of a cDNA for the type II iodothyronine deiodinase. J. Biol. Chem. 270, 26786–26789.

D'Avino, P. P. and Thummel, C. S. (1998). Crooked legs encodes a family of zinc finger proteins required for leg morphogenesis and ecdysone-regulated gene expression during Drosophila metamorphosis. Development 125, 1733–1745.

D'Avino, P. P. and Thummel, C. S. (1998). Crooked legs encodes a family of zinc finger proteins required for leg morphogenesis and ecdysone-regulated gene expression during Drosophila metamorphosis. Development 125, 1733–1745.

David, M., Petricoin II, E. F. I., Igarashi, K. I., Feldman, G. M., Finbloom, D. S., and Larner, A. C. (1994). Prolactin activates the interferon-regulated p91 transcription factor and the Jak2 kinase by tyrosine phosphorylation. Proc. Natl. Acad. Sci. USA 91, 7174–7178.

Davis, P. J. and Davis, F. B. (1996). Nongenomic actions of thyroid hormone. Thyroid 6, 497–504.

Dawid, I. B. (1991). Mesoderm induction. Methods Cell Biol. 36, 311–328.

Debruyne, F., Vanderschueren-Lodeweyckx, M., and Bastijns, P. (1983). Hearing in congenital hypothyroidism. Audiology 22, 404–409.

DeGroot, L. J., Larsen, P. R., Refetoff, S., and Stanbury, J. B. (1984). Hypothyroidism in infants and children, and developmental abnormalities of the thyroid gland. In *The Thyroid and Its Diseases*, Wiley Medical, New York, pp. 610–641.

DeLong, G. R., Stanbury, J. B., and Ficro-Benitz, R. (1985). Neurological signs in congenital iodine-deficiency disorder (endemic cretinism). Develop. Med. Child Neurol. 27, 317–324.

De Luze, A., Sachs, L., and Demeneix, B. (1993). Thyroid hormone-dependent transcriptional regulation of exogenous genes transferred into Xenopus tadpole muscle in vivo. Proc. Natl. Acad. Sci. USA 90, 7322–7326.

Dent, J. N. (1968). Survey of amphibian metamorphosis. In *Metamorphosis: A Problem in Developmental Biology* (W. Etkin, and L. I. Gilbert, eds.), Appleton-Crofts, New York, pp. 271–311.

Denver, R. J. (1988). Several hypothalamic peptides stimulate in vitro thyrotropin secretion by pituitaries of anuran amphibians. Gen. Comp. Endocrinol. 72, 383–393.

Denver, R. J. (1993). Acceleration of anuran amphibian metamorphosis by corticotropin releasing hormone-like peptides. Gen. Comp. Endocrinol. 91, 38–51.

Denver, R. J. (1996). Neuroendocrine control of amphibian metamorphosis. In *Metamorphosis: Postembryonic Reprogramming of Gene Expression in Amphibian and Insect Cells.* (I. L. Gilbert, J. R. Tata, and B. G. Atkinson, eds.), Academic Press, New York, pp. 433–464.

Denver, R. J. (1997a). Proximate mechanisms of phenotypic plasticity in amphibian metamorphosis. Am. Zool. 37, 172–184.

Denver, R. J. (1997b). Environmental stress as a developmental cue: Corticotropin-releasing hormone is a proximate mediator of adaptive phenotypic plasticity in amphibian metamorphosis. Hormones Behav. 31, 169–179.

Denver, R. J. (1998). Hormonal correlates of environmentally induced metamorphosis in the western spadefoot toad scaphiopus hammondii. Gen. and Comp. Endocrinol. 110, 326-336

Denver, R. J. and Licht, P. (1989). Neuropeptide stimulation of thyrotropin secretion in the larval bullfrog: Evidence for a common neuroregulator of thyroid and interrenal activity during metamorphosis. J. Exp. Zool. 252, 101–104.

Denver, R. J., Pavgi, S., and Shi, Y.-B. (1997). Thyroid hormone-dependent gene expression program for Xenopus neural development. J. Biol. Chem. 272, 8179–8188.

Denver, R. J., Mirhadi, N., and Phillips, M. (1998). Adaptive plasticity in amphibian-metamorphosis: Response of Scaphiopus hammondii tadpoles to habitat desiccation. Ecology 79, 1859–1872.

Dhanarajan, Z. C. and Atkinson, B. G. (1981). Thyroid-hormone-induced differentiation and development of anuran tadpole hind limbs: Detection and quantitation of M-line protein and α-actinin synthesis. Develop. Biol. 82, 317–328.

Dhanarajan, Z. C., Merrifield, P. A., and Atkinson, B. G. (1988). Thyroid hormone induces synthesis and accumulation of tropomyosin and myosin heavy chain in limb buds of premetamorphic tadpoles. Biochem. Cell Biol. 66, 724–734.

Dillman, W. H. (1996). Thyroid hormone and the heart: basic mechanistic and clinical issues. Thyroid 9, 1–11.

Dmytrenko, G. M. and Kirby, G. S. (1981). Mechanics of tail resorption in triiodothyronine-induced metamorphosing tadpoles. J. Exper. Zool. 215, 179–182.

Dodd, M. H. I. and Dodd, J. M. (1976). The biology of metamorphosis. In *Physiology of the Amphibia* (B. Lofts ed.), Academic Press, New York, pp. 467–599.

Doerr-Schott, J. (1980). Immunohistochemistry of the adenohypophysis of non-mammalian vertebrates. Acta Histochem. Suppl. 22, 185–223.

Driscoll M (1992). Molecular genetics of cell death in the nematocle Caenorhabditis elegans. J. Neurobiol. 23, 1327–1351.

Du Pasquier, L. and Flajnik, M. F. (1990). Expression of MHC class II antigens during Xenopus development. Dev. Immunol. 1, 85–95.

Duckett, C. S., Nava, V. E., Gedrich, R. W., Clem, R. J., Dongen, J. L. V., Gilfillan, M. C., Shielss, H., Hardwich, J. M., and Thompson, C. B. (1996). A conserved family of cellular genes related to the baculovirus iap gene and encoding apoptosis inhibitors. EMBO J. 15, 2685–2694.

Duellman, W. E. and Trueb, L. (1986). *Biology of Amphibians*. McGraw Hill, New York.

Dusanter-Fourt, I., Muller, O., Ziemiecki, A., Mayeux, P., Drucker, B., Djiane, J., Wilks, A., Harpur, A. G., Fischer, S., and Gisselbrecht, S. (1994). Identification of JAK protein tyrosine kinases assignaling molecules for prolactin. Functional analysis of prolactin receptor and prolactin-erythropoietin receptor chimera expressed in lymphoid cells. EMBO J. 13, 2583–2591.

Dussault, J. H. and Ruel, J. (1987). Thyroid hormones and brain development. Ann. Rev. Physiol. 49, 321–34.

Duvall, E., Wyllie, A. H., and Morris, R. G. (1985). Macrophage recognition of cells undergoing programmed cell death (apoptosis). Immunology 56, 351–358.

Eddy, L. and Lipner, H. (1975). Acceleration of thyroxine-induced metamorphosis by prolactin antiserum. Gen. Comp. Endocrinol. 25, 462–466.

Edery, M., Jolicoeur, C., Levimeyrueis, C., Dusanter-Fourt, I., Petridou, B., Boutin, J.-M., Lesueur, L., Kelly, P. A., and Djiane, J. (1989). Identification and sequence analysis of a second form of prolactin receptor by molecular cloning of complementary DNA from rabbit mammary gland. Proc. Natl. Acad. Sci. USA 86, 2112–2116.

Eeckhout, Y. (1969). Etude biochmimique de la metamorphose caudale des amphibiens anoures. Acad. Roy. Bel. Cl. Sci. Mem. Collect, 38, 1–113.

Ekker, S. C., McGrew, L. L., Lai, C.-J., Lee, J. J., von Kessler, D. P., Moon, R. T., and Beachy, P. A. (1995). Distinct expression and shared activities of members of the hedgehog gene family of Xenopus laevis. *Development* 121, 2337–2347.

Eliceiri, B. P. and Brown, D. D. (1994). Quantitation of endogenous thyroid hormone receptors α and β during embryogenesis and metamorphosis in Xenopus laevis. J. Biol. Chem. 269, 24459–24465.

Ellis, H. M. and Horvitz, H. R. (1986). Genetic control of programmed cell death in the nematode C. elegans. Cell 44, 817–829.

Enari, M., Sakahira, H. J., Yokoyama, H., Okawa, K., Iwamatsu, A., and Nagata, S. (1998). Acaspase-activated DNase that degrades DNA during apoptosis, and its inhibitor ICAD. Nature 391, 43–50.

Etkin, W. (1964). Metamorphosis. In *Physiology of the Amphibia* (J. A. Moore, ed.), Academic Press, New York, pp. 427–468.

Etkin, W. (1968). Hormonal control of amphibian metamorphosis. In *Metamorphosis, a Problem in Developmental Biology* (W. Etkin, and L. I. Gilbert, eds.). Appleton, New York, pp. 313–348.

Etkin, W. and Gilbert, L. I. (1968) *Metamorphosis: A Problem in Developmental Biology*, Appleton, New York

Etkin, W. and Lehrer, R. (1960). Excess growth in tadpoles after transplantation of the adenohypophysis. Endocrinology (Baltimore) 67, 457–466.

Evans, R. M. (1988). The steroid and thyroid hormone receptor superfamily. Science 240, 889–895.

Fairclough L. and Tata J. R. (1997). An immunocytochemical analysis of the expression of thyroid hormone receptor α and β proteins during natural and thyroid hormone-induced metamorphosis in Xenopus. Develop. Growth Differ. 39, 273–283.

Feng, W., Ribeiro, R. C. J., Wagner, R. L., Nguyen, H., Apriletti, J. W., Fletterick, R. J., Baxter, J. D., Kushner, P. J., and West, B. L. (1998). Hormone-dependent coactivator binding to a hydrophobic cleft on nuclear receptors. Science 280, 1747–1749.

Fini, M. E., Plucinska, I. M., Mayer, A. S., Gross, R. H., and Brinckerhoff, C. E. (1987). A Gene of Rabbit Synovial Cell Collagenase: Member of a Family of Metalloproteinases that Degrade the Connective Tissue Matrix. Biochemistry 26, 6156–6165.

Fisk, G. J. and Thummel, C. S. (1998). The DHR78 nuclear receptor is required for ecdysteroid signaling during the onset of Drosophila metamorphosis. Cell 93, 543–555.

Flajnik, M. F., Hsu, E., Kaufman, J. F., and Du Pasquier, L. (1987). Changes in the immune system during metamorphosis of Xenopus. Immunol. Today 8, 58–64.

Fleischhauer, K., Park, J. H., DiSanto, J. P., Marks, M., Ozato, K., and Yang, S. Y. (1992). Isolation of a full-length cDNA clone encoding a N-terminally variant form of the human retinoid X receptor beta. Nucl. Acids Res. 20, 1801.

Fletcher, J. C. and Thummel, C. S. (1995). The ecdysone-inducible broad-complex and E74 early genes interact to regulate target gene transcription and Drosophila metamorphosis. Genetics 141, 1025–1035.

Fondell, J. D., Roy, A. L., and Roeder, R. G. (1993). Unliganded thyroid hormone receptor inhibits formation of a functional preinitiation complex: implications for active repression. Genes Develop. 7, 1400–1410.

Fondell, J. D., Ge, H., and Roeder, R. G. (1996). Ligand induction of a transcriptionally active thyroid hormone receptor coactivator complex. Proc. Natl. Acad. Sci. USA 93, 8329–8333.

Ford, L. S., and Cannatella (1993). The major clades of frogs. Herpetological monographs 7, 94–117.

Forman, B. M. and Samuels, H. H. (1990). Interaction among a subfamily of nuclear-hormone receptors: the regulatory zipper model. Mol. Endocrinol. 4, 1293–1626.

Forman, B. M., Yang, C. R., Au, M. Casanova, J., Ghysdael, J., and Samuels, H. H. (1989). A domain containing leucine-zipper-like motifs mediate novel in vivo interactions between the thyroid hormone and retinoic acid receptors. Mol. Endocrinol. 3, 1610–1626.

Forrest, D., Erway, L. C., Ng, L., Altschuler, R., and Curran, T. (1996a). Thyroid hormone receptor beta is essential for development of auditory function. Nature Genet. 13, 354–357.

Forrest, D., Hanebuth, E., Smeyne, R. J., Everds, N., Stewart, C. L., Wehner, J. M., and Curran, T. (1996b). Recessive resistance to thyroid hormone in mice lacking thyroid hormone receptor beta: evidence for tissue-specific modulation of receptor function. EMBO J. 15, 3006–3015.

Fox, H. (1983). *Amphibian morphogenesis.* Humana Press, Clifton, N.J.

Fraichard, A., Chassande, O., Plateroti, M., Roux, J. P., Trouillas, J., Dehay, C., Legrand, C., Gauthier, K., Kedinger, M., Malaval, L., et al. (1997). The T3Rα gene encoding a thyroid hormone receptor is essential for post-natal development and thyroid hormone production. EMBO J. 16, 4412–4420.

Franklyn, J. A. and Gammage, M. D. (1996). Thyroid disease: Effects on cardiovascular function. Trends Endocrinol. Metab. 7, 50–54.

Freake, H. C. and Oppenheimer, J. H. (1995). Thermogenesis and thyroid function. Ann. Rev. Nutr. 15, 263–291.

Frieden, E. (1961). Biochemical adaptation and anuran metamorphosis. Am. Zool. 1, 115–149.

Frieden, E. (1968). Biochemistry of amphibian metamorphosis. In *Metamorphosis: A Problem in Developmental Biology* (W. Etkin and L. I. Gilbert, eds.), Appleton, New York, pp. 349–398.

Frieden, E. and Just, J. J. (1970). Hormonal responses in amphibian metamorphosis. In *Action of Hormones on Molecular Processes* (G. Litwack, ed.), Vol. 1, Academic Press, New York, pp. 1–53.

Frieden, E. and Naile, B. (1955). Biochemistry of amphibian metamorphosis: I. Enhancement of induced metamorphosis by glucocorticoids. Science 121, 37–39.

Frieden, E. Wahlborg, A. and Howard, E. (1965). Temperature control of the response of tadpoles to triiodthyronine. Nature 205, 1173–1176.

Friedman, A. D., Landschulz, W. H., and McKnight, S. L. (1989). CCAAT/enhancer binding protein activates the promoter of the serum albumin gene in cultured hepatoma cells. Genes Develop. 3, 1314–1322.

Frisch, S. M. and Ruoslahti, E. (1997). Integrins and anoikis. Curr. Opinion Cell Biol. 9, 701–706.

Fristrom, D. and Fristrom, J. W. (1993). The metamorphic development of the adult epidermis. In *The Development of Drosophila Melanogaster* (M. Bate, and A. M. Arias, eds.), Cold Spring Harbor Laboratory Press, Cold Spring Harbor, pp. 843–897.

Fristrom, J. W. (1981). Drosophila imaginal discs as a model system for the study of metamorphosis. In *Metamorphosis: A Problem in Developmental Biology*, 2nd ed. (L. I. Gilbert, and E. Frieden, eds.), Plenum Press, New York, pp. 217–240.

Fristrom, J. W., Logan, W. R. and Murphy, C. (1973). The synthetic and minimal culture requirements for evagination of imaginal discs of Drosophila melanogaster in vitro. Develop. Biol. 33, 441–456.

Fristrom, J. W., Fristrom, D., Fekete, E., and Kuniyuki, A. H. (1977). The mechanism of evagination of imaginal discs of Drosophila melanogaster. Am. Zool. 17, 671–684.

Fryer, C. J. and Archer, T. K. (1998). Chromatin remodelling by the glucocorticoid receptor requires the BRG1 complex. Nature 393, 88–91.

Fu, X.-Y. (1995). Direct signal transduction by tyrosine phosphorylation of transcription factors with SH2 domains. In *Inducible Gene Expression*, (P. Bauerle, ed.), Vol. 2, Birkhauser, Boston, pp. 99–130.

Fu, X.-Y. (1992). A transcription factor with SH2 and SH3 domains is directly activated by aninterferon α-induced cytoplasmic protein tyrosine kinase(s). Cell 70, 323–335.

Fu, Y., Wang, Y., and Evans, S. M. (1998). Viral sequences enable efficient and tissue-specific expression of transgenes in Xenopus. Nature Biotechnology 16, 253–257.

Furlow, J. D., Berry, D. L., Wang, Z., and Brown, D. D. (1997). A set of novel tadpole specific genes expressed only in the epidermis are down-regulated by thyroid hormone during Xenopus laevis metamorphosis. Develop, Biol. 182, 284–298.

Gaire, M., Magbanua, Z., McDonnell, S., McNeil, L., Louvett, D. H., and Matrisian, L. M. (1994). Structure and expression of the human gene for the matrix metalloproteinase matrilysin. J. Biol. Chem. 269, 2032–2040.

Galton, V. A. (1983). Thyroid hormone action in amphibian metamorphosis. In (J. H. Oppenheimer, and H. H. Samuels, eds.), *Molecular Basis of Thyroid Hormone Action*. Academic Press, New York, pp. 445–483.

Galton, V. (1990). Mechanisms underlying the acceleration of thyroid hormone-induced tadpole metamorphosis by corticosterone. Endocrinology (Baltimore) 127, 2997–3002.

Galton, V. A., Morganelli, C. M., Schneider, M. J., and Yee, K. (1991) The role of thyroid hormone in the regulation of hepatic carbamyl phosphate synthetase activity in Rana catesbeiana. Endocrinology 129, 2298–2304.

Gancedo, B., Corpas, I., Alonso-Gomez, A. L., Delgado, M. J., Morreale de Escobar, G., and Alonso-Bedate, M. (1992). Corticotropin-releasing factor stimulates metamorphosis and increases thyroid hormone concentration in prometamorphic Rana perezi larvae. Gen. Comp. Endocrinol. 87, 6–13.

Gao, X., Kalkhoven, E. Peterson-Maduro, J., van der Burg, B., and Destree, O. H. (1994). Expression of the glucocorticoid receptor gene is regulated during early embryogenesis of Xenopus laevis. Biochim. Biophys. Acta 1218, 194–198.

Gauthier, K., Chassande, O., Plateroti, M. et al. (1999). Different functions for the thyroid hormone receptors TRα and TRβ in the control of thyroid hormone production and post-natal development. EMBO J. 18, 623–631.

Gavrieli, Y., Sherman, Y., and Ben-Sasson, S. A. (1992). Identification of programmed cell death in situ via specific labeling of nuclear DNA fragmentation. J. Cell Biol. 119, 493–501.

Geetha-Habib, M., Noiva, R., Kaplan, H. A., and Lennarz, W. J. (1988). Glycosylation site binding protein, a component of oligosaccharyl transferase, is highly similar to three other 57 kd luminal proteins of the ER. Cell 54, 1053–1060.

Giguere, V., Hollenberg, S. M., Rosenfeld, M. G., and Evans, R. M. (1986). Functional domains of human glucocorticoid receptor. Cell 46, 645–652.

Gilbert, L. I. and Frieden, E. (1981). *Metamorphosis: A Problem in Developmental Biology*, 2nd ed., Plenum Press, New York.

Gilbert, L. I. and Goodman, W. (1981). Chemistry, metabolism, and transport of hormones controlling insect metamorphosis. In *Metamorphosis: A Problem in Developmental Biology*, 2nd ed. (L. I. Gilbert, and E. Frieden, eds.), Plenum Press, New York, pp. 139–175.

Gilbert, L. I., Tata, J. R., Atkinson, B. G. (1996). Metamorphosis: Postembryonic reprogramming of gene expression in amphibian and insect cells. Academic Press, New York.

Gilmour, K. C. and Reich, N. C. (1994). Receptor to nucleus signaling by prolactin and interleukin 2 via activation of latent DNA-binding factors. Proc. Natl. Acad. Sci. USA 91, 6850–6854.

Glass, G. B. J. (1968). *Introduction to Gastrointestinal Physiology*. Prentice-Hall, Englewood Cliffs, N. J.

Glass, C. K., Lipkin, S. M., Devary, O. V., and Rosenfeld, M. G. (1989). Positive and negative regulation of gene transcription by a retinoic acid-thyroid hormone receptor heterodimer. Cell 59, 697–708.

Godowski, P. J., Picard, D., and Yamamoto, K. R. (1988). Signal transduction and transcriptional regulation by glucocorticoid receptor-LexA fusion proteins. Science 241, 812–816.

Gona, A. G. (1969). Light and electronmicroscopic study on thyroxine-induced in vitro resorption of the tadpole fin. Z. Zellforsch. Mikrosk. Anat. 95, 483–494.

Gona, A. G., Uray, N. J., and Hauser, K. F. (1988). Neurogenesis of the frog cerebellum. In *Developmental Neurobiology of the Frog* (E. D. Pollack, and H. D. Bibb, eds.), Alan R. Liss, New York, pp. 255–276.

Gonzalez, G. C. and Lederis, K. (1988). Sauvagine-like and corticotropin-releasing factor-like immunoreactivity in the brain of the bullfrog (Rana catesbeiana). Cell Tissue Res. 253, 29–37.

Gosner, K. L. (1960). A simplified table for staging anuran embryos and larvae with notes on identification. Herpetology 16, 183–190.

Gouilleux, F., Wakao, H., Mundt, M., and Groner, B. (1994). Prolactin induces phosphorylation of Tyr694 of Stat5 (MGF), a prerequisite for DNA binding and induction of transcription. EMBO J. 13, 4361–4369.

Gould, A. P., Brookman, J. J., Strutt, D. I., and White, R. A. H. (1990). Targets of homeotic gene control in Drosophila. Nature 348, 308–312.

Gracia-Navarro, F., Lamacz, M., Tonon, M.-C., and Vaudry, H. (1992). Pituitary adenylate cyclase-activating polypeptide stimulates calcium mobilization in amphibian pituitary cells. Endocrinology (Baltimore) 131, 1069–1074.

Granger, N. A. and Bollenbacher, W. E. (1981). Hormonal control of insect metamorphosis. In *Metamorphosis: A Problem in Developmental Biology*, 2nd ed. (L. I. Gilbert, and E. Frieden, eds.), Plenum Press, New York, pp. 105–137.

Graupner, G., Wills, K. N., Tzukerman, M., Zhang, X. K., and Pfahl, M. (1989). Dual regulatory role for thyroid-hormone receptors allows control of retinoic-acid receptor activity. Nature 340, 653–656.

Gray, K. M. and Janssens, P. A. (1990). Gonadal hormones inhibit the induction of metamorphosis by thyroid hormones in Xenopus laevis tadpoles in vivo, but not in vitro. Gen. Comp. Endocrinol. 77, 202–211.

Green, D. R. and Reed, J. C. (1998). Mitochondria and apoptosis. Science 281, 1309–1312.

Green, S. and Chambon, P. (1988). Nuclear receptors enhace our understanding of transcription regulation. Trends Genet. 4, 309–313.

Gronostajski, R. M., Knox, J., Berry, D., and Miyamoto, N. G. (1988). Stimulation of transcription in vitro by binding sites for nuclear factor I. Nucl. Acids Res. 5, 2087–1098.

Gross, J. (1964). Studies on the biology of connective tissues: remodelling of collagen in metamorphosis. Medicine 43, 291–303.

Gross, J. (1966). How tadpoles lose their tails. J. Invest. Dermatol. 47, 274–277.

Gross, J. and Lapiere, C. M. (1962). Collagenolytic activity in amphibian tissues: A tissue culture assay. Proc. Natl. Acad. Sci. USA 48, 1014–1022.

Grunstein, M. (1997). Histone acetylation in chromatin structure and transcription. Nature 389, 349–352.

Gu, W. and Roeder, R. G. (1997). Activation of p53 sequence specific DNA binding by acetylation of the p53 C-terminal domain. Cell 90, 595–606.

Gudernatsch, J. F. (1912). Feeding experiments on tadpoles. I. The influence of specific organs given as food on growth and differentiation: a contribution to the knowledge of organs with internal secretion. Arch. Entwicklungsmech. Org. 35, 457–483.

Guernsey, D. L. and Edelman, I. S. (1983). Regulation of thermogenesis by thyroid hormones. In *Molecular Basis of Thyroid Hormone Action* (J. Oppenheimer and H. Samuels, eds.), Academic Press, New York, pp. 293–324.

Guiochon-Mantel, A., Loosfelt, H., Lescop, P., Sar, S., Atger, M., Perrot-Applanat, M., and Milgrom, E. (1989) Mechanism of nuclear localization of the progesterone receptor: evidence for interaction between monomers. Cell 57, 1147–1154.

Gurdon, J. B. (1991). Nuclear transplantation in Xenopus. Methods Cell Biol. 36, 299–310.

Haffen, K., Lacroix, B., Kedinger, M., and Simon-Assmann, P. M. (1983). Inductive properties of fibroblastic cell cultures derived from rat intestinal mucosa on epithelial differentiation. Differentiation 23, 226–233.

Hager, G. L., Archer, T. K., Fragoso, G., Bresnick, E. H., Tsukagoshi, Y., John, S., and Smith, C. L. (1993). Influence of chromatin structure on the binding of transcription factors to DNA. Cold Spring Harbor Symp. Quant. Biol. 58, 63–72.

Hagiwara, M., Mamiya, S., and Hidaka, H. (1989). Selective binding of L-thyroxine by myosin light chain kinase. J. Biol. Chem. 264, 40–44.

Hakem, R., Hakem, A., Duncan, G. S., Henderson, J. T., Woo, M., Soengas, M. S., Elia, A., de laPompa, J. L., Kagi, D., Khoo, W., Potter, J., Yoshida, R., Kaufman, S. A., Lowe, S. W., Penninger, J. M., and Mak, T. W. (1998). Differential requirement for caspase 9 in apoptotic pathways in vivo. Cell 94, 339–352.

Halachmi, S., Marden, E., Martin, G., Mackay, H., Abbondanza, C., and Brown, M.

(1994). Estrogen receptor-associated proteins: possible mediators of hormone-induced transcription. Science 264, 1455–1458.

Han, Y., Watling, D., Rogers, N. C., and Stark, G. R. (1997). JAK2 and STAT5, but not JAK1 and STAT1, are required for prolactin-induced β-lactoglobulin transcription. Mol. Endocrinol 11, 1180–1188.

Hauser, P. (1994). The thyroid receptor Beta gene and resistance to thyroid hormone: Implications for behavioral and brain research. Psychoneuroendocrinol. 19, iii–vii.

Hay, E. D. (1991). *Cell Biology of Extracellular Matrix.* 2nd ed. Plenum Press, New York.

Hay, R. T. (1985). Origin of adenovirus DNA replication. Role of the nuclear factor I binding site in vivo. J. Mol. Biol. 1, 129–136.

Hayes, T. B. (1997). Steroids as potential modulators of thyroid hormone activity in anuran metamorphosis. Am. Zool. 37, 185–194.

Hayes, T. B. and Wu, T. H. (1995a). Interdependence of corticosterone and thyroid hormones in larval growth and development in the western toad (Bufo boreas). II. Regulation of corticosterone and thyroid hormones. J. Exp. Zool. 271, 103–111.

Hayes, T. B. and Wu, T. H. (1995b). The role of corticosterone in anuran metamorphosis and its potential role in stress-induced metamorphosis: In Proceedings of the 17th Conference of the European Society of Comparative Endocrinology. Neth. J. Zool. 45, 107–109.

Hayes, T. B., Chan, R., and Licht, P. (1993). Interactions of temperature and steroids on larval growth, development, and metamorphosis in a toad (Bufo boreas). J. Exp. Zool. 266, 206–215.

Hebbes, T. R., Thorne, A. W., and Crane-Robinson, C. (1988). A direct link between core histone acetylation and transcriptionally active chromatin. EMBO J. 7, 1395–1402.

Heiden, M. G., Chandel, N. S., Schumasker, P. J., and Thompson, C. B. (1999) Bcl-xL prevents cell death following growth factor withdrawal by facilitation mitochondrial ATP/ADP exchange. Mol. Cell 3, 159–167.

Heinzel, T., Lavinsky, R. M., Mullen, T. M., Solderstrom, M., Laherty, C. D., Torchia, J., Yang, W. M., Brard, G., Ngo, S. D., Davie, J. R., et al. (1997). A complex containing N-CoR, mSin3 and histone deacetylase mediates transcriptional repression. Nature 387, 43–48.

Helbing, C. C. and Atkinson, B. G. (1994) 3, 5, 3'-triiodothyronine-induced carbamylphosphate synthetase gene expression is stabilized in the liver of rana catesbeiana tadpoles during heat shock. J. Biol. Chem. 269, 11743–11750.

Helbing, C. C., Gergely, G., and Atkinson, B. G. (1992). Sequential up-regulation of thyroid hormone β receptor, ornithine transcarbamylase, and carbamyl phosphate synthetase mRNAs in the liver of Rana Catesbeiana tadpoles during spontaneous and thyroid hormone-induced metamorphosis. Develop. Genet. 13, 289–301.

Helff, O. M. (1926). Factors involved in the atrophy of the tail of anuran larvae during metamorphosis. Anat. Rec. 34, 129.

Hengartner, M. O. and Horvitz H. R. (1994a). C. Elegans cell survival gene ced-9 encodes a functional homolog of the mammalian proto-oncogene bcl-2. Cell 76, 665–676.

Hengartner, M. O. and Horvitz, H. R. (1994b). The ins and outs of programmed cell death during C. elegans development. Phil. Trans. R. Soc. 345, 243–246.

Hetzel, B. S. (1989). *The Story of Iodine Deficiency: An International Challenge in Nutrition.* Oxford University Press, Oxford.

Heyman, R. A., Mangelsdorf, D. J., Dyck, J. A., Stein, R. B., Eichele, G., Evans, R. M., and Thaller, C. (1992). 9-Cis retinoic acid is a high affinity ligand for the retinoid X receptor. Cell 68, 397–406.

Highnam, K. C. (1981). A survey of invertebrate metamorphosis. In *Metamorphosis: A Problem in Developmental Biology*, 2nd ed (L. I. Gilbert, and E. Frieden, eds.), Plenum Press, New York, pp. 43–73.

Hinuma, S., Habata, Y., Fujii, R., Kawamata, Y., Hosoya, M., Fukusumi, S., Kitada, C., Masuo, Y., Asano, T., Matsumoto, H., et al. (1998). A prolactin-releasing peptide in the brain. Nature 393, 272–276.

Hoch, M. and Pankratz, M. J. (1996). Control of gut development By fork head and cell signaling molecules in Drosophila. Mechanisms Develop. 58, 3–14.

Hollenberg, S. M. and Evans, R. M. (1988). Multiple and cooperative transactivation domains of human glucocorticoid receptor. Cell 55, 899–906.

Hong, L., Schroth, G. P., Matthews, H. R., Yau, P., and Bradbury, E. M. (1993). Studies of the DNA binding properties of histone H4 amino terminus. Thermal denaturation studies reveal that acetylation markedly reduces the binding constant of the H4 "tail" to DNA. J. Biol. Chem. 268, 305–314.

Horlein, A. J., Naar, A. M., Heinzel, T., Torchia, J., Gloss, B., Kurokawa, R., Ryan, A., Kamei, Y., Soderstrom, M., Glass, C. K., and Rosenfeld. M. G. (1995). Ligand-independent repression by the thyroid hormone receptor mediated by a nuclear receptor co-repressor. Nature 377, 397–404.

Horvitz, H., Ellis, H., and Sternberg, P. (1982). Programmed cell death in nematode development. Neurosci. Comment. 1, 56–65.

Hoskins, S. G. (1990). Metamorphosis of the amphibian eye. J. Neurobiol. 21, 970–989.

Hourdry, J. (1974). Dosage de l'activite de quelques hydrolases lysosomiques intestinales, au cours du development larvaire de Discoglossus pictus Otth, Amphibien Anoure. W. Roux Arch. Entwicklungsmech. Org. 174, 222–233.

Hourdry, J. and Dauca, M. (1977). Cytological and cytochemical changes in the intestinal epithelium during anuran metamorphosis. Int. Rev. Cytol. 5, 337–385.

Hourdry, J., Chabot, J. G., Menard, D., and Hugon, J. S. (1979). Intestinal brush border enzyme activities in developing amphibian Rana catesbeiana. Comp. Biochem. Physiol. 63A, 121–125.

Hsu, J.-H. and Brent, G. A. (1998). Thyroid hormone receptor gene knockouts. Trends Endocrinol. Metab. 9, 103–112.

Huang, M., Marsh-Armstrong, N., Brown, D. D. (1999). Metamorphosis is inhibited in transgenic Xenopus laevis tadpoles that overexpress type III deiodinase. Proc. Natl. Acad. Sci. USA 96, 962–967.

Huet, F., Ruiz, C., and Richards, G. (1993). Puffs and PCR: the in vivo dynamics of early gene expression during ecdysone responses in Drosophila. Development 118, 613-627

Huhtala, P, Chow, L. T., and Tryggvason, K. (1990). Structure of the human type IV collagenase gene. J. Biol. Chem. 265, 11077–11082.

Huhtala, P., Tuuttila, A., Chow, L. T., Lohi, J., Keski-Oja, J., and Tryggvason, K. (1991). Complete Structure of the Human Gene for 92-kDa Type IV Collaganase. J. Biol. Chem. 266, 16485–16490.

Hukins, D. W. L. (1984). *Connective Tissue Matrix*. Verlag Chemie, Basel.

Hurban, P. and Thummel, C. S. (1993). Isolation and characterization of fifteen ecdysone-inducible Drosophila genes reveal unexpected complexities in ecdysone regulation. Mol. Cell. Biol. 13, 7101–7111.

Huxley, J. S. (1929). Thyroid and temperature in cold-blooded vertebrates. Nature 123, 712.

Ichikawa, K., Hashizume, K., Furuta, S., Osumi, T., Miyamoto, T., Yamauchi, K., Takeda, T., and Yamada, T. (1990). Human c-erb A protein expressed in Escherichia coli: changes in hydrophobicity upon thyroid hormone binding. Mol. Cell. Endocrinol. 70, 175–184.

Imbalzano, A. M., Kwon, H., Green, M. R., and Kingston, R. E. (1994). Facilitated binding of TATA-binding protein to nucleosomal DNA. Nature 370, 481–485.

Imhof, A., Yang, X. J., Ogryzko, V. V., Nakatani, Y., Wolffe, A. P., and Ge, H. (1997). Acetylation of general transcription factors by histone acetyltransferases. Curr. Biol. 7, 689–692.

Inaba, T. and Frieden, E. (1967). Changes in ceruloplasmin during anuran metamorphosis. J. Biol. Chem. 242, 4789–4796.

Ingham, P. W. (1998). Transducting hedgehog: the story so far. EMBO J. 17, 3505–3511.

Inui, Y. and Miwa, S. (1985). Thyroid hormone induces metamorphosis of flounder larvae. Gen. Comp. Endocrinol. 60, 450–454.

Ishizuya-Oka, A. (1996). Apoptosis of larval cells during amphibian metamorphosis. Microscopy Res. Technol. 34, 228–235.

Ishizuya-Oka, A. and Mizuno, T. (1984). Intestinal cytodifferentiation in vitro of chick stomach endoderm induced by the duodenal mesenchyme. J. Embryol. Exp. Morphol. 82, 163–176.

Ishizuya-Oka, A. and Mizuno, T. (1985). Chronological analysis of the intestinalization of chick stomach endoderm induced in vitro by duodenal mesenchyme. Roux's Arch. Dev. Biol. 194, 301–305.

Ishizuya-Oka, A. and Shimozawa, A. (1987a). Development of the connective tissue in the digestive tract of the larval and metamorphosing Xenopus laevis. Ant. Anz. Jena 164, 81–93.

Ishizuya-Oka, A. and Shimozawa, A. (1987b). Ultrastructural changes in the intestinal connective tissue of Xenopus laevis during metamorphosis. J. Morphol. 193, 13–22.

Ishizuya-Oka, A. and Shimozawa, A. (1990). Changes in lectin-binding pattern in the digestive tract of Xenopus laevis during metamorphosis. II. Small intestine. J. Morphol. 205, 9–15.

Ishizuya-Oka, A. and Shimozawa, A. (1991). Induction of metamorphosis by thyroid hormone in anuran small intestine cultured organotypically in vitro. In Vitro Cell. Dev. Biol. 27A, 853–857.

Ishizuya-Oka, A., Shimozawa, A. (1992a). Programmed cell death and heterolysis of larval epithelial cells by macrophage-like cells in the anuran small intestine in vivo and in vitro. J. Morphol. 213, 185–195.

Ishizuya-Oka, A. and Shimozawa, A. (1992b). Connective tissue is involved in adult epithelial development of the small intestine during anuran metamorphosis in vitro. Roux's Arch. Dev. Biol. 201, 322–329.

Ishizuya-Oka, A., Shimozawa, A., Takeda, H., and Shi, Y.-B. (1994). Cell-specific and spatio-temporal expression of intestinal fatty acid-binding protein gene during amphibian metamorphosis. Roux's Arch. Dev. Biol. 204, 150–155.

Ishizuya-Oka, A., Ueda, S., and Shi, Y.-B. (1996). Transient expression of stromelysin-3 mRNA in the amphibian small intestine during metamorphosis. Cell Tissue Res. 283, 325–329.

Ishizuya-Oka, A., Ueda, S., and Shi, Y. B. (1997a). Temporal and spatial regulation of a putative transcriptional repressor implicates it as playing a role in thyroid hormone-dependent organ transformation. Develop. Genetics 20, 329–337.

Ishizuya-Oka, A., Stolow, M. A., Ueda, S., and Shi, Y.-B. (1997b). Temporal and spatial expression of an intestinal Na^+/PO_4^{3-} cotransporter correlates with epithelial transformation during thyroidhormone-dependent frog metamorphosis. Develop. Genet. 20, 53–66.

Ishizuya-Oka, A., Ueda, S., Damjanovski, S., Li, Q., Liang, V. C.-T., and Shi, Y.-B. (1997c). Anteroposterior gradient of epithelial transformation during amphibian intestinal remodeling: Immunohistochemical detection of intestinal fatty acid-binding protein. Develop. Biol. 192, 149–161.

Ishizuya-Oka, A. and Ueda, S. (1996). Apoptosis and cell proliferation in the Xenopus small intestine during metamorphosis. Cell Tissue Res. 286, 467–476.

Ishizuya-Oka, A., Inokuchi, T., and Ueda, S. (1998). Thyroid hormone-induced apoptosis of larval cells and differentiation of pepsinogen-producing cells in the stomach of Xenopus laevis in vitro. Differentiation 63, 59–68.

Iwamuro, S. and Tata, J. R. (1995). Contrasting patterns of expression of thyroid hormone and retinoid X receptor genes during hormonal manipulation of Xenopus tadpole tail regression in culture. Mol. Cell. Endocrinol. 113, 235–243.

Iwase, K., Yamauchi, K., and Ishikawa, K. (1995). Cloning of cDNAs encoding argininosuccinate lyase and arginase from Rana catesbeiana liver and regulation of their mRNAs during spontaneous and thyroid hormone-induced metamorphosis. Biochim. Biophys. Acta 1260, 139–146.

Izutsu Y. and Yoshizato K. (1993). Metamorphosis-dependent recognition of larval skin as non-self by inbred adult frogs (Xenopus laevis). J. Exp. Zool. 266, 163–167.

Izutsu, Y., Yoshizato, K., and Tochinai, S. (1996). Adult-type splenocytes of Xenopus induce apoptosis of histocompatible larval tail cells in vitro. Differentiation 60, 277–286.

Jackson, I. M. D. and Reichlin, S. (1977). Thyrotropin-releasing hormone: Abundance in the skin of the frog, Rana pipiens. Science 198, 414–415.

Jacobs, G. F. M. and Kuhn, E. R. (1992). Thyroid hormone feedback regulation of the secretion of bioactive thyrotropin in the frog. Gen. Comp. Endocrinol. 88, 415–423.

Jacobson, M. D., Weil, M., and Raff, M. C. (1997). Programmed cell death in animal development. Cell 88, 347–354.

Jaffe, R. C. (1981). Plasma concentration of corticosterone during Rana catesbeiana tadpole metamorphosis. Gen. Comp. Endocrinol. 44, 314–318.

Jameson, J. L. (1994). Mechanisms by which thyroid hormone receptor mutations cause clinical syndromes of resistance to thyroid hormone. Thyroid 4, 485–492.

Jennings, D. H. and Hanken, J. (1998). Mechanistic basis of life history evolution in anuran amphibians: thyroid gland development in the direct-development frog. Gen. Comp. Endocrinol. 111, 225–32.

Jiang, C., Baehrecke, E. H., and Thummel, C. S. (1997). Steroid regulated programmed cell death during Drosophila metamorphosis. Development 124, 4673–4683.

Jolivet-Jaudet, G. and Leloup-Hatey, J. (1984) Variations in aldosterone and corticosterone plasma levels during metamorphosis in Xenopus laevis tadpoles. Gen. Comp. Endocrinol. 56, 59–65.

Johnson, R. L. and Scott, M. P. (1998). New players and puzzles in the hedgehog signaling pathway. Curr. Opinion in Genet. Develop. 8, 450–456.

Jorgensen, E. C. (1978). Thyroid hormones and analogs. II. Structure-activity relationships. In *Hormonal Proteins and Peptides*. (C. H. Li, ed.), Academic Press, New York, pp. 107–199.

Just, J. J. and Atkinson, B. G. (1972). Hemoglobin transitions in the bullfrog Rana catesbeiana during spontaneous and induced metamorphosis. J. Exp. Zool. 182, 271–280.

Just, J. J., Kraus-Just, J., and Check, D. A. (1981). Survey of chordate metamorphosis. In *Metamorphosis: A Problem in Developmental Biology*, 2nd ed. (L. I. Gilbert, and E. Frieden, eds.), Plenum Press, New York, pp. 265–326.

Kaltenbach, J. C. (1953). Local action of thyroxin on amphibian metamorphosis. 1. Local metamorphosis in Rana pipiens larvae effected by thyroxin-cholesterol implants. J. Exp. Zool. 122, 21–39.

Kaltenbach, J. C. (1958). Direct steroid enhancement of induced metamorphosis in peripheral tissues. Anat. Rec. 131, 569–570 (abstract).

Kaltenbach, J. C. (1968). Nature of hormone action in amphibian metamorphosis. In *Metamorphosis: A Problem in Developmental Biology* (W. Etkin and L. I. Gilbert, eds.), Appleton, New York, pp. 399–441.

Kaltenbach, J. C. (1985). Amphibian metamorphosis: Influence of thyroid and steroid hormones. In *Current Trends in Comparative Endocrinology* (B. Lofts and W. N. Holmes, eds.), Hong Kong University Press, Hong Kong, pp. 533–534.

Kaltenbach, J. A. (1996). Endocrinology of amphibian metamorphosis. In *Metamorphosis: Postembryonic Reprogramming of Gene Expression in Amphibian and Insect Cells* (L. I. Gilbert, J. R. Tata, and B. G. Atkinson, eds.), Academic Press, New York, pp. 403–431.

Kaltenbach, J. C., Lipson, M. J., and Wang, C. H. K. (1977). Histochemical study of the amphibian digestive tract during normal and thyroxine-induced metamorphosis. J. Exp. Zool. 202, 103–120.

Kamei, Y., Xu, L., Heinzel, T., Trochia, J., Kurokawa, R., Gloss, B., Lin, S.-C., Heyman, R. A., Rose, D. W., Glass, C. K., and Rosenfeld, M. G. (1996). A CBP Integrator Complex Mediates Transcriptional Activation and AP-1 Inhibition by Nuclear Receptors. Cell 85, 403–414.

Kanamori, A. and Brown, D. D. (1992). The regulation of thyroid hormone receptor β genes by thyroid hormone in Xenopus laevis. J. Biol. Chem. 267, 739–745.

Kanamori, A. and Brown, D. D. (1993). Cultured cells as a model for amphibian metamorphosis. Proc. Natl. Acad. Sci. USA 90, 6013–6017.

Kanamori, A. and Brown, D. D. (1996). The analysis of complex developmental programmes: amphibian metamorphosis. Genes to Cells 1, 429–435.

Kandel, E. R. and Schwarts, J. H. eds. (1985). *Principles of Neural Science*. Elsevier, New York.

Kaplan, M. M., Swartz, S. L., and Larsen, P. R. (1981). Partial peripheral resistance to thyroid hormone. Am. J. Med. 70, 1115–1121.

Karim, F. D. and Thummel, C. S. (1992). Temporal coordination of regulatory gene expression by the steroid hormone ecdysone. EMBO J. 11, 4083–4093.

Karim, F. D., Guild, G. M., and Thummel, C. S. (1993). The Drosophila broad-complex plays a key role in controlling ecdysone-regulated gene expression at the onset of metamorphosis. Development 118, 977–988.

Kato, H., Fukuda, T., Parkison, C., McPhie, P., and Chen, S.-Y. (1989). Cytosolic thyroid hormone-binding protein is a monomer of pyruvate kinase. Proc. Natl. Acad. Sci. USA 86, 7861–7865.

Kawahara, A., Hikosaka, A., Sasado, T., and Hirota, K. (1997). Thyroid hormone-dependent repression of α 1-microglobulin/bikunin precursor (AMBP) gene expression during amphibian metamorphosis. Develop. Genes Evol. 206, 355–362.

Kawahara, A., Baker, B. S., and Tata, J. R. (1991). Developmental and regional expression of thyroid hormone receptor genes during Xenopus metamorphosis. Development 112, 933–943.

Kedinger, M., Simon-Assmann, P., Alexandre, E., and Haffen, K. (1987). Importance of a fibrablastic support for in vitro differentiation of intestinal endodermal cells and for their response to glucocorticoids. Cell Differ. 20, 171–182.

Kedinger, M., Bouziges, F., Simon-Assmann, P., and Haffen, K. (1989). Influence of cell interactions on intestinal brush border enzyme expression, in *Highlights of Modern Biochemistry*, Vol. 2 (A. Kotyk, J. Skoda, V. Paces, and V. Kostka, eds.), VSP International, Zeist, pp. 1103–1112.

Kemp, H. F., and Taylor, P. M. (1997). Interactions between thyroid hormone and tryptophan transport in rat liver are modulated by thyroid status. Am. J. Physiol. 272, E809–16.

Kendall, E. C. (1915). The isolation in crystalline form of the compound containing iodine which occurs in the thyroid: Its chemical nature and physiological activity. Trans. Assoc. Am. Phys. 30, 420–449.

Kendall, E. C. (1919). Physiological action of the thyroid hormone. Am. J. Physiol. 49, 136–137.

Kerr, J. F. R., Wyllie, A. H., and Currie, A. R. (1972). Apoptosis: A basic biological phenomenon with wide-ranging implication in tissue kinetics. Br. J. Cancer 26, 239–257.

Kerr, J. F. R., Harmon, B., and Searle, J. (1974). An electron-microscope study of cell eletion in theanuran tadpole tail during spontaneous metamorphosis with special reference poptosis of striated muscle fibres. J. Cell Sci. 14, 571–585.

Ketola-Pirie, C. A. and Atkinson, B. G. (1988). 3,3′,5-triiodothyronine-induced differences in water-insoluble protein synthesis in primary epidermal cell cultures from the hind limb of premetamorphic Rana catesbeiana tadpoles. Gen. Comp. Endocrinol. 69, 197–204.

Ketola-Pirie, C. A. and Atkinson, B. G. (1990). Thyroid hormone-induced differential synthesis of water-insoluble proteins in epidermal cell cultures from the hind limb of Rana catesbeiana tadpoles in Stages XII-XV and XVI-XIX. Gen. Comp. Endocrinol. 79, 275–282.

Kikuyama, S., Yamamoto, K., and Mayumi, M. (1980). Growth-promoting and antimetamorphic hormone in pituitary glands of bullfrogs. Gen. Comp. Endocrinol. 41, 212–216.

Kikuyama, S., Niki, K., Mayumi, M., and Kawamura, K. (1982). Retardation of thyroxine-induced metamorphosis by Amphenone B in toad tadpoles. Endocrinol. J. 29, 659–662.

Kikuyama, S., Niki, K., Mayumi, M., Shibayama, R., Nishikawa, M., and Shitake, N. (1983). Studies on corticoid action on the toad tadpoles in vitro. Gen. Comp. Endocrinol. 52, 395–399.

Kikuyama, S., Suzuki, M. R., and Iwamuro, S. (1986). Elevation of plasma aldosterone levels of tadpoles at metamorphic climax. Gen. Comp. Endocrinol. 63, 186–190.

Kikuyama, S., Kawamura, K., Tanaka, S., and Yamamoto, K. (1993). Aspects of amphibian metamorphosis: hormonal control. Int. Rev. Cytol. 145, 105–148.

King, J. A. and Miller, R. P. (1981). TRH, GH-RIH, and LH-RH in metamorphosing Xenopus laevis. Gen. Comp. Endocrinol. 44, 20–27.

Kinoshita, T., Sasaki, F., and Watanabe, K. (1985). Autolysis and heterolysis of the epidermal cells in anuran tadpole tail regression. J. Morphol. 185, 269–275.

Kinzler, K. and Vogelstein, B. (1989). Whole genome PCR: application to the identification of sequences bound by gene regulatory proteins. Nucl. Acids Res. 17, 3645–3653.

Kishimoto, T., Taga, T., and Akira, S. (1994). Cytokine signal transduction. Cell 76, 253–262.

Kleiner Jr, D. E. and Stetler-Stevenson, W. G. (1993). Structural biochemistry and activiation of matrix metalloproteases. Current Opinion in Cell Biol. 5, 891–897.

Kluck, R. M., Bossy-Wetzel, E., Green, D. R., and Newmeyer, D. D, (1997). The release of cytochrome c from mitochondria: A primary site for Bcl-2 regulation of apoptosis. Science 275, 1132–1236.

Kobayashi, H. (1958). Effect of desoxycorticosterone acetate on metamorphosis induced bythyroxine in anuran tadpoles. Endocrinology 62, 371–377.

Koelle, M. R., Talbot, W. S., Segraves, W. A., Bender, M. T., Cherbas, P., and Hogness, D. S. (1991). The *Drosophila* EcR gene encodes an ecdysone receptor, a new member of the steroid receptor superfamily. Cell 67, 59–77.

Koelle, M. R., Segraves, W. A., and Hogness, D. S. (1992). DHR3: A Drosophila steriod receptor homolog. Proc. Natl. Acad. Sci. USA 89, 6167–6171.

Koenig, R. J. (1998). Thyroid hormone receptor coactivators and corepressors. Thyroid 8, 703–713.

Kollros, J. J. (1981) Transitions in the nervous system during amphibian metamorphosis. In *Metamorphosis: A Problem in Developmental Biology*, 2nd ed, (L. I. Gilbert, and E. Frieden, eds.), Plenum Press, New York, pp. 445–459.

Kordylewski, L. (1983). Light and electron microscopic observations of the development of intestinal musculature in Xenopus. Anat. Forsch. 97, 719–734.

Kornberg, R. D. and Lorch, Y. (1995). Interplay between chromatin structure and transcription. Curr. Opinion Cell. Biol. 7, 371–375.

Kralli, A. and Yamamoto, K. R. (1996). An FK506-sensitive transporter selectively decreases intracellular levels and potency of steroid hormones. J. Biol. Chem. 271, 17152–17156.

Kralli, A., Bohen, S. P., and Yamamoto, K. R. (1995). LEM1, an ATP-binding-cassette transporter, selectively modulates the biological potency of steroid hormones. Proc. Natl. Acad. Sci. 92, 4701–4705.

Kroll, K. L. and Amaya, E. (1996). Transgenic Xenopus embryos from sperm nuclear transplantations reveal FGF signaling requirements during gastrulation. Development 122, 3173–3183.

Kroll, K. L. and Gerhart, J. C. (1994). Transgenic X. laevis embryos from eggs transplanted with nuclei of transfected cultured cells. Science 266, 650–653.

Krug, E. C., Honn, K. V., Battista, J., and Nicoll, C. S. (1983). Corticosteroids in serum of Rana catesbeiana during development and metamorphosis. Gen. Comp. Endocrinol. 52, 232–241.

Krust, A., Green, S., Argos, P., Kumar, Walter, Ph., Bornert, J.-M., and Chambon, P. (1986). The chicken oestrogen receptor sequence: homology with v-erbA and the human oestrogen and glucocorticoid receptors. EMBO J. 5, 891–897.

Kubler, H. and Frieden, E. (1964). The increase in β-glucuronidase of the tadpole tail during amphibian metamorphosis and its relation to lysosomes. Biochim. Biophys. Acta 93, 635–643.

Kuhn, E. R. and Jocobs, G. F. M. (1989). Metamorphosis. In *Developmental Biology of the Axolotl* (J. B. Armstrong, and G. M. Malacinski, eds.), Oxford Univ. Press, New York, pp. 187–197.

Kuida, K., Lippke, J. A., Ku, G., Harding, M. W., Livingston, D. J., Su, M. S.-S., and Flavell, R. A. (1995). Altered cytokine export and apoptosis in mice deficient in interleukin-1 beta converting enzyme. Science 267, 2000–2003.

Kuida, K., Zheng, T. S., Na, S., Kuan, C. K., Yang, D., Karasuyama, H., Rakic, P., and Flavell, R. A. (1996). Decreased apoptosis in the brain and premature lethality in CPP32-deficient mice. Nature 384, 368–372.

Kuida, K., Haydar, T. F., Kuan, C.-Y., Gu, Y., Taya, C., Karasuyama, H., Su, M. S.-S., Rakic, P., and Flavell, R. A. (1998). Reduced apoptosis and cytochrome c-mediated caspase activation in mice lacking caspase 9. Cell 94, 325–337.

Kurokawa, R., Yu, V. C. Naar, A., Kyakumoto, S., Han, Z., Silverman, S., Rosenfeld, M. G., and Glass, C. K. (1993). Differential orientations of the DNA-binding domain and carboxy-terminal dimerization interface regulate binding site selection by nuclear receptor heterodimers. Genes Develop. 7, 1423–1435.

Kwon, H., Imbalzano, A. N., Khavari, P. A., Kingston, R. E., and Green, M. R. (1994). Nucleosome disruption and enhancement of activator binding by a human SW1/SNF complex. Nature 370, 477–481.

Lam, G. T., Jiang, C., and Thummel, C. S. (1997). Coordination of larval and prepupal gene expression by the DHR3 orphan receptor during Drosophila metamorphosis. Development 124, 1757–1769.

Landschulz, W. H., Johnson, P. F., and McKnight, S. L. (1988). The leucine zipper: A hypothetical structure common to a new class of DNA binding proteins. Science 240, 1759–1764.

Lapiere, C. M. and Gross, J. (1963). Animal collagenase and collagen metabolism. In *Mechanisms of Hard Tissue Destruction* (R. F. Sognnaes, ed.), American Association for the Advancement of Science, Washington, DC, pp. 663–694.

Larsen, P. R. (1989). Maternal thyroxine and congenital hypothyroidism. New Engl. J. Med. 321, 44–46.

Laudet, V., Begue, A., Henry-Duthout, C., Joubel, A., Martin, P., Stehelin, D., and Saule, S. (1991). Genomic organization of the human thyroid hormone receptor (c-erbA-1) gene. Nucleic Acids Res. 19, 1105–1112.

Lavorgna, G., Ueda, H., Clos, J., and Wu, C. (1991). FTZ-F1, a steroid hormone receptor-like protein implicated in the activation of fushi tarazu. Science 252, 848–851.

Lavorgna, G., Karim, F. D., Thummel, C. S., and Wu, C. (1993). Potential role for a FTZ-F1 steroid receptor superfamily member in the control of Drosophila metamorphosis. Proc. Natl. Acad. Sci. USA 90, 3004–3008.

Lazar, M. A. (1993). Thyroid hormone receptors: Multiple forms, multiple possibilities. Endocrine Rev. 14, 184–193.

Lebrun, J. J., Ali, S., Sofer, L., Ullrich, A., and Kelly, P. A. (1994). Prolactin-induced proliferation of Nb2 cells involves tyrosine phosphorylation of the prolactin receptor and its associated tyrone kinase JAK2. J. Biol. Chem. 269, 14021–14026.

Le Douarin, B., Zechel, C., Garnier, J.-M., Lutz, Y., Tora, L., Pierrat, B., Heery, D., Gronemeyer, H., Chambon, P., and Losson, R. (1995). The N-terminal part of TIF1, a putative mediator of the ligand-dependent activation function (AF-2) of nuclear receptors, is fused to B-raf in the oncogenic protein T18. EMBO J. 14, 2020–2033.

Lee, D. Y., Hayes, J. J., Pruss, D., and Wolffe, A. P. (1993). A positive role for histone acetylation in transcription factor access to nucleosomal DNA. Cell 72, 73–84.

Lee, H. H. and Archer, T. K. (1994). Nucleosome-mediated disruption of transcription factor-chromatin initiation complexes at the mouse mammary virus long terminal repeat in vivo. Mol. Cell. Biol. 14, 32–41.

Lee, J. J., von Kessler, D. P., Parks, S., and Beachy, P. A. (1992). Secretion and localized transcription suggest a role in positional signaling for products of the segmentation gene hedgehog. Cell 71, 33–50.

Lee, J. J., Ekker, S. C., von Kessler, D. P., Porter, J. A., Sun, B. I., and Beachy, P. A. (1994a). Autoproteolysis in hedgehog protein biogenesis. Science 266, 1528–1537.

Lee, J. W., Moore, D. D., and Heyman, R. A. (1994b). A chimeric thyroid hormone receptor constitutively bound to DNA requires retinoid X receptor for hormone-dependent transcriptional activation in yeast. Mol. Endocrinol. 8, 1245–1252.

Lee, J. W., Choi, H.-S., Gyuris, J., Brent, R., and Moore, D. D. (1995a). Two classes of proteins dependent on either the presence or absence of thyroid hormone for interaction with the thyroid hormone receptor. Mol. Endocrinol. 9, 243–254.

Lee, J. W., Ryan, F., Swaffield, J. C., Johnston, S. A., and Moore, D. D. (1995b). Interaction of thyroid-hormone receptor with a conserved transcriptional mediator. Nature 374, 91–94.

Lee, Y. and Mahdavi, V. (1993). The D domain of the thyroid hormone receptor α1 specifies positive and negative transcriptional regulation functions. J. Biol. Chem. 268, 2021–2028.

Lefebvre, O., Wolf, C., Limacher, J.-M., Hutin, P., Wendling, C., LeMeur, M., Basset, P., and Rio, M.-C. (1992). The breast cancer-associated stromelysin-3 gene is expressed during mouse mammary gland apoptosis. J. Cell. Biol. 119, 997–1002.

Leid, M., Kastner, P., Lyons, R., Nakshatri, H., Saunders, M., Zachrewski, T., Chen, J. Y., Staub, A., Garnier, J. M., Mader, S., and Chambon, P. (1992). Purification, cloning, and RXR identity of the HeLa cell factor with which RAR or TR heterodimerizes to bind target sequences efficiently. Cell 68, 377–395.

Leloup, J. and Buscaglia, M. (1977). La triiodothyronine: hormone de la métamorphose des amphibiens. C. R. Acad. Sci. 284, 2261–2263.

Levi, G., Broders, F., Dunon, D., Edelman, G. M. and Thiery, J. P. (1990). Thyroxine-dependent modulations of the expression of the neural cell adhesion molecule N-CAM during Xenopus laevis metamorphosis. Development 109, 681–692.

Lewin, B. (1994). Chromatin and gene expression: constant questions, but changing answers. Cell 79, 397–406.

Lezoualch, F., Hassan, A. H. S., Giraud, P., Loeffler, J.-P., Lee, S. L., and Demeneix, B. A. (1992). Assignment of the β-thyroid hormone receptor to 3, 5, 3′-triiodothyronine-dependent inhibition of transcription from the thyrotropin-releasing hormone promoter in chick hypothalamus neurons. Mol. Endocrinol. 6, 1797–1804.

Li, H., Gomes, P. J., and Chen, J. D. (1997a). RAC3, a steroid/nuclear receptor-associated coactivator that is related to SRC-1 and TIF2 . Proc. Natl. Acad. Sci. USA 94, 8479–8484.

Li, P., Nijhawan, D., Budihardjo, I., Srinivasula, S. M., Ahmad, M., Alnemri, E. S., and Wang, X. (1997b). Cytochrome c and dATP-dependent formation of apaf-1/Caspase-9 complex initiates an apoptotic protease cascade. Cell 91, 479–489.

Li, Y. C., Bergwitz, C., Juppner, H., and Demay, M. B. (1997c). Cloning and characterization of the vitamin D receptor from Xenopus laevis. Endocrinology 138, 2347–2653.

Li, H., Zhu, H., Xu, C.-J. and Yuan, J. (1998a). Cleavage of BID by Caspase 8 mediates the Mitochondrial Damage in the Fas Pathway of Apoptosis. Cell 94, 491–501.

Li, J., Liang, V. C. T., Sedgwick, T., Wong, J., and Shi, Y.-B. (1998b). Unique organization and involvement of GAGA factors in transcriptional regulation of the Xenopus stromelysin-3 gene. Nucl. Acids Res. 26, 3018–3025.

Liang, P. and Pardee, A. B. (1992). Differential display of eukaryotic messenger RNA by means of the polymerase chain reaction. Science 257, 967–971.

Liang, V. C.-T., Sedgwick, T., and Shi, Y.-B. (1997). Characterization of the Xenopus homolog of an immediate early gene associated with cell activation: Sequence analysis and regulation of its expression by thyroid hormone during amphibian metamorphosis. Cell Res. 7, 179–193.

Little, G. H. and Flores, A. (1992). Inhibition of programmed cell death by cyclosporin. Comp. Biochem. Physiol. 103C, 463–467.

Liu, J., Farmer, J. D., Lane, W. S., Friedman, J., Weissman, I., and Schreiber, S. L. (1991). Calcineurin is a common target of cyclophilin-cyclosporin A and FKBP-FK506 complexes. Cell 66, 807–815.

Liu, X., Kim, C. N., Yang, J., Jemmerson, R., and Wang, X. (1996). Induction of apoptotic program in cell-free extracts: Requirement for dATP and cytochrome c. Cell 86, 147–157.

Liu, X., Zou, H., Slaughter, C., and Wang, X. (1997). DFF, a heterodimeric protein that functions downstream of caspase-3 to trigger DNA fragmentation during apoptosis. Cell 89, 175–184.

Liu, X., Li, P., Widlak, P., Zou, H., Luo, X., Garrard, W. T., and Wang, X. (1998). The 40-kDa subunit of DNA fragmentation factor induces DNA fragmentation and chromatin condensation during apoptosis. Proc. Natl. Acad. Sci. USA 95, 8461–8466.

Louvard, D., Kedinger, M., and Hauri, H. P. (1992). The differentiating intestinal epithelial cell: Establishment and maintenance of functions through interactions between cellular structures. Ann. Rev. Cell. Biol. 8, 157–195.

Lund, L. R., Romer, J., Thomasset, N., Solberg, H., Pyke, C., Bissell, M. J., Dono, K., and Werb, Z. (1996). Two distinct phases of apoptosis in mammary gland involution: proteinase-independent and -dependent pathways. *Development* 122, 181–193.

Luo, X., Budihardjo, I., Zou, H., Slaughter, C., and Wang, X. (1998). Bid, a Bcl2 interacting protein, mediates cytochrome c release from mitochondria in response to activation of cell surface death receptors. Cell 94, 481–490.

Macchi, I. A. and Phillips, J. G. (1966). In vitro effect of adrenocorticotropin on corticoid secretion in the turtle, snake, and bullfrog. General and Comp. Endocrinol. 6, 170–182.

Macchia, E., Gurnell, M., Agostini, M., Giorgilli, G., Marcocci, C., Valenti, T. M. L., Martino, E., Chatterjee, K. K., and Pinchera, A. (1997). Identification and characterization of a novel de novo mutation (L346V) in the thyroid hormone receptor Beta gene in a family with generalized thyroid hormone resistance. Eur. J. Endocrinol. 137, 370–376.

Machuca, I., Esslemont, G., Fairclough, L., and Tata, J. R. (1995). Analysis of structure and expression of the *Xenopus* thyroid hormone receptor b gene to explain its autoregulation. Mol. Endocrinol. 9, 96–107.

Madhavan, M. M. and Schneiderman, H. A. (1977). Histological analysis of the dynamics of growth of imaginal discs and histoblast nests during the larval development of Drosophila melanogaster. Roux's Arch. Dev. Biol. 183, 269–305.

Malagon, M. M., Garcia-Navarro, S., Ruiz-Navarro, A., and Gracia-Navarro, F. (1989). Morphometric evaluation of subcellular changes induced by in vitro TRH treatment in the pituitary gland of Rana perezi: Effects on prolactin and thyrotropic cells. Cell Tissue Res. 256, 391–398.

Malagon, M. M., Ruiz-Navarro, A., Torronteras, R., and Gracia-Navarro, F. (1991). Effects of ovine CRF on amphibian pituitary ACTH and TSH cells in vivo: A quantiative ultrastructural study. Gen. Comp. Endocrinol. 83, 487–497.

Mamajiwalla, S. N., Fath, K. R., and Burgess, D. R. (1992). Development of the chicken intestinal epithelium. Curr. Topics in Develop. Biol. 26, 123–143.

Mandaron, P., Guillermet, C., and Sengei, P. (1977). In vitro development of Drosophila imaginal discs: Hormonal control and mechanism of evagination. Am. Zool. 17, 661–670.

Mandel, S. J., Brent, G. A., and Larsen, P. R. (1993). Levothyroxine therapy in patients with thyroid disease. Ann. Intern. Med. 119, 492–502.

Mangelsdorf, D. J. and Evans, R. M. (1995). The RXR heterodimers and orphan receptors. Cell 83, 841–850.

Mangelsdorf, D. J., Ong, E. S., Dyck, J. A., and Evans, R. M. (1990). Nuclear receptor that identifies a novel retinoic acid response pathway. Nature 345, 224–229.

Mangelsdorf, D. J., Borgmeyer, U., Heyman, R. A., Zhou, J. Y., Ong, E. S., Oro, A. E., Kakizuka, A., and Evans, R. M. (1992). Characterization of three RXR genes that mediate the action of 9-cis retinoic acid. Genes Dev. 6, 329–344.

Mangelsdorf, D. J., Thummel, C., Beato, M., Herrlich, P., Schutz, G., Umesono, K., Blumberg, B., Kastner, P., Mark, M., Chambon, P., and Evans, R. M. (1995). The nuclear receptor superfamily: The second decade. Cell 83, 835–839.

Marigo, V., Davey, R. A., Zuo, Y., Cunningham, J. M., and Tabin, C. J. (1996). Biochemical evidence that patched is the hedgehog receptor. Nature 384, 176–179.

Marklew, S., Smith, D. P., Mason, C. S., and Old, R. W. (1994). Isolation of a novel RXR from Xenopus that most closely resembles mammalian RXRβ and is expressed throughout early development. Biochim. Biophys. Acta 1218, 267–272.

Marks, M. S., Hallenbeck, P. L., Nagata, T., Segars, J. H., Appella, E., Nikodem, V. M., and Ozato, K. (1992). H-2RIIBP (RXR-β) heterodimerization provides a mechanism for combinatorial diversity in the regulation of retinoic acid and thyroid hormone responsive genes. EMBO J. 11, 1419–1435.

Marsh, M. N. and Trier, J. S. (1974). Morphology and cell proliferation of subepithelial fibroblasts in adult mouse jejunum. I. structural features. Gastroenterology 67, 622–635.

Marshall, J. A. and Dixon, K. E. (1978). Cell specialization in the epithelium of the small intestine of feeding Xenopus laevis. J. Anat. 126, 133–144.

Masson, R., Lefebvre, O., Noel, A., El Fahime, M., Chenard, M.-P., Wendling, C., Kebers, F., LeMeur, M., Dierich, A., Foidart, J.-M., et al. (1998). In vivo evidence that the Stromelysin-3 metalloproteinase contributes in a paracrine manner to epithelial cell malignancy. J. Cell Biol. 140, 1535–1541.

Mathisen, P. M. and Miller, L. (1989). Thyroid hormone induces constitutive keratin gene expression during *Xenopus laevis* development. Mol. Cell. Biol. 9, 1823–1831.

Matrisian, L. M. (1992). The matrix-degrading metalloproteinases. BioEssays 14, 455–463.

Matrisian, L. M. and Hogan, B. L. M. (1990). Growth factor-regulated proteases and extracellular matrix remodeling during mammalian development. Curr. Topics Develop. Biol. 24, 219–259.

McAvoy, J. W. and Dixon, K. E. (1977). Cell proliferation and renewal in the small intestinal epithelium of metamorphosing and adult Xenopus laevis. J. Exp. Zool. 202, 129–138.

McAvoy, J. W. and Dixon, K. E. (1978a). Cell specialization in the small intestinal epithelium of adult Xenopus laevis: structural aspects. J. Anat. 125, 155–169.

McAvoy, J. W. and Dixon, K. E. (1978b). Cell specialization in the small intestinal epithelium of adult Xenopus laevis: functional aspects. J. Anat. 125, 237–245.

McCollum, S. A. and Van Buskirk, J. (1996). Costs and benefits of a predator-induced polyphenism in the gray tree frog HyLa chrysoscelis. Evolution 50, 583–593.

McCutcheon, F. H. (1936). Hemoglobin function during the life history of the bullfrog. J. Cellular Comp. Physiol. 8, 63–81.

McEwen, B. S., Coirini, H., Danielson, A., Frankfurt, M., Gould, E., Mendelson, S., Schumacher, M., Segarra, A., and Woolley, C. (1991). Steroid and thyroid hormones modulate a changing brain. J. Ster. Biochem. Mol. Biol. 40, 1–14.

McInerney, E. M., Rose, D. W., Flynn, S. E., Westin, S., Mullen, T.-M., Krones, A., Inostroza, J., Torchia, J., Nolte, R. T., Assa-Munt, N., et al. (1998). Determinants of coactivator LXXLL motif specificity in nuclear receptor transcriptional activation. Genes Develop. 12, 3357–3368.

McKenna, N. J., Nawaz, Z., Tsai, S. Y., Tsai, M.-J., and O'Malley, B. W. (1998). Distinct steady—state nuclear receptor coregulator complexes exist in vivo. Proc. Natl. Acad. Sci. USA 95, 11697–11702.

McKerrow, J. H., Pino-Heiss, S., Lindquist, R., and Werb, Z. (1985). Purification and characterization of an elastinolytic proteinase secreted by cercariae of Schistosoma mansoni. J. Biol. Chem. 260, 3703–3707.

Mellstrom, B., Naranjo, J. R., Santos, A., Gonzalez, A. M., and Bernal, J. (1991). Independent expression of the alpha and beta c-erbA genes in developing rat brain. Mol. Endocrinol. 5, 1339–1350.

Meredith Jr. J. E. and Schwartz, M. A. (1997). Integrins, adhesion and apoptosis. Trends Cell Biol. 7, 146–150.

Metzenberg, R. L., Marshall, M., Paik, W. K., and Cohen, P. P. (1961). The synthesis of carbamyl phosphate synthetase in thyroxin-treated tadpoles. J. Biol. Chem. 236, 162–165.

Millar, R. P., Nicolson, S., King, J. A., and Louw, G. N. (1983). Functional significance of TRH in metamorphosing and adult anurans. In *Thyrotropin-Releasing Hormone* (E. C. Griffiths and G. W. Bennett, eds.), Raven, New York, pp. 217–227.

Miller, L. (1996). Hormone-induced changes in keratin gene expression during amphibian skin metamorphosis. In *Metamorphosis: Postembryonic Reprogramming of Gene Expression in Amphibian and Insect Cells.* (L. I. Gilbert, J. R. Tata, B. G. Atkinson, eds.), Academic Press, New York, pp. 599–624.

Milner, M. J. (1977a). The time during which beta-ecdysone is required for the differentiation in vitro and in situ of wing imaginal discs of Drosophila melanogaster. Develop. Biol. 56, 206–212.

Milner, M. J. (1977b). The eversion and differentiation of Drosophila melanogaster leg and wing imaginal discs cultured in vitro with an optimal concentration of beta-ecdysone. J. Embryol. Exp. Morphol. 37, 105–117.

Mimnagh, K. M., Bolaffi, J. L., Montgomery, N. M., and Kaltenbach, J. C. (1987). Thyrotropin-releasing hormone (TRH): Immunohistochemical distribution in tadpole and frog brain. Gen. Comp. Endocrinol. 66, 394–404.

Minn, A. J. ., Kettlum, C. S., Liang, H., Kelekar, A., Van der Heiden, M. G., Chang, B. S., Fesik, S. W., Fill, M., and Thompson, C. B. (1999). Bcl-xL regulates apoptosis by heterodimerization-dependent and -independent mechanisms. EMBO J 18, 632–643.

Miyajima, N., Horiuchi, R., Shibuya, Y., Fukushige, S. I., Matsubara, K.-I., Toyoshima, K., and Yamamoto, I. (1989). Two erbA homologs encoding proteins with different T-3 binding capacities are transcribed from opposite DNA strands of the same genetic locus. Cell 57, 31–39.

Miyoshi, Y., Nakamura, H., Tagami, T., Sasaki, S., Dorin, R. I., Taniyama, M., and Nakao, K. (1998). Comparison of the functional properties of three different truncated thyroid hormone receptors identified in subjects with resistance to thyroid hormone. Mol. Cell. Endocrinol. 137, 169–176.

Montague, J. W., Gaido, M. L., Frye, C., and Cidlowski, J. A. (1994). A calcium-dependent nuclease from apoptotic rat thyromocytes is homologous with cyclophilin: Recombination cyclophilins A, B, and C have nuclease activity. J. Biol. Chem. 269, 18877–18880.

Montague, J. W., Hughes Jr., F. M., and Cidlowski, J. A. (1997). Native recombinant cyclophilins A, B, and C degrade DNA independently of peptidylprolyl cis-trans-isomerase activity. J. Biol. Chem. 272, 6677–6684.

Montgomery, A. M. P., Reisfeld, R. A., and Cheresh, D. A. (1994). Integrin $\alpha v \beta 3$ rescues melanoma cells from apoptosis in three dimensional dermal collagen. Proc. Natl. Acad. Sci. USA 91, 8856–8860.

Moore, J. T., Fristrom, D., Hammonds, A. S., and Fristrom, J. W. (1990). Characterization of IMP-E3, a gene active during imaginal disc morphogenesis in Drosophila melanogaster. Develop. Genet. 11, 299–309.

Moore, R. C. and Oka, T. (1993). Cloning and sequencing of the cDNA encoding the murine mammary gland long-form prolactin receptor. Gene 134, 263–265.

Moras, D. and Gronemeyer, H. (1998). The nuclear receptor ligand-binding domain: structure and function. Curr. Opinion Cell Biol. 10, 384–391.

Morley, J. E. (1981). Neuroendocrine control of thyrotropin secretion. Endocrin. Rev. 2, 396–436.

Morris, S. M. Jr. (1987). Thyroxine elicits divergent changes in mRNA levels of two urea cycle enzymes and one gluconeogenic enzyme in tadpole liver. Arch. Biochem. Biophys. 259, 144–148.

Moskaitis, J. E., Sargen, T. D., Smith Jr., L. H., Pastori, R. L., and Schoenberg, D. R. (1989). Xenopus laevis serum albumin: sequence of the complementary deoxyribonucleic acids encoding the 68- and 74-kilodalton peptides and the regulation of albumin gene expression by thyroid hormone during development. Mol. Endocrinol. 3, 464–473.

Muchardt, C. and Yaniv, M. (1993). A human homologue of Saccharomyces cerevisiae SNF2/SWI2 and Drosophila brm genes potentiates transcriptional activation by the glucocorticoid receptor. EMBO J. 12, 4279–4290.

Mukae, N., Enari, M., Sakahira, H., Fukuda, Y., Inazawa, J., Toh, H., and Nagata, S. (1998). Molecular cloning and characterization of human caspase-activated DNase. Biochemistry. 95, 9123–9128.

Muller D., Wolf C., Abecassis J., Millon R., Engelmann A., Bronner G., Rouyer N., Rio M. C., Eber M., Methlin G., et al. (1993). Increased stromelysin 3 gene expression is associated with increased local invasiveness in head and neck squamous cell carcinomas. Cancer Res. 53, 165–169.

Munro, A. F. (1939). Nitrogen excretion and arginase activity during amphibian development. Biochem. J. 33, 1957–1965.

Munro, A. F. (1953). The ammonia and urea excretion of different species of amphibian during their development and metamorphosis. Biochem. J. 54, 29–36.

Murata, E. and Merker, H. J. (1991). Morphologic changes of the basal lamina in the small intestine of Xenopus laevis during metamorphosis. Acta Anat. 140, 60–69.

Murphy, G., Willenbrock, R., Crabbe, T., O'Shea, M., Ward, R., Atkinson, S., O'Connell, J., and Docherty, A. (1994). Regulation of matrix metalloproteinase activity. Ann. N. Y. Acad. Sci. 732, 31–41.

Mymryk, J. S. and Archer, T. K. (1995). Dissection of progesterone receptor-mediated chromatin remodeling and transcriptional activation in vivo. Genes Develop. 9, 1366–1376.

Naar, A. M., Boutin, J.-M., Lipkin, S. M., Yu, V. C., Holloway, J. M., Glass, C. K., and Rosenfeld, M. G. (1991). The orientation and spacing of core DNA-binding motifs dictate selective transcriptional responses to three nuclear receptors. Cell 65, 1267–1279.

Nagasawa, T., Takeda, T., Minemura, K., and DeGroot, L. J. (1997). Oct-1, silencer sequence, and GC box regulate thyroid hormone receptor $\beta1$ promoter. Mol. Cell Endocrinol. 130, 153–165.

Nagase, H. (1998). Cell surface activation of progelatinase A (pro MMP-2) and cell migration. Cell Res. 8, 179–186.

Nagase, J., Suzuki, K., Morodomi, T., Enghild, J. J., and Salvesen, G. (1992). Activation mechanisms of the precursors of matrix metalloproteinases 1, 2, and 3. MATRIX Suppl. 1, 237–244.

Nagata, K., Guggenheimer, R. A., Enomoto, T., Lichy, J. H., and Hurwitz, J. (1982). Adenovirus DNA replication in vitro: Identification of a host factor that stimulates synthesis of the preterminal protein-dCMP complex. Proc. Natl. Acad. Sci. USA 21, 6438–6442.

Nagy L., Kao H. Y., Chakravarti D., Lin R. J., Hassig C. A., Ayer D. E., Schreinber S. L., and Evans R. M. (1997). Nuclear receptor repression mediated by a complex containing SMRT, mSin3A, and histone deacetylase. Cell 89, 373–380.

Nakagawa, H., Kim, K., and Cohen, P. P. (1967). Studies on ribonucleic acid synthesis in tadpole liver during metamorphosis induced by thyroxine. I. Relation of synthesis of ribonucleic acid and of carbamylphosphate synthetase. J. Biol. Chem. 242–635.

Nakai, A., Sakurai, A., Bell, G. I., and DeGroot, L. J. (1988a). Characterization of a third human thyroid hormone receptor coexpressed with other thyroid hormone receptors in several tissues. Mol. Endocrinol. 2, 1087–1092.

Nakai, A., Seino, S., Sakurai, A., Szilak, I., Bell, G. I., and DeGroot, L. J. (1988b). Characterization of a thyroid hormone receptor expressed in human kidney and other tissues. Proc. Natl. Acad. Sci. USA 85, 2781–2785.

Narita, M., Shimizu, S., Ito, T., Chittenden, T., Lutz, R. J., Matsuda, H., and Tsujimoto, Y. (1998). Bax interacts with the permeability transition pore to induce permeability transition and cytochrome c release in isolated mitochondria. Proc. Natl. Acad. Sci. USA 95, 14681–14686.

Natzle, J. E. (1993). Temporal regulation of Drosophila imaginal disc morphogenesis: A hierarchy of primary and secondary 20-hydroxyecdysone-responsive loci. Develop. Biol. 155, 516–532.

Natzle, J. E., Hammonds, A. S., and Fristrom, J. W. (1986). Isolation of genes active during hormone-induced morphogenesis in Drosophila imaginal discs. J. Biol. Chem. 261, 5575–5583.

Natzle, J. E., Fristrom, D. K., and Fristrom, J. W. (1988). Genes expressed during imaginal disc morphogenesis: IMP-E1, a gene associated with epithelial cell rearrangement. Develop. Biol. 129, 428–438.

Naylor, S. L., Bishop, D. T. (1989). Report of the committee on the genetic constitution of chromosome 3 (HGH10). Cytogenet. Cell Genet. 51, 106–120.

Newman, R. A. (1989). Developmental plasticity of Scaphiopus couchii tadpoles in an unpredictable environment. Ecology 70, 1775–1787.

Newman, R. A. (1992). Adaptive plasticity in amphibian metamorphosis. Bioscience 42, 671–678.

Nicholson, D. W. and Thornberry, N. A. (1997). Caspases: killer proteases. TIBS 22, 299–306.

Nicoll, C. S. and Nichols Jr., C. W. (1971). Evolutionary biology of prolactins and somatotropins. I. Electrophoretic comparison of tetrapod prolactins. Gen. Comp. Endocrinol. 17, 300–310.

Nicoll, C. S., Bern, H. A., Dunlop, D., and Strohman, R. C. (1965). Prolactin, growth hormone, thyroxine and growth of tadpoles of Rana catesbeiana. Am. Zool. 5, 738–739.

Nieuwkoop, P. D. and Faber, J. (1956). *Normal Table of Xenopus laevis*. North Holland Publishing, Amsterdam.

Niki, K. and Yoshizato, K. (1986). An epidermal factor which induces thyroid hormone-dependent regression of mesenchymal tissues of the tadpole tail. Develop. Biol. 118, 306–308.

Niki, K., Namiki, H., Kikuyama, S., and Yoshizato, K. (1982). Epidermal tissue requirement for tadpole tail regression induced by thyroid hormone. Develop. Biol. 94, 116–120.

Nishikawa, A. and Hayashi, H. (1995). Spatial, temporal and hormonal regulation of programmed muscle cell death during metamorphosis of the frog Xenopus laevis. Differentiation 59, 207–214.

Nishikawa, A. and Yoshizato, K. (1986). Hormonal regulation of growth and life span of bullfrog tadpole tail epidermal cells cultured in vitro. J. Experimental Zool. 237, 221-230

Nishikawa, A., Kaiho, M., and Yoshizato, K. (1989). Cell death in the anuran tadpole tail: Thyroid hormone induces keratinization and tail-specific growth inhibition of epidermal cells. Develop. Biol. 131, 337–344.

Nishikawa, A., Shimizu-Nishikawa, K., and Miller, L. (1992). Spatial, temporal, and hormonal regulation of epidermal keratin expression during development of the frog, Xenopus laevis. Develop. Biol. 151, 145–153.

Nishikawa, A., Murata, E., Akita, M., Kaneko, K., Moriya, O., Tomita, M., and Hayashi, H. (1998). Roles of macrophages in programmed cell death and remodeling of tail and body muscle of Xenopus laevis during metamorphosis. Histochem. Cell Biol. 109, 11–17.

Noel, A. C., Lefebvre, O., Maquoi, E., VanHoorde, L., Chenard, M. P., Mareel, M., Foidart, J.-M., Basset, P., and Rio, M.-C. (1997). Stromelysin-3 expression promotes tumor take in nude mice. J. Clin. Invest. 97, 1924–2930.

Noiva, R. and Lennarz W. J. (1992). Protein disulfide isomerase. A multifunctional protein resident in the lumen of the endoplasmic reticulum. J. Biol. Chem. 267, 3553–3556.

Nolte, R. T., Wisely, G. B., Westin, S., Cobb, J. E., Lambert, M. H., Kurokawa, R., Rosenfeld, M. G., Willson, T. M., Glass, C. K., and Milburn, M. V. (1998). Ligand binding and co-activator assembly of the peroxisome proliferator-activated receptor-γ. Nature 395, 137–143.

Norton, V. G., Imai, B. S., Yau, P., and Bradbury, E. M. (1989). Histone acetylation reduces nucleosome core particle linking number change. Cell 57, 449–457.

O'Donnell, A. L., Rosen, E. D., Darling, D. S., and Koenig, R. J. (1991). Thyroid hormone receptor mutations that interfere with transcriptional activation also interfere with receptor interaction with a nuclear protein. Mol. Endocrinol. 5, 94–99.

Ogryzko V. V., Schiltz R. L., Russanova V., Horward B. H., and Nakatani Y. (1996). The transcriptional coactivators p300 and CBP are histone acetyltransferases. Cell 87, 953–959.

Olivereau, M., Vandesande, F., Boucique, E., Ollevier, F., and Oliveau, J. M. (1987). Immunocytochemical localization and spatial relation to the adenohypophysis of a somatostatin-like and a corticotropin-releasing factor-like peptide in the brain of four amphibian species. Cell Tissue. Res. 247, 317–324.

Onate, S. A., Tsai, S. Y., Tsai, M.-J., and O'Malley, B. W. (1995). Sequence and characterization of a coactivator for the steroid hormone receptor superfamily. Science 270, 1354–1357.

Onigata, K., Yagi, H., Sakuri, A., Nagashima, T., Nomura, Y., Nagashima, K., Hashizume, K., and Morikawa, A. (1995). A novel point mutation (R243Q) in exon 7 of the c-erbA beta thyroid hormone receptor gene in a family with resistance to thyroid hormone. Thyroid 5, 355–358.

Oofusa, K., Yomori, S., and Yoshizato, K. (1994). Regionally and hormonally regulated expression of genes of collagen and collagenase in the anuran larval skin. Int. J. Develop. Biol. 38, 345–350.

Oofusa, K., and Yoshizato, K. (1991) Biochemical and immunological characterization of collagenase in tissue of metamorphosing bullfrog tadpoles. Develop. Growth of Differ., 33, 329–339.

Oofusa, K. and Yoshizato, K. (1996). Thyroid hormone-dependent expression of bullfrog tadpole collagenase gene. Roux's Arch Develop. Biol. 205, 241–251.

Oppenheimer, J. H. (1979). Thyroid hormone action at the cellular level. Science 203, 971–979.

Oppenheimer, J. H., Schwartz, H. L., Mariash, C. N., Kinlaw, W. B., Wong, N. C. W., and Freake, H. C. (1987). Advances in our understanding of thyroid hormone action at the cellular level. Endocr. Rev. 8, 288–308.

Oro, A. and Evans, R. M. (1990). Relationship between the product of the Drosophila ultraspiracle locus and the vertebrate retinoid X receptor. Nature 347, 298–301.

Orth, K., O'Rourke, K., Salvesen, G. S., and Dixit, V. M. (1996). Molecular ordering of apoptotic mammalian CED-3/ICE-like proteases. J. Biol. Chem 271, 20977–20980.

Ostlund Farrangs, A.-K., Blomquist, P., Kwon, H., and Wrange, O. (1997). Glucocorticoid receptor-glucocorticoid response element binding stimulates nucleosome disruption by the SWI/SNF complex. Mol. Cell Biol. 17, 895–905.

Ouatas, T., Mevel, S. L., Demenleix, B. H., and De Luze, A. (1998). T_3-dependent physiological regulation of transcription in the Xenopus tadpoles brain studied by polyethylenimine based in vivo gene transfer. Int. J. Dev. Biol. 42; 1159–1164.

Owen-Hughes, T., Utley, R. T., Cote, J., Peterson, C. L., and Workman, J. L. (1996). Persistent site-specific remodeling of a nucleosome array by transient action of the SWI/SNF complex. Science 273, 513–516.

Paik, W. and Cohen, P. P. (1960). Biochemical studies on amphibian metamorphosis. I. The effect of thyroxine on protein synthesis in the tadpole. J. Gen. Physiol. 43, 683–696.

Paine-Saunders S, Fristrom D, and Fristrom J. W. (1990). The *Drosophila* IMP-E2 gene encodes an apically secreted protein expressed during imaginal disc morphogenesis. Dev. Biol. 140, 337–351.

Parisi, M. J. and Lin, H. (1998). The role of the hedgehog/patched signaling pathway in epithelial stem cell proliferation: from fly to human. Cell Res. 8, 15–21.

Parkison, C., Ashizawa, K., McPhie, P., Lin, K-H., and Cheng, S-Y. (1991). The monomer of pyruvate kinase, subtype M1, is both a kinase and a cytosolic thyroid hormone binding protein. Biochem. Biophys. Res. Commun. 179, 668–674.

Parks, W. C. and Mecham, R. P. (1998). *Matrix Metalloproteinases*. Academic Press, New York.

Parrilla, R., Mixson, A. J., McPherson, J. A., McClaskey, J. H., and Weintraub, B. D. (1991). Characterization of seven novel mutations of the c-erbA gene in unrelated kindreds with generalized thyroid hormone resistance. Evidence for two "hot spot" regions of the ligand binding domain. J. Clin. Invest. 88, 2123–2130.

Patterton, D. and Shi, Y.-B. (1994). Thyroid hormone-dependent differential regulation of multiple arginase genes during amphibian metamorphosis. J. Biol. Chem. 269, 25328–25334.

Patterton, D., Hayes, W. P., and Shi, Y.-B. (1995). Transcriptional activation of the matrix metalloproteinase gene stromelysin-3 coincides with thyroid hormone-induced cell death during frog metamorphosis. Develop. Biol. 167, 252–262.

Pazin, M. J. and Kadonaga, J. T. (1997). What's up and down with histone deacetylation and transcription? Cell 89, 325–328.

Pei, D. and Weiss, S. J. (1995). Furin-dependent intracellular activation of the human stromelysin-3 zymogen. Nature 375, 244–247.

Pei, D., Majmudar, G., and Weiss, S. J. (1994). Hydrolytic inactivation of a breast carcinoma cell-derived serpin by human stromelysin-3. J. Biol. Chem. 269, 25849–25855.

Perlman, A. J., Stanley, F., and Samuels, H. H. (1982). Thyroid hormone nuclear receptor. J. Biol. Chem. 257, 930–938.

Perlmann, T., Rangarajan, P. N., Umesono, K., and Evans, R. M. (1993). Determinants for selective RAR and TR recognition of direct repeat HREs. Genes Dev. 7, 1411–1422.

Perrimon, N. (1995). Hedgehog and beyond. Cell 80, 517–520.

Phillips, C. R. (1991). Neural Induction. Methods Cell Biol. 36, 329–346.

Picard, D. and Yamamoto, K. R. (1987). Two signals mediate hormone-dependent nuclear localization of the glucocorticoid receptor. EMBO J. 6, 3333–3340.

Pierce, J. G. and Parsons, T. F. (1981). Glycoprotein hormones: structure and function. Ann. Rev. Physiol. 50, 465–495.

Pihlajaniemi, T., Helaakoshi, T., Tasanen, K., Myllyla, R., Huhtala, M.-L., Koivu, J., and I. Kivirikko, K. (1987). Molecular cloning of the β-subunit of human prolyl

4-hydroxylase. This subunit and protein disulphide isomerase are products of the same gene. EMBO J. 6, 643–649.

Pina, B., Bruggemeier, U., and Beato, M. (1990). Nucleosome positioning modulates accessibility of regulatory proteins to the mouse mammary tumor virus promoter. Cell 60, 719–731.

Platt, N., Silva, R. P. D., and Gordon, S. (1998). Recognizing death: The phagocytosis of apoptotic cells. Trends Cell Biol. 8, 365–372.

Pogo, B. G. T., Allfrey, V. G., and Mirsky, A. E. (1966). RNA synthesis and histone acetylation during the course of gene activation in lymphocytes. Proc. Natl. Acad. Sci. USA 55, 805–812.

Porter, J. A., von Kessler, D. P., Ekker, S. C., Yong, K. E., Lee, J. J., Moses, K., Beachy, P. A. (1995). The product of hedgehog autoproteolytic cleavage active in local and long-range signalling. Nature 374, 363–366.

Porter, J. A., Young, K., and Beachy, P. (1996a). Cholesterol modification of hedgehog signaling proteins in animal development. Science 274, 255–259. Porter, J. A., et al. (1996b). Hedgehog pattering activity-role of a lipophilic modification mediated by the carboxy-terminal auto-processing domain. Cell 86, 21–34.

Porterfield, S. P. and Hendrich, C. E. (1993). The role of thyroid hormones in prenatal and neonatal neurological development: current perspectives. Endocr. Rev. 14, 94–106.

Pouyet, J.-C. and Hourdry, J. (1977). Effet de la thyroxine sur la structure des epitheliocytes intestinaux en culture organotypique, chez la larve de Discoglossus pictus otth (Amphibien Anoure). Biol. Cell. 29, 123–134.

Powell, W. C. and Matrisian, L. M. (1997). Mechanisms by which matrix metalloproteinases may influence apoptosis. In *Programmed Cell Death* (Y. B. Shi, Y. Shi, Y. Xu, and D. W. Scott, eds.), Plenum Press, New York, pp. 27–34.

Prahlad, K. V. and DeLanney, L. E. (1965). A study of induced metamorphosis in the Axolotl. J. Exp. Zool. 160, 137–146.

Price, S. and Frieden, E. (1963). β-Glucuronidase from amphibian tails. Comp. Biochem. Physiol. 10, 245–251.

Puzianowska-Kuznicka, M. and Shi, Y.-B. (1996). Nuclear factor I as a potential regulator during post-embryonic organ development. J. Biol. Chem. 271, 6273–6282.

Puzianowska-Kuznicka, M., Wing, J., Kanamori, A., and Shi, Y.-B. (1996). Functional characterization of a mutant thyroid hormone receptor in Xenopus laevis. J. Biol. Chem. 271, 33394–33403.

Puzianowska-Kuznicka, M., Damjanovski, S., and Shi, Y.-B. (1997). Both thyroid hormone and 9-cis retinoic acid receptors are required to efficiently mediate the effects of thyroid hormone on embryonic development and specific gene regulation in Xenopus laevis. Mol. Cell Biol. 17, 4738–4749.

Rachez, C., Suldan, Z., Ward, J., Betty Chang, C.-P., Burakov, D., Erdjument-Bromage, H., Tempst, P., and Freedman, L. P. (1998). A novel protein complex that interacts with the vitamin D_3 receptor in a ligand-dependent manner and enhances VDR transactivation in a cell-free system. Genes Develop. 12, 1787–1800.

Ranjan, M., Wong, J. and Shi, Y.-B. (1994). Transcriptional repression of Xenopuf TRβ gene is mediated by a thyroid hormone response element located near the start site. J. Biol. Chem. 269, 24699–24705.

Rao, L. and White, E. (1997). Bcl-2 and the ICE family of apoptotic regulators: making a connection. Curr. Opinion Genet. Dev. 7, 52–58.

Rastinejad, F., Perlmann, T., Evans, R. M., and Sigler, P. B. (1995). Structural determinants of nuclear receptor assembly on DNA direct repeats. Nature 375, 203–211.

Raz, Y. and Kelley, M. W. (1997). Effects of retinoid and thyroid receptors during development of the inner ear. Semin. Cell Develop. Biol. 8, 257–264.

Reeder, W. G. (1964). The digestive system. In *Physiology of the Amphibia* (J. A. Moore, ed.), Academic Press, New York, pp. 99–149.

Refetoff, S., DeWind, L. T., and DeGroot, L. J. (1967). Familial syndrome combining deaf mutism, stippled epiphyses, goiter, and abnormally high PB1: Possible target organ refractoriness to thyroid hormone. J. Clin. Endocrinol. Metab. 27, 279–294.

Refetoff, S., DeGroot, L. J., Bernard, B., and DeWind, L. T. (1972). Studies of a sibship with apparent hereditary resistance to the intracellular action of thyroid hormone. Metabolism 21, 723–756.

Refetoff, S., Weiss, R. E., and Usala, S. J. (1993). The syndromes of resistance to thyroid hormone. Endocrinol. Rev. 14, 348–399.

Regard, E. (1978). Cytophysiology of the amphibian thyroid gland through larval development and metamorphosis. Int. Rev. Cytol. 52, 81–118.

Reilley, D. S., Tomassini, N., and Zasloff, M. (1994). Expression of magainin antimicrobial peptidegenes in the developing granular glands of Xenopus skin and induction by thyroid hormone. Develop. Biol. 162, 123–133.

Renaud, J.-P., Rochel, N., Ruff, M., Vivat, V., Chambon, P., Gronemeyer, H., and Moras, D. (1995). Crystal structure of the RAR-γ ligand-binding domain bound to all-trans retinoic acid. Nature 378, 681–689.

Ribeiro, R. C. J., Cavalieri, R. R., Lomri, N., Rahmaoui, C. M., Baxter, J. D., and Scharschmidt, B. F. (1996) Thyroid hormone export regulates cellular hormone content and response. J. Biol. Chem. 271, 17147–17151.

Richard-Foy, H. and Hager, G. L. (1987). Sequence-specific positioning of nucleosomes over the steroid-inducible MMTV promoter. EMBO J. 6, 2321–2328.

Richards, G. (1981a). Insect hormones in development. Biol. Rev. 56, 501–549.

Richards, G. (1981b). The radioimmune assay of ecdysteroid titres in *Drosophila melanogaster*. Mol. Cell Endocrinol. 21, 181–197.

Richards, C. M. and Nace, G. W. (1978). Gynogenetic and hormonal sex reversal used in tests of the XX-XY hypothesis of sex determination in Rana pipiens. Growth 42, 319–331.

Riddiford, L. M. (1993a). Hormones and Drosophila development. In *The Development of Drosophila melanogaster* (M. Bate, and A. M. Arias, eds.), Vol. II, Cold Spring Harbor Laboratory Press, Cold Spring Harbor, NY, pp. 899–939.

Riddiford, L. M. (1993b). Hormone receptors and the regulation of in sect metamorphosis. Receptor 3, 203–209.

Riddiford, L. M. (1996). Molecular aspects of juvenile hormone action in insect metamorphosis. In *Metamorphosis: Postembryonic Reprogramming of Gene Expression in Amphibian and Insect Cells.* (L. I. Gilbert, J. R. Tata, and B. G. Atkinson, eds.), Academic Press, New York, pp. 223–251.

Riddiford, L. M. and Truman, J. W. (1978). Biochemistry of insect hormones and insect growth regulators. In *Biochemistry of Insects*, (M. Rockstein, ed), Academic Press, New York, pp. 307–356.

Riggs, A. (1951). The metamorphosis of hemoglobin in the bullfrog. J. Gen. Physiol. 35, 23–40.

Riggs, A. (1960). The nature and significance of the Bohr effect in mammalian hemoglobins. J. Gen. Physiol. 43, 737–752.

Rivier, J., Rivier, C., and Vale, W. (1984). Synthetic competitive antagonists of corticotropin-releasing factor: Effect on ACTH secretion in the rat. Science 224, 889–891.

Rivkees, S. A., Bode, H. H., and Crawford, J. D. (1988). Long-term growth in juvenile acquired hypothyroidism. New Engl. J. Med. 318, 599–602.

Robbins, J. (1992). Thyroxine transport and the free hormone hypothesis. Endocrinology 131, 546–547.

Roberts, D. J., Johnson, R. L., Burke, A. C., Nelson, C. E., Morgan, B. A., and Tabin, C. (1995). Sonic hedgehog is an endodermal signal inducing Bmp-4 and Hox genes during induction and regionalization of the chick hindgut. Development 121, 3163–3174.

Robinow, S., Talbot, W. S., Hogness, D. S., and Truman, J. W. (1993). Programmed cell death in the Drosophila CNS is ecdysone-regulated and coupled with a specific ecdysone receptor isoform. Development 119, 1251–1259.

Robinson, H. (1970). Acid phosphatase in the tail of Xenopus laevis during development and metamorphosis. J. Exp. Zool. 173, 215–224.

Robinson, H. (1972). An electrophoretic and biochemical analysis of acid phosphatase in the tail of Xenopus laevis during development and metamorphosis. J. Exp. Zool. 180, 127–140.

Roseland, C. R. and Schneiderman, H. A. (1979). Regulation and metamorphosis of the abdominal histoblasts of Drosophila melanogaster. Roux's Arch. Dev. Biol. 186, 235–265.

Rosenkilde, P. and Ussing, A. P. (1996). What mechanisms control neoteny and regulate induced metamorphosis in urodeles. Int. J. Dev. Biol. 40, 665–673.

Roskelley, C. D., Srebrow, A., and Bissell, M. J. (1995). A hierarchy of ECM-mediated signalling regulates tissue-specific gene expression. Curr. Biol. 7, 736–747.

Rossi, A. (1959). Tavole cronologiche dello svilluppo embryionale e larvale del "Bufo bufo" Mont. Zool. Ital. 66, 133.

Roth, P. (1941). Action antagoniste du propionate de testosterone dans la metamorphose experimentale des batraciens provoquee par thyroxine. Bull. Mus. Natl. Hist. Nat. (Paris) 13, 500–502.

Roth, P. (1948). Sur l'action antagoniste des substances oestrogenes dans la metamorphose experimentale des amphibiens (3e note). Bull. Mus. Natl. Hist. Natl. (Paris) 20, 408–415.

Roth, S. Y. and Allis, C. D. (1996). Histone acetylation and chromatin assembly: A single escort, multiple dances? Cell 87, 5–8.

Roy, N., Mahadevan, M. S., McLean, M., Shutler, G., Yaraghi, Z., Farahani, R., Baird, S., Besner-Johnson, A., Lefebvre, V, Kang X, et al. (1995). The gene for neuronal apoptotis inhibitory proteinis partially deleted in individuals with spinal muscular atrophy. Cell 80, 167–178.

Ruoslahti, E. and Reed, J. C. (1994). Anchorage dependence, integrins, and apoptosis. Cell 77, 477–478.

Rui, H., Kirken, R. A., and Farrar, W. L. (1994). Activation of receptor-associated tyrosine kinase JAK2 by prolactin. J. Biol. Chem. 269, 5364–5368.

Ruiz i Altaba, A. (1997). Catching a Glimpse of hedgehog. Cell 90, 193–196.

Russell, S., and Ashburner, M. C. (1996). Ecdysone-regulated chromosome puffing in Drosphila melanogaster. In Metamorphosis: Postembryonic reprogramming of gene expression during amphibian and insect cells. (L. J. Gilbert, J. R. Tata, and B. G. Atkinson, eds.), Academic Press, New York, pp. 109–144.

Sachs, L. M., Abdallah, B., Hassan, A., Levi, G., De Luze, A., Reed, J. C., and Demeneix, B. A. (1997a). Apoptosis in Xenopus tadpole tail muscles involves Bax-dependent pathways. FASEB J. 11, 801–808.

Sachs, L. M., Lebrun, J. J., de Luze, A., Kelly, P. A., and Demeneix, B. A. (1997b). Tail regression, apoptosis and thyroid hormone regulation of myosin heavy chain isoforms in Xenopus tadpoles. Mol. Cell Endocrinol. 131, 211–219.

Sachs, L. M., De Luze, A., and Demeneix, B. A. (1998). Studying amphibian metamorphosis by in vivo gene transfer. Ann. N. Y. Acad. Sci. 839, 152–156.

Safer, J. D., Langlois, M. F., Cohen, R., et al. (1997). Isoform variable action among thyroid hormone receptor mutants provides insight into pituitary resistance to thyroid hormone. Mol. Endocrinol. 11, 16–26.

Safi, R., Begue, A., Haenni, C., Stehelin, D., Tata, J. R., and Laudet, V. (1997a). Thyroid hormone receptor genes of neotenic amphibians. J. Mol. Evol. 44, 595–604.

Safi, R., Deprez, A., and Laudet, V. (1997b). Thyroid hormone receptors in perennibranchiate amphibians. Int. J. Dev. Biol. 41, 533–535.

Sakahira, H., Enari, M., and Nagata, S. (1998). Cleavage of CAD inhibitor in CAD activation and DNA degradation during apoptosis. Nature 391, 96–99.

Sakai, M., Hanaoka, Y., Tanaka, S., Hayashi, H., and Kikuyama, S. (1991). Thyrotropic activity of various adenohypophyseal hormones of the bullfrog. Zool. Sci. 8, 929–934.

Sakurai, A., Takeda, K., Ain, K., Ceccarelli, P., Nakai, A., Seino, S., Bell, G. I., Refetoff, S., and DeGroot, L. J. (1989). Generalized resistance to thyroid hormone associated with a mutation in the ligand-binding domain of the human thyroid hormone receptor β. Proc. Natl. Acad. Sci. USA 86, 8977–8981.

Sakurai, A., Nakai, A., and DeGroot, L. J. (1990). Structural analysis of human thyroid hormone receptor β gene. Mol. Cell. Endocrinol. 71, 83–91.

Salamonsen, L. A. (1996). Matrix metalloproteinases and their tissue inhibitors in endometrial remodelling and menstruation. Reproductive Med. Rev. 5, 185–203.

Salvesen, G. S. and Dixit, V. M. (1997). Caspases: Intracellular signaling by proteolysis. Cell 91, 443–446.

Sande, S. and Privalsky, M. L. (1994). Reconstitution of thyroid hormone receptor and retinoic acid receptor function in the fission yeast schizosaccharomyces pombe. Mol. Endocrinol. 8, 1455–1464.

Sandhofer, C. R., Forrest, D., Schwartz, H. L., and Oppenheimer, J. H. (1996). Apparent lack of effect of thyroid hormone receptor β (TRβ) knockout on Purkinje cell-specific protein (PCP2) and myelin basic protein (MBP) (abstr). Thyroid 6 (suppl 1):S32.

Sang, Q. A. and Douglas, D. A. (1996). Computational sequence analysis of matrix metalloproteinases. J. Protein Chem. 15, 137–160.

Sang, Q. X. A. (1998). Complex role of matrix metalloproteinases in angiogenesis. Cell Res. 8, 171–177.

Sap, J., Munoz, A., Damm, K., Goldberg, Y., Ghysdael, J., Leutz, A., Berg, H., and Vennstrom, B. (1986). The C-erb-A protein is a high affinity receptor for thyroid hormone. Nature 324, 635–640.

Sap, J., Munoz, A., Schmitt, J., Stunnenberg, H., and Vennstrom, B. (1989). Repression of transcription mediated at a thyroid hormone response element by the v-erb-A oncogene product. Nature 340, 242–244.

Sargent, T. D. (1987). Isolation of differentially expressed genes. Methods Enzymol. 152, 423–432.

Sargent, T. D. and Dawid, L. B. (1983). Differential gene expression in the gastrula of Xenopus laevis. Science 222, 135–139.

Sargent, T. D. and Mathers, P. H. (1991). Analysis of class II gene regulation. Methods Cell Biol. 36, 347–367.

Sato, H. and Seiki M. (1996). Membrane-Type Matrix Metallproteinases (MT-MMPs) in Tumor Metastasis. J. Biochem. 119, 209–215.

Savage, J. M. (1973) The geographic distribution of frogs: Patterns and predictions. In *Evolutionary Biology of the Anurans: Contemporary Research on Major Problems* (J. L. Vial, ed.), University of Missouri Press, Columbia, MO, pp. 351–445.

Savill, J. (1998). Phagocytic docking without shocking. Nature 392, 442–509.

Savill, J., Fadok, V., Henson, P., and Haslett, C. (1993). Phagocyte recognition of cells undergoing apoptosis. Immunol. Today, 14, 131–136.

Saxen, L., Saxen, E., Toivonen, S., and Salimaki, K. (1957a). Quantitative investigation on the anterior pituitary-thyroid mechanism during frog metamorphosis. Endocrinology. 61, 35–44.

Saxen, L., Saxen, E., Toivonen, S., and Salimaki, K. (1957b). The anterior pituitary and the thyroid function during normal and abnormal development of the frog. Ann. Zool. Soc. Zool. Bot. Fenn. Varamo 18, 1–44.

Schlaepfer, D. D. and Hunter, T. (1998). Integrin signalling and tyrosine phosphorylation: just the FAKs? Trends Cell. Biol. 8, 151–157.

Schmidt, J. W., Piepenhagen, P. A., and Nelson, W. J. (1993). Modulation of epithelial morphogenesis and cell fate by cell-to-cell signals and regulated cell adhesion. Semin. Cell Biol. 4, 161–173.

Schneider, M. J. and Galton, V. A. (1991). Regulation of c-erbA-α messenger RNA species in tadpole erythrocytes by thyroid hormone. Mol. Endocrinol. 5, 201–208.

Schneider, M. J. and Galton, V. A. (1995). Effect of glucocorticoids on thyroid hormone action in cultured red blood cells from Rana catesbeiana tadpoles. Endocrinology. 136, 1435–1440.

Schueler, P. A., Schwartz, H. L., Strait, K. A., Mariash, C. N., and Oppenheimer, J. H. (1990). Binding of 3, 5, 3′-triiodothyronine (T_3) and its analogs to the in vitro translational products of c-erbA protooncogenes: differences in the affinity of the α- and β-forms for the acetic acid analog and failure of the human testis and kidney α-2 products to bind T_3. Mol. Endocrinol. 4, 227–234.

Schuler, L. A., Nagel, R. J., Gao, J., Horseman, N. D., and Kessler, M. A. (1997). Prolactin receptor heterogeneity in bovine fetal and maternal tissues. Endocrinology. 138, 3187–3194.

Schultz, J. J., Price, M. P., and Frieden, E. (1988). Triiodothyronine increases translatable albumin messenger RNA in *Rana catesbeiana* tadpole liver. J. Exp. Zool. 247, 69–76.

Schwartz, H. L. (1983). Effects of thyroid hormone on growth and development. In *Molecular Basis of Thyroid Hormone Action*. (J. Oppenheimer and H. Samuels, eds.), Academic Press, NY, pp. 413–444.

Schwartzman R. A., and Cidlowski, J. A. (1993). Apoptosis: The biochemistry and molecular biology of programmed cell death. Endocrine Rev. 14, 133–151.

Scott, P., Kessler, M. A., and Schuler, L. A. (1992). Molecular cloning of the bovine prolactin receptor and distribution of prolactin and growth hormone receptor transcripts in fetal and uteroplacental tissues. Mol. Cell. Endocrinol. 89, 47–58.

Sealy, L. and Chalkley, R. (1978). DNA associated with hyperacetylated histone is preferentially digested by DNase I. Nucl. Acids Res. 5, 1863–1876.

Segal, G. H. and Petras, R. E. 1992. Small intestine. In *Histology for Pathologists* (S. S. Sternberg, ed), Raven Press, New York, pp. 547–571.

Segraves, W. A. and Hogness, D. S. (1990). The E75 ecdysone-inducible gene responsible for the 75B early puff in Drosophila encodes two new members of the steroid receptor superfamily. Genes Dev. 4, 204–219.

Seidman, S. and Soreq, H. (1997). Transgenic Xenopus: microinjection methods and developmental neurobiology. Neuromethods 28, 1–165.

Shaffer, H. B., and Voss, S. R. (1996). Phylogenetic and mechanistic analysis of a developmentally integrated character complex: Alternate life history modes in ambystomatid salamanders. Am. Zool. 36, 24–35.

Shapiro, S. D. (1998). Matrix metalloproteinase degradation of extracellular matrix: Biological consequences. Curr. Opinion Cell Biol. 10, 602–608.

Shi, Y.-B. (1994). Molecular biology of amphibian metamorphosis: A new approach to an old problem. Trends Endocrinol. Metab. 5, 14–20.

Shi Y.-B. (1996). Thyroid hormone-regulated early and late genes during amphibian metamorphosis. In *Metamorphosis: Postembryonic Reprogramming of Gene Expression in Amphibian and Insect Cells*. (L. I. Gilbert, J. R. Tata, and B. G. Atkinson eds.). Academic Press, New York, pp. 505–538.

Shi, Y.-B. (1997). Cell-cell and cell-ECM interactions in epithelial apoptosis and cell renewal during frog intestinal development. Cell Biochem. Biophys. 27, 179–202.

Shi, Y.-B. and Brown, D. D. (1990). Developmental and thyroid hormone dependent regulation of pancreatic genes in Xenopus laevis. Genes Develop. 4, 1107–1113.

Shi, Y.-B. and Brown, D. D. (1993). The earliest changes in gene expression in tadpole intestine induced by thyroid hormone. J. Biol. Chem. 268, 20312–20317.

Shi, Y.-B. and Hayes, W. P. (1994). Thyroid hormone-dependent regulation of the intestinal fatty acid-binding protein gene during amphibian metamorphosis. Develop. Biol. 161, 48–58.

Shi, Y.-B. and Ishizuya-Oka, A. (1996). Biphasic intestinal development in amphibians: Embryogensis and remodeling during metamorphosis. Curr. Topics Develop. Biol. 32, 205–235.

Shi Y.-B. and Ishizuya-Oka, A. (1997a). Autoactivation of Xenopus Thyroid Hormone Receptor β GENES Correlates with Larval Epithelial Apoptosis and Adult Cell Proliferation. J. Biomed. Sci. 4, 9–18.

Shi, Y.-B. and Ishizuya-Oka, A. (1997b). Activation of matrix metalloproteine genes during thyroid hormone-induced apoptotic tissue remodeling. In *Programmed Cell Death* (Y.-B Shi, Y. Shi, Y. Xu, and D. W. Scott, eds.), Plenum Press, New York, pp. 13–26.

Shi, Y.-B. and Liang, V. C.-T. (1994). Cloning and characterization of the ribosomal protein L8 gene from Xenopus laevis. Biochim. Biophys. Acta 1217, 227–228.

Shi, Y.-B., Yaoita, Y., and Brown, D. D. (1992). Genomic organization and alternative promoter usage of the two thyroid hormone receptor genes in Xenopus laevis. J. Biol. Chem. 267, 733–788.

Shi, Y.-B., Shi, Y., Xu, Y., and Scott, D. W. (1997). *Programmed Cell Death*. Plenum Press, New York

Shi, Y.-B., Liang, V. C.-T., Parkison, C., and Cheng, S.-Y. (1994). Tissue-dependent developmental expression of a cytosolic thyroid hormone protein gene in Xenopus: its role in the regulation of amphibian metamorphosis. FEBS Letters 355, 61–64.

Shi, Y.-B., Su, Y., Li, Q., and Damjanovski, S. (1998) Auto-regulation of thyroid hormone receptor genes during metamorphosis: Roles in apoptosis and cell proliferation. Int. J. Dev. Biol. 42, 107–116.

Shi, Y. F., Sahai, B. M., and Green, D. R. (1989). Cyclosporin A inhibits activation-induced cell death in T-cell hybridomas and thymocytes. Nature 339, 625–626.

Shimizu-Nishikawa, K. and Miller, L. (1992). Hormonal regulation of adult type keratin gene expression in larval epidermal cells of the frog Xenopus laevis. Differentiation 49, 77–83.

Shirota, M., Banville, D. L., Ali, S., Jolicoeur, C., Boutin, J.-M., Edery, M., Djiane, J., and Kelly, P. A. (1990). Expression of two forms of prolactin receptor in rat ovary and liver. Mol. Endocrinol. 4, 1136–1143.

Silva, J. E. (1995). Thyroid hormone control of thermogenesis and energy balance. Thyroid 5, 481–492.

Simo, P., Simon-Assmann, P., Bouzigos, F., Leberquier, C., Kedinger, M., Ekblom, P., and Sorokin, L. (1991). Changes in the expression of laminin during intestinal development. Development 112, 477–487.

Simo, P., Simon-Assmann, P., Arnold, C., and Kedinger, M. (1992). Mesenchyme-mediated effect of dexamethasone on laminin in cocultures of embryonic gut epithelial cells and mesenchyme-derived cells. J. Cell. Sci. 101, 161–171.

Simon, T. C. and Gordon, J. I. (1995). Intestinal epithelial cell differentiation: New insights from mice, flies and nematodes. Curr. Biol. 5, 577–586.

Simon-Assmann, P., Bouziges, F., Arnold, C., Haffen, K., and Kedinger, M. (1988). Epithelial-mesenchymal interactions in the production of the basement membrane components in the gut. Development 102, 339–347.

Simon-Assmann, P. and Kedinger, M. (1993). Heterotypic cellular cooperation in gut morphogenesis and differentiation. Semin. Cell Biol. 4, 221–230.

Simon-Assmann, P., Duclos, B., Orian-Rousseau, V., Arnold, C., Mathelin, C., Engvali, E., and Kedinger, M. (1994). Differential expression of laminin isoforms and $\alpha 6$-$\beta 4$ integrin subunits in the developing human and mouse intestine. Dev. Dynamics 201, 71–85.

Skaer, H. (1993) The alimentary canal. In *The Development of Drosophila Melanogaster* (M. Bate, A. M. Arias, eds.), Vol. II, Cold Spring Harbor Laboratory Press, Cold Spring Harbor, NY, pp. 941–1012.

Sladek, F. M. and Darnell, J. E. (1992). Mechanisms of liver-specific gene expression. Curr. Opinion Gene Develop. 2, 256–259.

Smith, J. C. (1994). Hedgehog, the floor plate, and the zone of polarizing activity. Cell 76, 193–196.

Smith, P. E. (1916). Experimental ablation of the hypophysis in the frog embryo. Science 44, 280–282.

Smith-Gill, S. J. and Carver, V. (1981). *Biochemical Characterization of Organ Differentiation and Maturation.* In *Metamorphosis: A Problem in Developmental Biology.* (L. I. Gilbert, and E. Frieden eds.), New York, Plenum Press, pp. 491–544.

Solomon, E. and Barker, D. F. (1989). Report of the committee on the genetic constitution of chromosome 17 (HGM10) Cytogenet. Cell Genet. 51, 319–337.

Spanjaard, R. A., Darling, D. S., and Chin, W. W. (1991). Ligand binding and heterodimerization activities of a conserved region in the ligand-binding domain of the thyroid hormone receptor. Proc. Natl. Acad. Sci. USA 88, 8587–8591.

Spector, M. S., Desnoyers, S., Hoeppner, D. J., and Hengartner, M. O. (1997). Interaction between the C. elegans cell-death regulators CED-9 and CED-4. Nature 385, 653–656.

Spencer T. E., Jenster G., Burcin M. M., Allis D., Zhou J., Mizzen C. A., McKenna N. J., Onate S. A., Tsai S. Y., Tsai M.-J., and O'Malley B. W. (1997). Steroid receptor coactivor-1 is a histone acetyltransferase. Nature 389, 194–198.

Stallmach, A., Hahn, U., Merker, H. J., Hahn, E. G., and Riecken, E. O. (1989). Differentiation of rat intestinal epithelial cells is induced by organotypic mesenchymal cells in vitro. Gut 30, 959–970.

Stearns, S. C. (1989). The evolutionary significance of phenotypic plasticity. Bioscience 39, 436–445.

Steger, D. J. and Workman, J. L. (1996). Remodeling chromatin structures for transcription: what happens to the histones? Bio Essays 18, 875–884.

Stenzel-Poore, M. P., Heldwein, K. A., Stenzel, P., Lee, S., and Vale, W. W. (1992). Characterization of the genomic corticotropin-releasing factor (CRF) gene from Xenopus laevis: Two members of the CRF family exist in amphibians. Mol. Endocrinol. 6, 1716–1724.

St. Germain, D. L. (1994). Iodothyronine deiodinases. Trends Endocrinol. Metab. 5, 36:42.

St. Germain, D. L., Schwartzman, R. A., Croteau, W., Kanamori, A., Wang, Z., Brown, D. D., and Galton, V. A. (1994). A thyroid hormone-regulated gene in Xenopus laevis encodes a type III iodothyronine 5-deiodinase. Proc. Natl. Acad. Sci. USA 91, 7767–7771.

St. Germain, D. L. and Galton, V. A. (1997). The deiodinase family of selenoproteins. Thyroid 7, 655–668.

Stetler-Stevenson, W. G., Aznavoorian, S., and Liotta, L. A. (1993). Tumor cell interactions with the extracellular matrix during invasion and metastasis. Ann. Rev. Cell Biol. 9, 541–573.

Stoecklin, E., Wissler, M., Gouilleux, F., and Groner, B. (1996). Functional interactions between Stat5 and the glucocorticoid receptor. Nature 383, 726–728.

Stoecklin, E., Wissler, M., Moriggl, R., and Groner, B. (1997). Specific DNA binding of Stat5, but not of glucocorticoid receptor, is required for their functional cooperation in the regulation of gene transcription. Mol. Cell. Biol. 17, 6708–6716.

Stolow, M. A. and Shi, Y.-B. (1995). Xenopus sonic hedgehog as a potential morphogen during embryogenesis and thyroid hormone-dependent metamorphosis. Nucl. Acids Res. 23, 2555–2562.

Stolow, M. A., Bauzon, D. D., Li, J., Segwick, T., Liang, V. C-T., Sang, Q. A., and Shi, Y.-B. (1996). Identification and characterization of a novel ollagenase in Xenopus laevis: Possible roles during frog development. Mol. Biol. Cell 7, 1471–1483.

Stolow, M. A., Ishizuya-Oka, A., Su, Y., and Shi, Y.-B. (1997). Gene regulation by thyroid hormone during amphibian metamorphosis: implications on the role of cell-cell and cell-extracellular matrix interactions. Amer. Zool. 37, 195–207.

Stone, D. M., Hynes, M., Armanini, M., Swanson, T. A., Ku, O., Johnson, R. L., Scott, M. P., Pennica, D., Goddard, A., Phillips, H., et al. (1996). The tumor-suppressor gene patched encodes a candidate receptor for sonic hedgehog. Nature 384, 129–134.

Stromblad, S., Brooks, P. C., Becker, J., Rosenfeld, M., and Cheresch, D. A. (1997). The role of integrin $\alpha v \beta 3$ in cell survival and angiogenesis. In *Programmed Cell Death* (Y.-B. Shi, Y. Shi, Y. Xu, and D. W. Scott, eds.), Plenum Press, New York, pp. 35–42.

Struhl, K. (1998). Histone acetylation and transcriptional regulatory mechanisms. Genes Develop. 12, 599–606.

Struhl, K. (1996). Chromatin structure and RNA polymerase II connection: Implications for transcription. Cell 84, 179–182.

Su, Y., Shi, Y., and Shi, Y.-B. (1997a). Cyclosporin A but not FK506 inhibits thyroid hormone-induced apoptosis in Xenopus tadpole intestinal epithelium. FASEB J. 11, 559–565.

Su, Y., Shi, Y., Stolow, M., and Shi, Y.-B. (1997b). Thyroid hormone induces apoptosis in primary cell cultures of tadpole intestine: cell type specificity and effects of extracellular matrix. J. Cell Biol. 139, 1533–1543.

Sulston, J. (1998). Cell lineage. In *The Nematade Caenorhabditis Elegans.* (edited by Wood and the community of C. elegans researchers). Cold Spring Harbor Laboratory, Cold Spring Harbor, NY, pp 123–155.

Suzuki, S., Miyamoto, T., Opsahl, A., Sakurai, A., and DeGroot, L. J. (1994). Two thyroid hormone response elements are present in the promoter of human thyroid hormone receptor $\beta 1$. Mol. Endocrinol. 8, 305–314.

Suzuki, S., Takeda, T., Liu, R.-T., Hashizume, K., and DeGroot, L. J. (1995). Importance of the most proximal GC box for activity of the promoter of human thyroid hormone receptor $\beta 1$. Mol. Endocrinol. 9, 1288–1296.

Svaren, J. and Horz, W. (1993). Histones, nucleosomes and transcription. Curr. Opinion Genet. Develop. 3, 219–225.

Svaren, J. and Horz, W. (1996). Regulation of gene expression by nucleosomes. Curr. Opinion Genet. Develop., 6, 164–170.

Svaren, J., Schmitz, J., and Horz, W. (1994). The transactivation domain of Pho4 is required for nucleosome disruption at the PHO5 promoter. EMBO J. 13, 4856–4862.

Schwartzman, R. A. and Cidlowski, J. A. (1993). Apoptosis: the biochemistry and molecular biology of programmed cell death. 14, 133–151.

Sympson, C. J., Talhouk, R. S., Alexander, C. M., Chin, J. R., Clift, S. M., Bissell, M. J., and Werb, Z. (1994). Targeted expression of stromelysin-1 in mammary gland provides evidence for a role of proteinases in branching morphogenesis and the requirement for an intact basement membrane for tissue-specific gene expression. J. Cell Biol. 125, 681–693.

Takahashi, N., Yoshihama, K., Kikuyama, S., Yamamoto, K., Wakabayashi, K., and Kato, Y. (1990). Molecular cloning and nucleotide sequence analysis of complementary DNA for bullfrog prolactin. J. Mol. Endocrinol. 5, 281–287.

Takeda, K., Sakurai, A., DeGroot, L. J., and Refetoff, S. (1992a). Recessive inheritance of thyroid hormone resistance caused by complete deletion of the protein-coding region of the thyroid hormone receptor-β gene. J. Clin. Endocrinol. Metab. 74, 49–55.

Takeda, K., Weiss, R. E., and Refetoff, S. (1992b). Rapid localisation of mutations in the thyroid hormone β receptor gene by denaturing gradient gel electrophoresis in 18 families with thyroid hormone resistance. J. Clin Endocrinol. Metab. 74, 712–719.

Takeshita, A., Cardona, G. R., Koibuchi, N., Suen, C.-S., and Chin, W. W. (1997). TRAM-1, a novel 160-kDa thyroid hormone receptor activator molecule, exhibits distinct properties from steroid receptor coactivator-1. J. Biol. Chem. 272, 27629–27634.

Takeshita, A., Yen, P. M., Ikeda, M., Cardona, G. R., Liu, Y., Koibuchi, N., Norwitz, E. R., and Chin, W. W. (1998). Thyroid hormone response elements differentially modulate the interactions of thyroid hormone receptors with two receptor binding domains in the steroid receptor coactivator-1. J. Biol. Chem. 273, 21554–21562.

Talbot, N. S., Swyryd, E. A., and Hogness, D. S. (1993). Drosophila tissues with different metamorphic responses to ecdysone express different ecdysone receptor isoforms. Cell 73, 1323–1337.

Talhouk, R. S., Bissell, M. J., and Werb, Z. (1992). Coordinated expression of extracellular matrix-degrading proteinases and their inhibitors regulates mammary epithelial function during involution. J. Cell Biol. 118, 1271–1282.

Tanaka, M., Maeda, K., Okubo, T., and Nakashima, K. (1992). Double antenna structure of chicken prolactin receptor deduced from the cDNA sequence. Biochem. Biophys. Res. Commun. 188, 490–496.

Tanaka, S., Sakai, M., Park, M. K., and Kurosumi, K. (1991). Differential appearance of the subunits of glycoprotein hormones (LH, FSH, and TSH) in the pituitary of bullfrog (Rana catesbeiana) larvae during metamorphosis. Gen. Comp. Endocrinol. 84, 318–327.

Tanaka-Hall, T. M., Porter, J. A., Beachy, P. A., and Leahy, D. J. (1995). A potential catalytic site revealed by the 1.7-Å crystal structure of the amino-terminal signalling domain of sonic hedgehog. Nature 378, 212–216.

Tanenbaum, D. M., Wang, Y., Williams, S. P., and Sigler, P. B. (1998). Crystallographic comparison of the estrogen and progesterone receptor's ligand binding domains. Proc. Natl. Acad. Sci. USA 95, 5998–6003.

Tata, J. R. (1965). Turnover of nuclear and cytoplasmic ribonucleic acid at the onset of induced metamorphosis. Nature (London) 207, 378–381.

Tata, J. R. (1966). Requirement for RNA and protein synthesis for induced regression of the tadpole tail in organ culture. Develop. Biol. 13, 77–94.

Tata, J. R. (1967). The formation and distribution of ribosomes during hormone-induced growth and development. Biochem. J. 104, 1.

Tata, J. R. (1968). Early metamorphic competence of Xenopus larvae. Develop. Biol. 18, 415–440.

Tata, J. R. (1972). Hormonal regulation of metamorphosis. Symp. Soc. Exp. Biol. 25, 163–181.

Tata, J. R., Kawahara, A., and Baker, B. S. (1991). Prolactin inhibits both thyroid hormone-induced morphogenesis and cell death in cultured amphibian larval tissues. Develop. Biol. 146, 72–80.

Tata, J. R. (1993). Gene expression during metamorphosis: An ideal model for post-embryonic development. BioEssays 15, 239–248.

Tata, J. R. (1996). Hormonal interplay and thyroid hormone receptor expression during amphibian metamorphosis. In *Metamorphosis: Postembryonic Reprogramming of Gene Expression in Amphibian and Insect Cells* (L. I. Gilbert, J. R. Tata, B. G. and Atkinson, eds.), Academic Press, New York, pp. 465–503.

Tata, J. R. (1997). How hormones regulate programmed cell death during amphibian-metamorphosis. In *Programmed Cell Death* (Y.-B. Shi, Y. Shi, Y. Xu, and D. W. Scott, eds.), Plenum Press, New York, pp. 1–11.

Tata, J. R. and Widnell, C. C. (1966). Ribonucleic acid synthesis during the early action of thyroid hormones. Biochem. J. 88, 604.

Tata, J. R. (1991). Brain development and molecular genetics. In *Dolentium Hominum: The human mind*. Proceedings of the Fifth International Conference Organised by the Pontifical Council, Vatican City, pp. 28–36.

Taylor, A. C. and Kollros, J. J. (1946). Stages in the normal development of Rana pipies larvae. Anat. Rec. 93, 7–23.

Thompson, C. B. (1995). Apoptosis in the pathogenesis and treatment of diseases. Science 267, 1456–1462.

Thornberry, N. A. and Lazebnik, Y. (1998). Caspases: enemies within. Science 281, 1312–1316.

Thummel, C. S. (1990). Puffs and gene regulation—molecular insights into the Drosophila ecdysone regulatory hierarchy. BioEssays 12, 561–568.

Thummel, C. S. (1995). From embryogenesis to metamorphosis: The regulation and function of Drosophila nuclear receptor superfamily members. Cell 83, 871–877.

Thummel, C. S. (1996). Flies on steroids—Drosophila metamorphois and the mechanisms of steroid hormone action. Trends Genet. 12, 306–310.

Thummel, C. S. (1997). Dueling orphans—interacting nuclear receptors coordinate Drosophila metamorphosis. BioEssays 19, 669–672.

Timpl, R. and Brown, J. C. (1996). Supramolecular assembly of basement membranes. BioEssays 18, 123–132.

Tone, Y., Collingwood, T. N., Adams, M., and Chatterjee, V. K. (1994). Functional analysis of a transactivation domain in the thyroid hormone β receptor. J. Biol. Chem. 269, 31157–31161.

Toney, J. H., Ling, W., Summerfield, A. E., Sanyal, G., Forman, B. M., Zhu, J., and Samuels, H. H. (1993). Conformational changes in chicken thyroid hormone receptor $\alpha 1$ induced by binding to ligand or to DNA. Biochemistry 32, 2–6.

Tonon, M.-C., Cuet, P., Lamacz, M., Jegou, S., Cote, J., Gouteux, L., Ling, N., Pelletier, G., and Vaudry, H. (1986). Comparative effects of corticotropin-releasing factor, arginien vasopressin, and related neuropeptides on the secretion of ACTH and α-MSH by frog anterior pituitary cells and neurointermediate lobes in vitro. Gen. Comp. Endocrinol. 61, 438–445.

Torchia, J., Glass, C., and Rosenfeld, M. G. (1998). Co-activators and co-repressors in the integration of transcriptional responses. Current Opinion Cell Biol. 10, 373–383.

Torihashi, S. (1990). Morphological changes of the myenteric plexus neurons in the bullfrog (Rana catesbeiana) duodenum during metamorphosis. J. Comp. Neurol. 302, 54–65.

Torrents, D., Estevez, R., Pineda, M, Fernandez, E., Lloberas, J., Shi, Y.-B., Zorzano, A., and Palacin, M. (1998) Identification of a new cDNA (y^+LAT-1) that co-expression system y^+L activity with 4F2hc in oocytes: A candidate gene for Lysinuric Protein Intolerance. I. Biol. Chem. 273, 32437–32445.

Truman, J. W. (1990). Metamorphosis of the central nervous system of Drosophila. J. Neurobiol. 21, 1072–1084.

Truman, J. W., Taylor, B. J., and Awad, T. A. (1993). Formation of the adult nervous system. In *The Development of Drosophila melanogaster* (M. Bate, A. M. Arias, eds.), Vol. II, Cold Spring Harbor Laboratory Press, Cold Spring Harbor, NY, pp. 1245–1275.

Truman, J. W., Talbot, W. S., Fahrbach, S. E., and Hogness, D. S. (1994). Ecdysone receptor expression in the CNS correlates with stage-specific responses to ecdysteroids during Drosophila and Manduca development. Development 120, 219–234.

Truss, M., Bartsch, J., Schelbert, A., Hach, R. J. G., and Beato, M. (1995). Hormone induces binding of receptors and transcription factors to a rearranged nucleosome on the MMTV promoter in vivo. EMBO J. 14, 1737–1751.

Tryggvason, K., Hoyhtya, M., and Salo, T. (1987). Proteolytic degradation of extracellular matrix in tumor invasion. Biochim. Biophys. Acta 907, 191–217.

Tsai, M.-J. and O'Malley, B. W. (1994). Molecular mechanisms of action of steroid/thyroid receptor superfamily members. Ann. Rev. Biochem. 63, 451–486.

Tsukiyama, T. and Wu, C. (1995). Purification and properties of an ATP-dependent nucleosome remodeling factor. Cell 83, 1011–1020.

Tsukiyama, T., Becker, P. B., and Wu, C. (1994). ATP-dependent nucleosome disruption at a heat-shock promoter mediated by binding of GAGA transcription factor. Nature 367, 525–532.

Tsukiyama, T., Daniel, C., Tamkun, J., and Wu, C. (1995). ISWI, a member of the SWI2/SNF2 ATPase family, encodes the 140 kDa subunit of the nucleosome remodeling factor. Cell 83, 1021–1026.

Ulisse, S., Esslemont, G., Baker, B. S., Chatterjee, V. K. K., and Tata, J. R. (1996). Dominant-negative mutant thyroid hormone receptors prevent transcription from Xenopus thyroid hormone receptor β gene promoter in response to thyroid hormone in Xenopus tadpoles in vivo. Proc. Natl. Acad. Sci. USA 93, 1205–1209.

Umek, R. M., Friedman, A. D., and McKnight, S. L. (1991). CCAAT-enhancer binding protein: A component of a differentiation switch. Science 251, 288–292.

Umesono, K., Murakami, K. K., Thompson, C. C., and Evans, R. M. (1991). Direct repeats as selective response elements for the thyroid hormone, retinoic acid, and vitamin D3 receptors. Cell 65, 1255–1266.

Underhay, E. E. and Baldwin, W. (1955). Nitrogen excretion in the tadpoles of Xenopus laevis daudin. Biochem. J. 61, 544–547.

Uppaluri, R. and Towle, H. C. (1995). Genetic dissection of thyroid hormone receptor β: identification of mutations that separate hormone binding and transcriptional activation. Mol. Cell Biol. 15, 1499–1512.

Uren, A. G., Pakusch, M., Hawkins, C. J., Puls, K. L., and Vaux, D. L. (1996). Cloning and expression of apoptosis inhibitory protein homologs that function to inhibit apoptosis and/or bind tumor necrosis factor receptor-associated factors. Proc. Natl. Acad. Sci. 93, 4974–4978.

Uria, J. A. and Werb, Z. (1998). Matrix metalloproteinases and their expression in mammary gland. Cell Res. 8, 187–194.

Urness, L. D. and Thummel, C. S. (1990). Molecular interactions within the ecdysone regulatory hierarchy: DNA binding properties of the Drosophila ecdysone-inducible E74A protein. Cell 63, 47–61.

Urness, L. D. and Thummel, C. S. (1995). Molecular analysis of a steroid-induced regulatory hierarchy: the Drosophila E74A protein directly regulates L71-6 transcription. EMBO J. 14, 6239–6246.

Usala, S. J. and Weintraub, B. D. (1991). Thyroid hormone resistance syndromes. Trends Endocrinol. Metab. 2, 140–144.

Usuku, G. and Gross, J. (1965). Morphologic studies of connective tissue resorption in the tail fin of metamorphosing bullfrog tadpole. Develop. Biol. 11, 352–370.

Utley, R. T., Ikeda, K., Grant, P. A., Cote, J., Steger, D. J., Eberharter, A., John, S., and Workman, J. L. (1998). Transcriptional activators direct histone acetyltransferase complexes to nucleosomes. Nature 394, 498–502.

Vale, W., Speiss, J., Rivier, C., and Rivier, J. (1981). Characterization of a 41-residue ovine hypothalamic peptide that stimulates corticotropin and β-endorphin. Science 213, 1394–1397.

Van den Heuvel, M. and Ingham, P. W. (1996). Smoothened encodes a receptor-like serpentine protein required for hedgehog signaling. Nature 382, 547–551.

van Wart, H. E. and Birkedal-Hansen, H. (1990). The cysteine switch: A principle of regulation of metalloproteinase activity with potential applicability to the entire matrix metalloproteinase gene family. Proc. Natl. Acad. Sci. USA 87, 5578–5582.

Van der Heiden, M. G., Chandel, N. S., Williamson, E. K., Schumacker, P. T., and Thompson, C. B. (1997). Bcl-xL regulates the membrane potential and volume homeostasis of mitochondria. Cell 91, 627–637.

Varga-Weisz, P., Blank, T. A., and Becker, P. B. (1995). Energy-dependent chromatin accessibility and nucleosome mobility in a cell free system. EMBO J. 14, 2209–2216.

Vaux, D. L. (1997). CED-4 The third horseman of apoptosis. Cell 90, 389–390.

Verhaert, P., Marivoet, S., Vandesande, F., and De Loof, A. (1984). Localization of CRF immunoreactivity in the central nervous system of three vertebrate and one insect species. Cell Tissue Res. 238, 49–53.

Vettese-Dadey, M., Grant, P. A., Hebbes, T. R., Crane-Robinson, C., Allis, C. D., and Workman, J. L. (1996). Acetylation of histone H4 plays a primary role in enhancing transcription factor binding to nucleosomal DNA in vitro. EMBO J. 15, 2508–2518.

Vidali, G., Boffa, L. C., Bradbury, E. M., and Allfrey, V. G. (1978). Butyrate suppression of histone deacetylation leads to accumulation of multiacetylated forms of histones

H3 and H4 and increased DNase I sensitivity of the associated DNA sequences. Proc. Natl. Acad. Sci. 75, 2239–2243.

Vize, P. D., Hemmati-Brivanlou, A., Harland, R. M., and Melton, D. A. (1991). Assays for gene function in developing Xenopus embryos. Methods Cell Biol. 36, 368–388.

von Kalm, L., Fristrom, D., and Fristrom, J. (1995). The making of a fly leg: A model for epithelial morphogenesis. BioEssays 17, 693–702.

Vu, T. H., Shipley, M. J., Bergers, G., Berger, J. E., Helms, J. A., Hanahan, D., Shapiro, S. D., Senior, R. M., and Werb, Z. (1998). MMP-9/gelatinase B is a key regulatory of growth plate angiogenesis and apoptosis of hypertrophic chondrocytes. Cell 93, 411–422.

Vukicevic, S., Kleinman, H. K., Luyten, F. P., Roberts, A. B., Roche, N. S., Reddi, A. H. (1992). Identification of multiple active growth factors in basement membrane matrigel suggests caution in interpretation of cellular activity related to extracellular matrix components. Exp. Cell Res. 202, 1–8.

Wade, P. A., and Wolffe, A. P. (1997). Histone acetyltransferases in control. Current Biology 7:R82–R84.

Wagner, R. L., Apriletti, J. W., McGrath, M. E., West, B. L., Baxter, J. D., and Fletterick, R. J. (1995). A structural role for hormone in the thyroid hormone receptor. Nature 378, 690–697.

Wakao, H., Gouilleux, F., and Groner, B. (1994). Mammary gland factor (MGF) is a novel member of the cytokine regulated transcription factor gene family and confers the PRL response. EMBO J. 13, 2182–2191.

Wake, D. B. (1966). Comparative osteology and evolution of the lungless salamanders, family plethodontidae. Mem. South. Calif. Acad. Sci. 4, 1–111.

Wang, S., Miura, M., Jung, Y.-K., Zhu, H., Li, E., and Yuan, J. (1998). Murine caspase-11, an ICE-interaction protease, is essential for the activation of ICE. Cell 92, 501–509.

Wang, Z. and Brown, D. D. (1991). A gene expression screen. Proc. Natl. Acad. Sci. USA 88, 11505–11509.

Wang, Z. and Brown, D. D. (1993). Thyroid hormone-induced gene expression program for amphibian tail resorption. J. Biol. Chem. 268, 16270–16278.

Wang, Z. and Reed, J. C. (1998). Mechanisms of Bcl-2 protein function. Histol. Histopathol. 13, 521–530.

Weber, R. (1964). Ultrastructural changes in regressing tail muscles of Xenopus laevis at metamorphosis. J. Cell Biol. 22, 481–487.

Weber, R. (1965). Inhibitory effect of actinomycin D on tail atrophy in Xenopus larvae at metamorphosis. Experientia 21, 665–666.

Weber, R. (1967). Biochemistry of amphibian metamorphosis. In *The Biochemistry of Animal Development* (R. Weber, ed.), Vol. 2, Academic Press, New York, pp. 227–301.

Weber, R. (1969). The isolated tadpole tail as a model system for studies on the mechanism of hormone-dependent tissue involution. Gen. Comp. Endocrinol. Suppl. 2, 408–416.

Weber, R. (1995). Inhibitory effect of actinomycin D on tail atrophy in Xenopus larvae at metamorphosis. Experientia 21, 655–666.

Weber, R. (1996). Switching of global genes during anuran metamorphosis. In *Metamorphosis: Postembryonic Reprogramming of Gene Expression in Amphibian and Insect Cells.* (L. I. Gilbert, R. R. Tata, and B. G. Atkinson, eds.). Academic Press, New York, pp. 567–597.

Weber, R., Geiser, M., Muller, P., Sandmeier, E., and Wyler, T. (1989). The metamorphic switch in hemoglobin phenotype of Xenopus laevis involves erythroid cell replacement. Wilhelm Roux's Arch. Dev. Biol. 198, 57–64.

Weber, R., Blum, B., and Muller, P. R. (1991). The switch from larval to adult globin gene expression in Xenopus laevis is mediated by erythroid cells from distinct compartments. Development 112, 1021–1029.

Weinberger, C., Thompson, C. C., Ong, E. S., Lebo, R., Gruol, D. J., and Evans, R. M. (1986). The c-erb-A gene encodes a thyroid hormone receptor. Nature 324, 641–646.

Weinberger, C., Giguere, V., Hollenberg, S., Rosenfeld, M. G., and Evans, R. M. (1988). Human steroid receptors and erbA proto-oncogene products: Members of a new superfamily of enhancer binding proteins. Cold Spring Harbor Symp. Quant. Biol. 51, 759–772.

Werb, Z., Sympson, C. J., Alexander, C. M., Thomasset, N., Lund, L. R., MacAuley, A., Ashenase, J., and Bissell, M. J. (1996). Extracellular matrix remodeling and the regulation of epithelial stromal interactions during differentiation and involution. Kidney International Suppl. 54, S68–S74.

White, B. A. and Nicoll, C. S. (1981). Hormonal control of amphibian metamorphosis. In *Metamorphosis: A Problem in Developmental Biology.* (L. I. Gilbert and E. Frieden, eds.), Plenum Press, New York, pp. 363–396.

White, E. (1993). Death-defying acts: a meeting review on apoptosis. Genes Develop. 7, 2277–2284.

White, E. (1996). Life, death, and the pursuit of apoptosis. Genes Develop. 10, 1–15.

White, K. P., Hurban, P., Watanabe, T., and Hogness, D. S. (1997). Coordination of Drosophila metamorphosis by two ecdysone-induced nuclear receptors. Science 276, 114–117.

Wikstrom, L., Johansson, C., Salto, Ca., Barlow, C., Barros, A. C., Baas, F., Forrest, D., Thoren, P., and Veenstrom, B. (1998). Abnormal heart rate and body temperature in mice lacking thyroid hormone receptor α1. EMBO J. 17, 455–461.

Wilhelm, S. M., Collier, I. E., Kronberger, A., Eisen, A. Z., Marmer, B. L., Grant, G. A., Bauer, E. A., and Goldberg, G. I. (1987). Human skin fibroblast stromelysin: Structure, glycosylation, substrate specificity, and differential expression in normal and tumorigenic cells. Proc. Natl. Acad. Sci. USA 84, 6725–6729.

Wise, R. W. (1970). An immunochemical comparison of tadpole and frog hemoglobins. Comp. Biochem. Physiol. 32, 89–95.

Witty, J. P., Lempka, T., Coffey, R. J., Jr., and Matrisian, L. M. (1995b). Decreased tumor formation in 7, 12-dimethylbenzanthracene-treated stromelysin-1 transgenic mice is associated with alterations in mammary epithelial cell apoptosis. Cancer Res. 55, 1401–1406.

Witty, J. P., Wright, J. H., and Matrisian, L. M. (1995a). Matrix metalloproteinases are expressed during ductal and alveolar mammary morphogenesis, and misregulation of stromelysin-1 intransigenic mice induces unscheduled alveolar development. Mol. Biol. Cell 6, 1287–1303.

Woessner Jr., J. F. (1991). Matrix metalloproteinases and their inhibitors in connective tissue remodeling. FASEB J. 5, 2145–2154.

Wolffe, A. P. (1994). Transcription: In tune with histones. Cell 77, 13–16.

Wolffe, A. P. (1998). *Chromatin: Structure and Function.* 3rd ed. Academic Press, London.

Wolffe, A. P. (1996). Histone deacetylase: a regulator of transcription. Science 272, 371–372.

Wolffe, A. P. (1997). Chromatin remodeling regulated by steroid and nuclear receptors. Cell Res. 7, 127–142.

Wolffe, A. P. and Pruss, D. (1996). Targeting chromatin disruption: Transcription regulators that acetylate histones. Cell 84, 817–819.

Wong, J. and Shi, Y.-B. (1995). Coordinated regulation of and transcriptional activation by *Xenopus* thyroid hormone and retinoid X receptors. J. Biol. Chem. 270, 18479–18483.

Wong, J., Shi, Y.-B., and Wolffe, A. P. (1995). A role for nucleosome assembly in both silencing and activation of the Xenopus TRβA gene by the thyroid Hormone receptor. Genes Develop. 9, 2696–2711.

Wong, J., Shi, Y.-B., and Wolffe, A. P. (1997a). Determinants of chromatin disruption and transcriptional regulation instigated by the thyroid hormone receptor: hormone-regulated chromatin disruption is not sufficient for transcriptional activation. EMBO J. 16, 3158–3171.

Wong, J., Li, Q., Levi, B.-Z, Shi, Y.-B., and Wolffe, A. P. (1997b). Structural and functional features of a specific nucleosome containing a recognition element for the thyroid hormone receptor. EMBO J. 16, 7130–7145.

Wong, J., Liang, V. C.-T., Sachs, L. M., and Shi, Y.-B. (1998a). Transcription from the thyroid hormone-dependent promoter of the Xenopus laevis thyroid hormone receptor βA gene requires a novel upstream element and the initiator, but not a TATA box. J. Biol. Chem. 273, 14186–14193.

Wong, J., Patterton, D., Imhof, A., Guschin, D., Shi, Y.-B., and Wolffe, A. P. (1998b). Distinct requirements for chromatin assembly in transcriptional repression by thyroid hormone receptor and histone deacetylase. EMBO J. 17, 520–534.

Wong, R., Vasilyev, V. V., Ting, Y. T., Kutler, D. I., Willingham, M. C., Weintraub, B. D., and Cheng, S. (1997c). Transgenic mice bearing a human mutant thyroid hormone beta 1 receptor manifest thyroid function anomalies, weight reduction and hyperactivity. Mol. Med. 3, 303–314.

Woo, M., Hakem, R., Soengas, M. S., Duncan, G. S., Shahinian, A., Kagi, D., Hakem, A., McCurrach, M., Khoo, W., Kaufman, S. A., et al., (1988). Essential contribution of caspase 3/CPP32 to apoptosis and its associated nuclear changes. Genes Develop. 12, 806–819.

Woodard, C. T., Baehrecke, E. H., and Thummel, C. S. (1994). A molecular mechanism for the stage specificity of the Drosophila prepupal genetic response to ecdysone. Cell 79, 607–615.

Woody, C. J. and Jaffe, R. C. (1984). Binding of dexamethasone by hepatic, intestine, and tail fin cytosol in Rana catesbeiana tadpoles during spontaneous and triiodothyronine-induced metamorphosis. Gen. Comp. Endocrinol. 54, 194–202.

Workman, J. L. and Kingston, R. E. (1998). Alteration of nucleosome structure as a mechanism of transcriptional regulation. Ann. Rev. Biochem. 67, 545–579.

Wright, L. G., Chen, T., Thummel, C. S., and Guild, G. M. (1996a). Molecular characterization of the 71E late puff in Drosophila melanogaster reveals a family of novel genes. J. Mol. Biol. 225, 387–400.

Wright, M. L., Blanchard, L. S., Jorey, S. T., Basso, C. A., Myers, Y. M., and Paquette, C. M. (1990). Metamorphic rate as a function of the light/dark cycle in Rana pipiens larvae. Comp. Biochem. Physiol. A. 96, 215–220.

Wright, M. L., Cykowski, L. J., Lundrigan, L., Hemond, K. L., Kochan, D. M., Faszewski, E. E., and Anuszewski, C. M. (1994). Anterior pituitary and adrenal cortical hormones accelerate or inhibit tadpole hindlimb growth and development depending on stage of spontaneous development or thyroxine concentration in induced metamorphosis. J. Exp. Zool. 270, 175–188.

Wright, M. L., Pikula, A., Cykowski, L. J., and Kuliga, K. (1996b). Effect of melatonin on the anuran thyroid gland: follicle cell proliferation, morphometry, and subsequent thyroid hormone secretion in vitro after melatonin treatment in vivo. Gen. Comp. Endocrinol. 103, 182–191.

Wright, M. L., Pikula, A., Babski, A. M., Labieniec, K. E., and Wolan, R. B. (1997). Effect of melatonin on the response of the thryroid to thyrotropin stimulation in vitro. Gen. Comp. Endocrinol. 108, 298–305.

Wu, D., Herschell, D. W., and Nunez, G. (1997). Interaction and Regulation of Subcellular Localization of CED-4 by CED-9. Science 275, 1126–1129.

Wu, Y.-C. and Horvitz, H. R. (1998). C. Elegans phagocytosis and cell-migration protein CED-5 is similar to human DOCK 180. Nature 392, 501–504.

Wyllie, A. H., Kerr, J. F. R., and Curie, A. R. (1980). Cell death: The significance of apoptosis. Int. Rev. Cytol. 68, 251–306.

Xing, H. and Shapiro, D. J. (1993). An estrogen receptor mutant exhibiting hormone-independent transactivation and enhanced affinity for the estrogen response element. J. Biol. Chem. 268, 23227–23233.

Xu, J., Qiu, Y., DeMayo, F. J., Tsai, S. Y., Tsai, M-J., and O'Malley, B. W. (1998). Partial hormone resistance in mice with disruption of the steroid receptor coactivator-1 (SRC-1) gene. Science 279, 1922–1925.

Xu, Q. and Tata, J. R. (1992). Characterization and developmental expression of Xenopus C/EBP gene. Mechanisms Develop. 38, 69–81.

Xu, Q., Baker, B. S., and Tata, J. R. (1993). Developmental and hormonal regulation of the Xenopus liver-type arginase gene. Eur. J. Biochem. 211, 891–898.

Xue, D. and Horvitz, H. R. (1997). Caenorhabditis elegans CED-9 protein is a bifunctional cell-death inhibitor. Nature 390, 305–308.

Xue-Yi, C., Xin-Min, J., Zhi-Hong, D., et al. (1994). Timing of vulnerability of the brain to iodine deficiency in endemic cretinism. New Engl. J. Med. 331, 1739–1744.

Yagi, H., Pohlenz, J., Hayashi, Y., Sakurai, A., and Refetoff, S. (1997). Resistance to thyroid hormone caused by two mutant thyroid hormone receptor Beta, R243Q and R243W, with marked impairment of function that cannot be explained by altered in vitro 3, 5, 3-triiodothyroinine binding affinity. J. Clin. Endocrinol. Metab. 82, 1608–1614.

Yamamoto, K. and Kikuyama, S. (1981). Purification and properties of bullfrog prolactin. Endocrinol. Jpn. 28, 59–64.

Yamamoto, K. and Kikuyama, S. (1982). Radioimmunoassay of prolactin in plasma of bullfrog tadpoles. Endocrinol. Jpn. 29, 159–167.

Yamamoto, K. and Kikuyama, S. (1993). Binding of aldosterone by epidermal cytosol in the tail of bullfrog larvae. Gen. Comp. Endrocrinol. 89, 283–290.

Yamamoto, K., Kanski, D., and Frieden, E. (1966). The uptake and excretion of thyroxine, triiodothyronine, and iodide in bull frog tadpoles after immersion or injection at 25EC and 6EC. Gen. Comp. Endocrinol. 6, 312–324.

Yamamoto, K., Niinuma, K., and Kikuyama, S. (1986). Synthesis and storage of prolactin in the pituitary gland of bullfrog tadpoles during metamorphosis. Gen. Comp. Endocrinol. 62, 247–253.

Yamauchi, K. and Tata, J. R. (1994). Purification and characterization of a cytosolic thyroid-hormone-binding protein (CTBP) in Xenopus liver. Eur. J. Biochem. 225, 1105–1112.

Yamauchi, K. and Tata, J. R. (1997). Tissue-dependent and developmentally regulated-cytosolic thyroid-hormone-binding proteins (CTBPs) in Xenopus. Comp. Biochem. Physiol. 118C, 27–32.

Yamauchi, K., Kasahara, T., Hayashi, H., and Horiuchi, R. (1993). Purification and characterization of a 3, 5, 3'-L-triiodothyronine-specific binding protein from bullfrog tadpole plasma: A homolog of mammalian transthyretin. Endocrinology 132, 2254–2261.

Yang, J., Liu, X., Bhalla, K., Kim, C. N., Ibrado, A. M., Cai, J., Peng, T.-I., Jones, D. P., and Wang, X. (1997). Prevention of apoptosis by Bcl-2: Release of cytochrome c from mitochondria blocked. Science 275, 1129–1132.

Yang, X., Chang, H. Y., and Baltimore, D. (1998). Essential role of CED-4 oligomerization in CED-3 activation and apoptosis. Science 281, 1355–1357.

Yao, T. P., Segraves, W. A., Oro, A. E., McKeown, M., and Evans, R. M. (1992). Drosophila ultraspiracle modulates ecdysone receptor function via heterodimer formation. Cell 71, 63–72.

Yaoita, Y. and Brown, D. D. (1990). A correlation of thyroid hormone receptor gene expression with amphibian metamorphosis. Gene Develop. 4, 1917–1924.

Yaoita, Y. and Nakajima, K. (1997). Induction of apoptosis and CPP32 expression by thyroid hormone in a myoblastic cell line derived from tadpole tail. J. Biol. Chem. 272, 5122–5127.

Yaoita, Y., Shi, Y.-B., and Brown, D. D. (1990). Xenopus α laevis and β thyroid hormone receptors. Proc. Natl. Acad. Sci. USA 87, 7090–7094.

Yasuda, A., Yamaguchi, K., Kobayashi, T., Yamamoto, K., Kikuyama, S., and Kawauchi, H. (1991). The complete amino acid sequence of prolactin from the bullfrog, Rana catesbeiana. Gen. Comp. Endocrinol. 83, 218–226.

Yen, P. M. and Chin, W. W. (1994). New advances in understanding the molecular-mechanisms of thyroid hormone action. Trends Endocrinol. Metab. 5, 65–72.

Yen, P. M., Sunday, M. E., Darling, D. S., and Chin, W. W. (1992). Isoform specific thyroid hormone receptor antibodies detect multiple thyroid hormone receptors in rat and human pituitaries. Endocrinology 130, 1539–1546.

Yen, P. M., Sugawara, A., Forgione, M., Spanjaard, R. A., Macchia, E., Cheng, S. Y., and Chin, W. W. (1993). Region-specific anti-thyroid hormone receptor (TR) antibodies detect changes in TR structure due to ligand-binding and dimerization. Mol. Cell. Endocrinol. 97, 93–99.

Yoshida, H., Kong, Y.-Y., Yoshida, R., Elia, A. J., Hakem, A., Hakom, R., Penninger, J. M., and Mak, T. W. (1998). Apaf1 is Required for mitochondrial pathways of apoptosis and brain development. Cell 94, 739–750.

Yoshinaga, S. K., Peterson, C. L., Herskowitz, I., and Yamamoto, K. R. (1992). Roles of SWI1, SWI2, and SWI3 proteins for transcriptional enhancement by steroid receptors. Science 258, 1598–1604.

Yoshizato, K. (1989). Biochemistry and cell biology of amphibian metamorphosis with a special emphasis on the mechanism of removal of larval organs. Int. Rev. Cytol. 119, 97–149.

Yoshizato, K. (1996). Cell death and histolysis in amphibian tail during metamorphosis. In *Metamorphosis Postembryonic Reprogramming of Gene Expression in Amphibian and Insect Cells.* (L. I. Gilbert, J. R. Tata, B. G. Atkinson, eds.), Academic Press, New York, pp. 647–671.

Yoshizato, K., Kikuyama, S., and Yasumasu, I. (1972). Presence of growth promoting factor in the frog pituitary gland. Gen. Comp. Endocrinol. 19, 247–252.

Yoh, S. M., Chatterjee, V. K. K., and Privalsky, M. L. (1997). Thyroid hormone resistance syndrome manifests as an aberrant interaction between mutant T_3 receptors and transcriptional corepressors. Mol. Endocrinol. 11, 470–480.

Yu, Y. C., Delsert, C., Andersen, B., Holloway, J. M., Devary, O. V., Naar, A. M., Kim, S. Y., Boutin, J. M., Glass, C. K., and Rosenfeld, M. G. (1991). RXRβ: A coregulator that enhances binding of retinoic acid, thyroid hormone, and vitamin D receptors to their cognate response elements. Cell 67, 1251–1266.

Yuan, C.-X., Ito, M., Fondell, J. D., Fu, Z.-Y., and Roeder, R. G. (1998). The TRAP220 component of a thyroid hormone receptor-associated protein (TRAP) coactivator complex interacts directly with nuclear receptors in a ligand-dependent fashion. Proc. Natl. Acad. Sci. USA 95, 7939–7944.

Yuan, J., Shaham, S., Ledoux, S., Elliss, H.M., and Horvitz, H. R. (1993). The C. elegans cell death gene ced-3 encodes a proteins similar to mammalian interleukin-1 beta-converting enzyme. Cell 75, 641–652.

Zamir, I., Harding, H. P., Atkins, G. B., Horlein, A., Glass, G. K., Rosenfeld, M. G., and Lazar, M. A. (1996). A nuclear hormone receptor corepressor mediates transcriptional silencing by receptors with distinct repression domains. Mol. Cell. Biol. 16, 5458–5465.

Zamir, I., Zhang, J., and Lazar, M. A. (1997a). Stoichiometric and steric principles governing repression by nuclear hormone receptors. Genes Develop. 11, 835–846.

Zamir, I., Dawson, J., Lavinsky, R. M., Glass, C. K., Rosenfeld, M. G., and Lazar, M. A. (1997b). Cloning and characterization of a corepressor and potential component of the nuclear hormone receptor repression complex. Proc. Natl. Acad. Sci. USA 94, 14400–14405.

Zaret, K. S. and Yamamoto, K. R. (1984). Reversible and persistent changes in chromatin structure accompany activation of a glucocorticoid dependent enhancer element. Cell 38, 29–38.

Zavacki, A. M., Harney, J. W., Brent, G. A., and Larsen, P. R. (1993). Dominant negative inhibition by mutant thyroid hormone receptors is thyroid hormone response element and receptor isoform specific. Mol. Endocrinol. 7, 1319–1330.

Zhang, X-K., Hoffmann, B., Tran, P. B.-V., Graupner, G., and Pfahl, M. (1992). Retinoid X receptor is an auxiliary protein for thyroid hormone and retinoic acid receptors. Nature 355, 441–446.

Zhang, Z., Jones, S., Hagood, J. S., Fuentes, N. L., and Fuller, G. M. (1997). STAT3 acts as a co-activator of glucocorticoid receptor signaling. J. Biol. Chem. 272, 30607–30610.

Zechel, C., Shen, X-Q., Chen J-Y., Chen, Z-P., Chambon, P., and Gronemeyer, H. (1994). The dimerization interfaces formed between the DNA binding domains of RXR, RAR and TR determine the binding specificity and polarity of the full-length receptors to direct repeats. EMBO J. 13, 1425–1433.

Zhuang, J., Ren, Y., Snowden, R. T., Zhu, H., Gogvadze, V., Savill, J. S., and Cohen, G. M. (1998). Dissociation of phagocyte recognition of cells undergoing apoptosis from other features of the apoptotic program. J. of Biol. Chem. 273, 15628–15632.

Zou, H., Henzel, W. J., Liu, X., Lutschg, A., and Wang, X. (1997). Apaf-1, a human protein homologous to C. elegans CED-4, participates in cytochrome c-dependent activation of cappase-3. Cell 90, 405–413.

Zucker, S. N., Catillo, G., and Horwitz, S. B. (1997). Down-regulation of the mdr gene by thyroid hormone during Xenopus laevis development. Mol. Cell. Endocrinol. 129, 73–81.

Index

Acid hydrolases, in tail resorption, 66–67
Acid phosphatase
  during intestinal remodeling, 67–68
  during tail resorption, 66–67
Activation of transcription, 90–91
  by TH, 88–90
Adrenocorticotropin (ACTH)
  corticoids and, 26–27
  CRF and, 24
Adult hemoglobin, 14
Albumin. See Serum albumin
Aldosterone, influence on metamorphosis, 25–26
Alytes obstetricans, environmental effects on metamorphosis in, 13
Ametabolous insects, 190
Amino acid sequences, of thyroid receptors, 213–215
Ammonotelism, 71–73
  transition to ureotelism from, 117–118
Amphibians
  classification of, 2–3
  intestinal development and metamorphosis in, 43–45
Anura, 2. See also Frogs; Tadpoles
  apoptosis during metamorphosis in, 50–65
  biochemical changes during metamorphosis in, 65–73
  metamorphosis in, 5–6
  metamorphosis of gills in, 39–40
  metamorphosis of tail in, 37–39
  morphological changes during metamorphosis in, 36–49
  nitrogenous waste excretion in, 71–73
  prolactin inhibition of metamorphosis in, 27–34
  thyroid gland development in, 18–19
Apoda, 2
Apoptosis, 51
  in cell cultures, 180–183
  in *Drosophila* salivary gland histolysis, 200–202
  inhibitors of, 55
  during intestinal remodeling, 55–61, 159
  mechanisms of, 50–55
  necrosis versus, 51–52
  stromelysin-3 gene and, 166–168
  during tail resorption, 61–64
  during tissue resorption, 50–65
  TRβ genes and, 140
Apoptotic bodies, 51–52, 53, 62, 63
Arginine synthetase, 73
Ashburner model, for ecdysone-induced gene regulation, 196–197

Basal lamina, 155, 169
  in intestinal metamorphosis, 156–159
Binding domains. See TR domains
Binding proteins, for thyroid hormone, 74–78

Blood
  metamorphosis and protein changes in, 68–71
  thyroid hormone action via, 76–77
Bone morphogenetic protein (BMP) genes, as signaling molecules, 160–162
Brain
  early TH response genes in frog, 110, 113–116, 126
  in frog metamorphosis, 42–43, 121
  TH role in development of human, 206
Bufo boreas, effects of corticoids on, 25–26
Bufo bufo, larval stage of, 7
Bufo bufo bufo, gonadal steroids in, 27
bZip domain, 128–130. See also TH/bZip gene

Caecilia, 2, 3
  metamorphosis in, 5
  thyroid hormone and, 5
Caenorhabditis elegans, apoptosis genes of, 52–55
Carbamyl phosphate synthetase (CPS) gene, 133–134
Cascade model, of gene regulation during metamorphosis, 102–104, 123–134, 202–204
Caudata, 2–3
Cell–cell interactions, 177–179
  extracellular matrix and, 154–155
  during intestinal metamorphosis, 155–159
  regulation of, 175–176
  regulation of intestinal, 159–163
Cell cultures, TH regulation of cell fate in, 180–183
Cell death. See Apoptosis; Necrosis
Cell proliferation, TRβ genes and, 140
Chemical inhibitors, effects on metamorphosis of, 18
Chordates, metamorphosis in, 4
Chromatin, 89
Chromatin fragmentation
  in apoptosis, 58–59
  in transcription, 91–96
Coactivation, of transcription, 90–91
Coactivator proteins, 90–91
  as histone acetyltransferases, 97–99
Collagen, during tail resorption, 67
Collagenase genes, 116, 171, 175
Collagenases, 164, 166, 170–175
  during tail resorption, 67
Competence
  molecular basis of, 150–153
  in TH response during metamorphosis, 135–136

Condensation, in tail resorption, 38–39
Connective tissue
  extracellular matrix in, 155
  in intestinal metamorphosis, 156–159
  in metamorphosis, 44, 47–48, 49, 118
  TRβ gene expression and metamorphosis of, 140, 141
Corepression, of transcription, 90–91
Corepressor proteins, 90–91
  in histone acetylation, 97–99
Cornufer hazelae, direct development of, 9
Corticoids, influence on metamorphosis, 25–27
Corticosterone, influence on metamorphosis, 25–26
Corticotropes, 22
Corticotropin-releasing factor (CRF)
  cloning of genes encoding, 25
  corticoids and, 26–27
  hypothalamus and, 24–25
Corticotropin-releasing hormone, effects on metamorphosis, 13
Cretinism, 207
CTHBP genes, 152
Cubitus interruptus factor, 162, 177–178
Cyclophilins, in apoptosis inhibition, 60
Cyclosporin A (CsA)
  as apoptosis inhibitor, 60
  in cell cultures, 181–182
Cytoplasm, thyroid hormone action in, 76–78
Cytosolic thyroid hormone binding proteins (CTHBPs), 145, 146–147, 148–149, 152

Degradative enzymes, during tail resorption, 65–67
Deiodinase genes, 152
Deiodinases
  families of, 147–149
  in TH biosynthesis, 147–150
Deoxycorticosterone, influence on metamorphosis of, 25–26
Developmental biology, central theme of, 1
Diet, effects on metamorphosis of, 11–13
Differential display, isolating TH response genes via, 107
Differential hybridization, isolating TH response genes via, 104–105
Differential screening
  isolating TH response genes via, 104–105, 105–107
  subtractive, 105–107
Digestive system, remodeling of frog, 116–117
Direct development, in anurans, 6–9
Direct Ec-response genes, 203–204

Direct TH response genes, 102–104, 203–204
DNA binding, by TRs, 84–87, 91–96
DNA injection experiments, 186
Double knockout, in mice, 222
Drosophila
  chromosomes of, 197
  ecdysone receptors in, 194–195
  ecdysone response genes in, 198–202
  hedgehog gene in, 159, 160–161
  hedgehog signaling in, 162
  hormonal regulation of metamorphosis in, 191–194
Drosophila melanogaster
  amphibian metamorphosis versus metamorphosis in, 202–204
  life cycle of, 193

Early puffs, 196–197, 198–200
Early TH response genes, 103–104, 108–116, 123, 204. *See also* TH/bZip gene
  table of, 112
Ecdysone receptors (EcRs), 194–196
Ecdysone response genes, 198–202, 202–204
  table of, 198
Ecdysones
  gene regulation cascade with, 196–202
  in insect metamorphosis, 191–194
  structure of, 192
Eggs. *See also* Oocytes
  in developmental biology, 1
  in direct development, 9
Eleutherodactylus coqui, direct development in, 9
Eleutherodactylus nubicola, direct development of, 9
Embryogenesis, postembryonic development and, 13–14
Embryo injection, 185–186
Endogenous corticoids, influence on metamorphosis of, 26
Environmental factors, in metamorphosis, 10–13
Enzymes. *See also* Acid hydrolases; Acid phosphatase; Arginine synthetase; Collagenases; Degradative enzymes; Deiodinases; Gelatinases; Histone acetyltransferases; Histone deacetylases; ICE-like proteases; Janus kinases (JAKs); Lysosomal enzymes; Matrix metalloproteinases (MMPs); RNA polymerase II; Urea cycle enzymes
  during liver metamorphosis, 71–73
  in tail resorption, 65–67, 109–112
  in transcriptional regulation, 96–99
Epidermal cells, metamorphosis of tail, 39, 63–64
Epithelium
  apoptosis during metamorphosis of, 55–61
  biochemical changes during metamorphosis of, 67–68
  cell cultures from, 181–183
  extracellular matrix in, 155
  in metamorphosis, 43–47, 47–49, 55–61, 136, 155–159, 169, 176–177
  TRβ gene expression and metamorphosis of, 140, 141
Estradiol, influence on metamorphosis of, 27
Exogenous corticoids, influence on metamorphosis of, 26
Extracellular matrix (ECM)
  in cell cultures, 183
  as cell-fate regulator, 176–177
  in frog embryos, 186
  functions of, 154–155
  interactions of cells with, 154–179
  during intestinal remodeling, 155–159
  matrix metalloproteinases and, 163–175
  in organ culture systems, 184
  regulation of cellular interactions with, 175–176

Fetal hemoglobin, 14
Forelimb development of, 41
Frog eggs, in developmental biology, 1
Frog embryos, studying gene regulation and function in, 185–186
Frog limbs
  apoptosis in, 65
  development of, 41
  early TH response genes in, 110, 113
Frogs. *See also* Anura
  apoptosis during intestinal remodeling in, 55–61
  biochemical changes during metamorphosis in, 65–73
  classification of, 2
  in developmental biology, 1–2
  environmental effects on metamorphosis in, 10–13
  enzymes during tail resorption in, 65–67
  intestinal metamorphosis in, 43–49, 55–61
  liver metamorphosis in, 42
  metamorphosis in, 5–6, 9–10
  nervous system metamorphosis in, 42–43
  nitrogenous waste excretion in, 71–73
Frog-specific organs, development of, 36, 41

INDEX

Gelatinases, 164, 170–175
Gene knockout, 187
   in mice, 218–222
Generalized resistance to thyroid hormone (GRTH) syndrome, 210–211
Gene regulation
   Ashburner model for, 196–197
   cascade model of, in metamorphosis, 102–104, 123–134
   in cell cultures, 180–183
   by ecdysones, 196–202
   in frog embryos, 185–186
   in organ cultures, 183–185
   by TH, 102–122
   in transgenic *Xenopus laevis*, 187–189
Gene transcription. *See* Transcription
Gills
   apoptosis in, 64–65
   degeneration of tadpole, 39–40
Glucocorticoid receptors (GRs), 26, 91
   prolactin function and, 34
   in transcription, 96
Glycoprotein, TSH as, 21
Goiter, 206–207
Goitrogens, effects on metamorphosis of, 18
Gonadal steroids, influence on metamorphosis of, 27
Gonadotropin-releasing hormone, effects on metamorphosis of, 35
Growth promotion, by prolactin, 27–28
Gymnophiona, 2

Hedgehog genes, 143–145
   during intestinal remodeling, 159–163
   molecular signaling via, 162–163, 177–178
Hemimetabolous insects, 190
Hemoglobins
   fetal versus adult, 14
   metamorphosis and changes in, 70–71, 206
Hind limb, early TH response genes in frog, 110, 113, 126
Hind limb development, 41, 118, 135–136, 139
   gene regulation during, 120–121
Histolysis, in tail resorption, 38–39
Histone acetylation, in transcriptional regulation, 96–97
Histone acetyltransferases, in transcriptional regulation, 96–99
Histone deacetylases, in transcriptional regulation, 96–99
Histone deacetylation, in TH regulation, 100–101
Holometabolous insects, 190–191

hormonal regulation of metamorphosis in, 191–194
Hormones. *See also* Thyroid hormone (TH); Thyroid-stimulating hormone (TSH)
   in metamorphosis regulation, 15–35
Human development, thyroid hormone in, 206–209
Human diseases
   thyroid hormone in, 206–209
   thyroid hormone receptor mutations in, 216–218
Human fetal plasma, thyroid hormones in, 206
Human, thyroid hormone and receptors in, 206–218
Hybridization
   differential, 104–105
   TH/bZip expression and, 127
Hydrolases, acid, 66–67
20-Hydroxyecdysone (Ec)
   in insect metamorphosis, 191–194
   structure of, 192
Hyperthyroidism, hypothyroidism versus, 210
Hypophysis. *See* Pituitary gland
Hypothalamus
   in metamorphosis, 24–25
   prolactin regulation by, 30
   in thyroid function, 19, 24–25
Hypothalectomy, effect of metamorphosis of, 24
Hypothyroidism, 208–209
   hyperthyroidism versus, 210

ICE-like proteases, 181–183
Immune cells, in metamorphosis, 48
Indirect TH response genes, 102–104
Inhibitors of apoptosis (IAPs), 55
Insect development, 190–191
Insect metamorphosis, amphibian metamorphosis versus, 190–204
Intestine
   early TH response genes in frog, 108–109, 110, 113, 116, 126
   late TH response genes in frog, 119
   MMP regulation during metamorphosis of, 170–173
   remodeling of frog, 43–49, 55–61, 67–68, 121, 136, 155–159, 159–163, 169, 205–206
   TRβ gene expression and metamorphosis of, 140, 141
Iodination, in thyroid hormone biosynthesis, 74–76
Iodine
   effects on metamorphosis of, 11–13
   goiter and, 207

Janus kinases (JAKs), prolactin function and, 34
Juvenile hormone (JH), in insect metamorphosis, 192–194

Lactotropes, 22
Laminin, 155, 156–158
Larval stage. *See also* Tadpoles
  biochemical changes during metamorphosis of, 65–73
  cellular changes during metamorphosis of, 50–65
  in direct development, 6–9
Late puffs, 196–197
Late TH response genes, 103–104, 108, 116–119, 133–134, 204
  table of, 114
Leptodactylidae, direct development in, 9
Leucine zipper region. *See* bZip domain
Light, effects on metamorphosis of, 13
Litoria glauerti, apoptosis during tail resorption in, 61
Liver
  metamorphosis and biochemical changes in, 71–73
  remodeling of frog, 42
  urea cycle enzymes in, 107–108, 117–118
Lungs, gill resorption and, 39–40
Lysosomal enzymes
  during intestinal remodeling, 67–68
  during tail resorption, 67

Macrophages, in epithelial apoptosis, 56–58
Mammals
  apoptosis genes of, 52–55
  postembryonic organogenesis in, 14
  thyroid hormone and receptors in, 205–206, 218–222
Manduca sexta, hormonal regulation of metamorphosis in, 191
Maternal TH, 207–208
Matrix metalloproteinases, 178–179
  differential regulation of, 168–175
  as ECM regulators, 163–175, 175–176
  in frog embryos, 185–186
  in organ culture systems, 184
Mauthner cells, death of, 42
Melatonin, effects on metamorphosis of, 35
Mesenchymal cells, in tail resorption, 39
Metabolism, of thyroid hormone, 74–78
Metalloproteinases. *See* Matrix metalloproteinases (MMPs)
Metamorphic climax, 6
  thyroid hormone concentrations during, 16, 17

of *Xenopus laevis*, 8
Metamorphosis, 4–6
  biochemical changes during, 50, 65–73
  blocking of, 17
  in caecilians, 5
  cascade model of gene regulation during, 102–104, 123–134
  cellular changes during, 50–65
  chemical inhibitors of, 18
  criteria of, 4
  developmental stages of, 135–136
  environmental effects on, 10–13
  extracellular matrix in regulation of, 154–179
  in frogs, 1–2, 5–6, 36–49
  hedgehog genes in, 159–163
  hormonal regulation of, 15–35
  hypothalamus in, 24–25
  influence of steroid hormones on, 25–27
  insect versus amphibian, 190–204
  as model for postembryonic vertebrate organ development, 13–14
  model systems for studying, 180–189
  morphological changes during, 36–49
  precocious, 16–17, 18
  prolactin inhibition of, 27–34
  of tadpole gills, 39–40
  of tadpole tail, 37–39
  temporal regulation of, 135–136
  thyroid gland and, 9–10
  thyroid hormone and, 15–18
  tissue-specific temporal regulation of, 135–153
  tissue-specific transformations during, 119–122
  in urodeles, 4–5
Mice, TR knockout studies in, 218–222
Muscle
  apoptosis during metamorphosis of, 61–64
  cell cultures from, 183
  extracellular matrix in, 155
  in metamorphosis, 48, 118
  TR-gene expression and metamorphosis of, 140, 141

Necrosis, apoptosis versus, 51–52
Necturus maculosus, thyroid hormone and, 5
Nervous system, remodeling of frog, 42–43
Neuroendocrine system
  effects on metamorphosis of, 13
  in synthesis of thyroid hormone, 19–25
Neurons, metamorphosis of myenteric, 48
Newts, 2. *See also* Urodela
Nitrogenous wastes, excretion of, 71–73
Northern blot hybridization, 104, 107

Notochord, apoptosis in, 64–65
Nuclear factor I (NFI) genes, 124–128, 129, 130, 131

Oocytes. See also Eggs
   histone deacetylases in, 97–99
Oocyte transcription system, 130
   in *Xenopus laevis*, 91–96
Organ cultures, TH regulation of metamorphosis in, 183–185
Organ development
   during metamorphosis, 36, 41–49
   in vertebrates, 13–14
Ornithine transcarbamoylase (OTC) gene, 133–134

Patched protein, 162, 177–178
PCR-based subtractive differential screening, 105–107
Peripheral tissue resistance to thyroid hormone (PTRTH) syndrome, 210–211
Pineal gland, effects on metamorphosis of, 35
Pituitary gland
   hypothalamus and, 24–25
   in thyroid function, 19–23, 24–25
Pituitary resistance to thyroid hormone (PRTH) syndrome, 210
Plasmid DNA, 187–188
Polytene puffs, 196–197, 198
Postembryonic organ development, in vertebrates, 13–14
Postembryonic organogenesis, 14
Premetamorphosis, 6
   intestinal length shrinkage during, 16, 19
   thyroid hormone concentrations during, 17
   of *Xenopus laevis*, 8
Prolactin (PRL)
   effects on TR genes of, 32–34
   growth promotion by, 27–28
   inhibition of metamorphosis by, 27–34
   models of action of, 31, 183
   regulation of, 28–30
   T_2 and, 30, 31
   TH and, 30–34
Prolactin receptors (PRL-Rs), prolactin function and, 32–34
Prometamorphosis, 6
   thyroid hormone concentrations during, 16, 17
   of *Xenopus laevis*, 8
Proteins. See Binding proteins; Coactivator proteins; Corepressor proteins; Enzymes; Glycoprotein; Serum proteins
Protein sequencing, screening with, 108
Purkinje cells, metamorphosis of, 42–43

Rana, phylogenetic relationship between *Xenopus* and, 6
Rana catesbeiana
   apoptosis during tail resorption in, 62–64
   cell cultures from, 181
   CRF in, 24–25
   deiodinases in development of, 149–150
   in developmental biology, 6
   early TH response genes in, 116
   gene regulation cascade in, 133
   gill resorption in, 39–40
   larval stage of, 9
   late response genes in, 118
   late TH response genes in (table), 114
   metamorphosis and serum protein changes in, 68
   metamorphosis of tadpole tail of, 38
   MMP expression in, 171, 174
   MMP genes in, 166
   prolactin growth promotion in, 28
   retinoic acid receptors in metamorphosis of, 136
   tail collagenase gene cloning in, 108
   thyroid hormone synthesis in, 76
   timing of tissue-specific transformations in, 153
   urea cycle enzymes of, 108
Rana esculenta, environmental effects on metamorphosis in, 13
Rana grylio, metamorphosis and serum protein changes in, 68
Rana heckscheri, metamorphosis and serum protein changes in, 68
Rana japonica
   apoptosis during tail resorption in, 62
   effects of corticoids on, 25–26
Rana perezi, CRF in, 24
Rana pipiens
   in developmental biology, 6
   larval stage of, 7, 9
   TRH in, 24
Rana temporaria
   environmental effects on metamorphosis in, 13
   gonadal steroids in, 27
RARβ (retinoic acid receptor) binding domain, 84, 90–91
Red blood cells, metamorphosis and changes in, 70–71
Remodeling
   apoptosis during, 50–65
   biochemical changes during, 65–73
   of organs in metamorphosis, 36, 41–49
Repression of transcription, 90–91
   by TRs, 88–90

Resistance to thyroid hormone (RTH)
syndromes, 209–211, 223
mutations causing, 216–218, 219
Retinoic acid (RA). *See also* RARβ (retinoic acid receptor) binding domain
effects on *Xenopus* tadpoles of, 29
9-cis-Retinoic acid receptors (RXRs). *See also* RXR binding domains
developmental function of, 136–145
prolactin function and, 32–34
RNA blot analysis, 31
RNA polymerase II, 88
Rohon–Beard neurons, death of, 42
RXR binding domains, 84, 85, 86–87
in transcription, 88–90, 92–101

Salamanders, 2. *See also* Urodela
thyroid hormone and, 5
Salamandridae, 2
Salientia, 2
Sarcolytes. *See* Apoptotic bodies
Scaphiopus hammondii
CRF in, 24
effects of water level on metamorphosis of, 11–13
Serum albumin, metamorphosis and changes in, 69–70, 71, 206
Serum proteins
metamorphosis and changes in, 68–70
TH binding by, 76–78
Sex steroids, influence on metamorphosis of, 27
Signaling molecules, hedgehog gene and, 159–163
Skin, development of frog, 118–119
Smoothened protein, 162–163
Somatostatin, effects on metamorphosis of, 35
Sonic hedgehog gene, during intestinal remodeling, 159–162
Southern blot hybridization, 104
Stats (signal transducer and activators of transcription), prolactin function and, 34
Steroid hormones, influence on metamorphosis of, 25–27
Stromelysin-3 (ST3) gene, 143–145, 165, 169, 175, 178–179
and apoptosis during metamorphosis, 166–168
tissue remodeling and, 173–175
Stromelysins, 164
Subtractive differential screening, isolating TH response genes via, 105–107

Tadpoles
apoptosis during metamorphosis of, 50–65
cell cultures from, 180–183
in direct development, 6–9
early TH response genes in, 108–116
effects of retinoic acid and prolactin on, 29
environmental effects on metamorphosis in, 10–13
limb morphogenesis in, 41
metamorphosis in, 1–2, 9–10, 13–14, 15–19, 36–49
nitrogenous waste excretion in, 71–73
transgenic, 187–189
of *Xenopus laevis*, 8
Tadpole-specific organs, resorption of, 36, 37–40
Tail
apoptosis during resorption of, 61–64
degradative enzymes during resorption of, 65–67
early TH response genes in frog, 109–112, 116, 126
MMP regulation during metamorphosis of, 170–173
resorption of tadpole, 37–39, 61–64, 65–67, 109–112, 119–121, 136, 139
Temperature, effects on metamorphosis of, 10–11
Testosterone, influence on metamorphosis of, 27
TH/bZip gene, 124–127, 129–130
sequencing of, 125
TH deficiency diseases, 206–209
TH-induced transcription factors, 128–134
biochemistry of, 128–130
target genes of, 130–134
TH response genes, 102–104
early, 103–104, 108–116, 123, 126, 204
ECM remodeling and, 175–176
late, 103–104, 108, 114, 116–119, 133–134, 204
methods of isolating, 104–108
3,5,3',5'-Tetraiodothyronine (T_4), 77, 207–208. *See also* Thyroxine; Thyroxine
biosynthesis of, 74–76
chemical structure of, 16
deiodinases and conversion to T3 of, 147–150
effects on metamorphosis of, 15–16
regulation of cellular levels of, 145–146
thyroid hormone receptor binding of, 84
TSH and, 21–23
Thyroidectomy, effect on metamorphosis of, 10, 17–18
Thyroid gland
biochemistry in, 74–78

Thyroid gland (*Continued*)
 development of, 18–19
 hypothalamic regulation of, 24–25
 metamorphosis and, 9–10
 pituitary regulation of, 20–23, 24–25
Thyroid hormone (TH)
 binding proteins of, 74–78
 biosynthesis of, 145–150, 151–153
 in cell cultures, 181–183
 competence in response to, 135–136
 in direct development, 9
 ECM remodeling and, 176–177, 183
 in frog embryos, 185–186
 gene regulation by, 102–122
 in humans, 206–218
 influence of corticoids on, 25–27
 in intestinal remodeling, 48, 58
 in limb morphogenesis, 41
 in mammals, 205–206
 mechanism of action of, 74–101
 in metamorphosis, 5–6, 13, 14, 15–18, 135–136
 in mice, 218–222
 in neural system metamorphosis, 43
 neuroendocrine control of synthesis of, 19–25
 in organ cultures, 183–185
 prolactin and, 30–34
 receptors for, 78–87
 regulation of cellular levels of, 145–150
 retinoic acid receptor expression and, 136–140
 stromelysin-3 regulation by, 166
 in tail resorption, 39
 and thyroid gland development, 18–19
 tissue-dependent variation in gene expression via, 119–122
 transcriptional regulation by, 87–101, 102
 in transgenic *Xenopus laevis*, 188–189
 TR binding of, 82–84
 in vertebrates, 205, 222–223
Thyroid hormone receptors (TRs), 74, 78–87
 cloning of, 180
 developmental function of, 136–145
 DNA binding by, 84–87
 domain structure of, 78–82, 79, 83
 ecdysone receptors and, 194
 in frog development, 140–145
 in humans, 209–218
 in metamorphosis, 14
 in mice, 218–222
 mutations in human, 216–218, 222–223
 in organ culture systems, 183–184
 sequence comparison among, 80–81, 213–215
 as target genes of TH-induced transcription factors, 130–134
 TH binding by, 82–84
 in transcription, 87–101
Thyroid hormone resistance syndromes (RTHs), 209–211, 223
 mutations causing, 216–218, 219
Thyroid hormone response elements (TREs), 140. See also TH response genes
 in DNA binding, 84–85, 87, 95–96
 in transcription, 88–90
Thyroid-stimulating hormone (TSH)
 chemical structure of, 21
 effects on metamorphosis of, 13
 homology of frog and mammalian, 23
 hypothalamus and, 24
 in thyroid gland function, 20–23
Thyrotropes, 22, 24
Thyrotropin-releasing hormone (TRH)
 hypothalamus and, 24–25
Thyroxine. See also 3,5,3′,5′-Tetraiodothyronine (T_4)
 effects on metamorphosis of, 15–16
Tissue inhibitors of MMPs (TIMPs), 178
 in organ culture systems, 184
Tissue resorption, gene regulation cascade for, 200–202
Tissues, metamorphosis of, 36, 45–49
Tissue-specific transformations
 during metamorphosis, 119–122, 135–153
 timing of, 150–153
Toads, 2. See also Anura
TRβ genes, 136–137, 142–145. See also TR genes
 in humans, 211
 knockout studies of, 218–220
Transcription, TH regulation of, 87–101, 102
Transcription factors (TFs)
 genes encoding, 124–128
 TH-induced, 128–130, 130–134
 in TH signal transduction cascade, 123–134
Transgenic *Xenopus laevis*, studying gene regulation and function in, 187–189
TRβA gene, 137–139. See also TR genes
TRβB gene, 136–140. See also TR genes
TRβ genes, 136–140. See also TR genes
 in humans, 211–216
 in intestinal apoptosis and cell proliferation, 140
 knockout studies of, 218–220, 220–222
 mutations in, 216–218
 in organ culture systems, 183–184

TR domains, 78–82, 83
  for DNA binding, 84–87
  for TH binding, 82–84
TR genes
  developmental function of, 136–145, 151–153
  in humans, 211–218
  prolactin function and, 32–34
  TR domain structure and, 78–82, 83
3,5,3'-Triiodothyronine (T_3), 77, 88, 103, 175, 207–208
  biosynthesis of, 74–76
  chemical structure of, 16
  chromatin disruption by, 92–96
  deiodinases and conversion of T4 to, 147–149
  early gene response to, 111, 113, 123
  effects on metamorphosis of, 15–17, 156
  and histone acetylation, 97–99
  in intestinal remodeling, 58–60, 61
  late gene response to, 117, 118
  NFI genes and, 127
  prolactin and, 30, 31
  regulation of cellular levels of, 145–146
  ST_3 gene and, 171
  stromelysin-3 regulation by, 167
  in subtractive differential screening, 105–107
  TH/bZip gene and, 124–126
  thyroid hormone receptor binding of, 82–84
  TR/RXR heterodimers and, 142–145
TR/RXR heterodimers
  developmental function of, 136–145
  in frog embryos, 185
  histone acetylation and, 97–99
  in humans, 212–216
  mechanism of action of, 100–101
  timing of tissue-specific transformations and, 151–152
  transcriptional regulation via, 92–96, 102–104
Typhlosole, in metamorphosis, 44–45, 46
Ultraspiracle (USP) gene, 194
Urea, excretion of, 71–73
Urea cycle, 73
  genes encoding enzymes of, 117–118
Urea cycle enzymes, genes encoding, 117–118
  isolation of, 107–108
Ureotelism, transition from ammonotelism to, 117–118
Urodela, 2–3. See also Newts; Salamanders
  metamorphosis in, 4–5

Vertebrates
  intestinal development in, 43–44
  postembryonic organ development in, 13–14
  thyroid hormone and receptors in, 205, 222–223

Water level, effects on frog metamorphosis, 11–13

Xenopus
  EcRs and TRs of, 194
  histone acetylation in, 97–99
  phylogenetic relationship between Rana and, 6
Xenopus laevis
  acid hydrolases during tail resorption in, 66–67
  apoptosis and stromelysin-3 in, 166–168
  apoptosis during tail resorption in, 62, 64
  cDNA encoding TSH in, 23
  cell cultures from, 181
  cloning of prolactin genes in, 28
  corticoids in, 26–27
  CRF in, 25
  cytosolic thyroid hormone binding proteins in, 146–147, 148–149
  deiodinases in development of, 147–150
  in developmental biology, 1, 6
  effects of retinoic acid and ovine prolactin on, 29
  effects of temperature on metamorphosis of, 10–11
  epithelial apoptosis in, 56–58, 59
  gill resorption in, 40
  gonadal steroids in, 27
  human TRs versus TRs of, 212–216
  insect metamorphosis versus metamorphosis in, 202–204
  intestinal metamorphosis in, 44, 46–47, 49, 56–58, 59, 157, 158
  isolating early TH response genes in, 108–116
  isolating late TH response genes in, 116–119
  isolating TH response genes in, 104–105, 107
  larval stage of, 7, 8, 9
  limb morphogenesis in, 41
  metamorphosis in, 9–10
  metamorphosis of muscle in, 48
  metamorphosis and serum protein changes in, 68
  metamorphosis of tadpole tail of, 37–39
  MMP expression in, 170–175
  MMP genes in, 165–166
  NFI genes of, 127–128, 129, 131

Xenopus laevis (Continued)
   nitrogenous waste excretion in, 72
   prolactin regulation in, 30–32
   regulation of cellular TH levels in, 145–150
   retinoic acid receptors in metamorphosis of, 136–145
   sonic hedgehog gene in, 159–162
   structure of TR ligand binding domain in, 83
   tail tip resorption in, 20
   temporal regulation of metamorphosis in, 135–136
   TH binding proteins of, 77–78
   TH/bZip gene in, 124–126, 129–130
   TH-induced transcription factors in, 130–134
   thyroid-deficient tadpole of, 21
   thyroid gland development in, 18–19
   thyroid hormone and, 5, 15–18
   thyroid hormone receptors in, 78–82
   thyroid hormone synthesis in, 76
   timing of tissue-specific transformations in, 150–153
   tissue-specific transformations in, 119–122
   transcription in oocytes of, 91–96
   transgenic, 187–189
   TR binding of DNA in, 86
   TRH in, 24
   TSH in prometamorphic, 22